AF773943

Other Titles in This Series

147 **S. G. Gindikin and L. R. Volevich,** Mixed problem for partial differential equations with quasihomogeneous principal part, 1996
146 **L. Ya. Adrianova,** Introduction to linear systems of differential equations, 1995
145 **A. N. Andrianov and V. G. Zhuravlev,** Modular forms and Hecke operators, 1995
144 **O. V. Troshkin,** Nontraditional methods in mathematical hydrodynamics, 1995
143 **V. A. Malyshev and R. A. Minlos,** Linear infinite-particle operators, 1995
142 **N. V. Krylov,** Introduction to the theory of diffusion processes, 1995
141 **A. A. Davydov,** Qualitative theory of control systems, 1994
140 **Aizik I. Volpert, Vitaly A. Volpert, and Vladimir A. Volpert,** Traveling wave solutions of parabolic systems, 1994
139 **I. V. Skrypnik,** Methods for analysis of nonlinear elliptic boundary value problems, 1994
138 **Yu. P. Razmyslov,** Identities of algebras and their representations, 1994
137 **F. I. Karpelevich and A. Ya. Kreinin,** Heavy traffic limits for multiphase queues, 1994
136 **Masayoshi Miyanishi,** Algebraic geometry, 1994
135 **Masaru Takeuchi,** Modern spherical functions, 1994
134 **V. V. Prasolov,** Problems and theorems in linear algebra, 1994
133 **P. I. Naumkin and I. A. Shishmarev,** Nonlinear nonlocal equations in the theory of waves, 1994
132 **Hajime Urakawa,** Calculus of variations and harmonic maps, 1993
131 **V. V. Sharko,** Functions on manifolds: Algebraic and topological aspects, 1993
130 **V. V. Vershinin,** Cobordisms and spectral sequences, 1993
129 **Mitsuo Morimoto,** An introduction to Sato's hyperfunctions, 1993
128 **V. P. Orevkov,** Complexity of proofs and their transformations in axiomatic theories, 1993
127 **F. L. Zak,** Tangents and secants of algebraic varieties, 1993
126 **M. L. Agranovskiĭ,** Invariant function spaces on homogeneous manifolds of Lie groups and applications, 1993
125 **Masayoshi Nagata,** Theory of commutative fields, 1993
124 **Masahisa Adachi,** Embeddings and immersions, 1993
123 **M. A. Akivis and B. A. Rosenfeld,** Élie Cartan (1869–1951), 1993
122 **Zhang Guan-Hou,** Theory of entire and meromorphic functions: Deficient and asymptotic values and singular directions, 1993
121 **I. B. Fesenko and S. V. Vostokov,** Local fields and their extensions: A constructive approach, 1993
120 **Takeyuki Hida and Masuyuki Hitsuda,** Gaussian processes, 1993
119 **M. V. Karasev and V. P. Maslov,** Nonlinear Poisson brackets. Geometry and quantization, 1993
118 **Kenkichi Iwasawa,** Algebraic functions, 1993
117 **Boris Zilber,** Uncountably categorical theories, 1993
116 **G. M. Fel′dman,** Arithmetic of probability distributions, and characterization problems on abelian groups, 1993
115 **Nikolai V. Ivanov,** Subgroups of Teichmüller modular groups, 1992
114 **Seizô Itô,** Diffusion equations, 1992
113 **Michail Zhitomirskiĭ,** Typical singularities of differential 1-forms and Pfaffian equations, 1992
112 **S. A. Lomov,** Introduction to the general theory of singular perturbations, 1992
111 **Simon Gindikin,** Tube domains and the Cauchy problem, 1992
110 **B. V. Shabat,** Introduction to complex analysis Part II. Functions of several variables, 1992
109 **Isao Miyadera,** Nonlinear semigroups, 1992
108 **Takeo Yokonuma,** Tensor spaces and exterior algebra, 1992
107 **B. M. Makarov, M. G. Goluzina, A. A. Lodkin, and A. N. Podkorytov,** Selected problems in real analysis, 1992
106 **G.-C. Wen,** Conformal mappings and boundary value problems, 1992

(*Continued in the back of this publication*)

Mixed Problem for Partial Differential Equations with Quasihomogeneous Principal Part

Translations of

MATHEMATICAL MONOGRAPHS

Volume 147

Mixed Problem for Partial Differential Equations with Quasihomogeneous Principal Part

S. G. Gindikin
L. R. Volevich

American Mathematical Society
Providence, Rhode Island

Л. Р. Волевич, С. Г. Гиндикин

СМЕШАННАЯ ЗАДАЧА ДЛЯ ДИФФЕРЕНЦИАЛЬНЫХ УРАВНЕНИЙ В ЧАСТНЫХ ПРОИЗВОДНЫХ С КВАЗИОДНОРОДНОЙ СТАРШЕЙ ЧАСТЬЮ

Translated by V. M. Volosov from an original Russian manuscript

1991 *Mathematics Subject Classification*. Primary 35Gxx

ABSTRACT. The book gives a uniform exposition of the mixed problem for hyperbolic and $2b$-parabolic differential equations with variable coefficients. The method makes it possible to consider simultaneously a nonclassical type of equations, q-hyperbolic equations.

The book is intended for researchers and graduate students working in partial differential equations.

Library of Congress Cataloging-in-Publication Data

Gindikin, S. G. (Semen Grigor′evich)

[Smeshannaia͡ zadacha dlia͡ different͡sial′nykh uravneniĭ v chastnykh proizvodnykh s kvaziodnorodnoĭ starsheĭ chast′i͡u. English]

Mixed problem for partial differential equations with quasihomogeneous principal part / S. G. Gindikin, L. R. Volevich; [translated by V. M. Volosov from an original Russian manuscript].

p. cm. — (Translations of mathematical monographs, ISSN 0065-9282; v. 147)

Includes bibliographical references (p. –) and index.

ISBN 0-8218-4617-5 (alk. paper)

1. Differential equations, Hyperbolic. 2. Differential equations, Parabolic. I. Volevich, L. R. (Leonid Romanovich), 1934– . II. Title. III. Series.

QA377.G52713 1996

515′.353—dc20

95-20515

CIP

Copying and reprinting. Individual readers of this publication, and nonprofit libraries acting for them, are permitted to make fair use of the material, such as to copy a chapter for use in teaching or research. Permission is granted to quote brief passages from this publication in reviews, provided the customary acknowledgment of the source is given.

Republication, systematic copying, or multiple reproduction of any material in this publication (including abstracts) is permitted only under license from the American Mathematical Society. Requests for such permission should be addressed to the Assistant to the Publisher, American Mathematical Society, P.O. Box 6248, Providence, Rhode Island 02940-6248. Requests can also be made by e-mail to `reprint-permission@math.ams.org`.

© Copyright 1996 by the American Mathematical Society. All rights reserved.
The American Mathematical Society retains all rights
except those granted to the United States Government.
Printed in the United States of America.

∞ The paper used in this book is acid-free and falls within the guidelines
established to ensure permanence and durability.

♻ Printed on recycled paper.

10 9 8 7 6 5 4 3 2 1 01 00 99 98 97 96

Contents

Preface

The prime objective of the present book is systematic presentation of the theory of mixed problem for hyperbolic differential equations with variable coefficients. In a sense, this theory completes the classical line of investigation in partial differential equations which began with studying the Cauchy problem and mixed problem for the wave equation (for related historical remarks see the survey by Gindikin and Volevich [**GV8**]). The main stages in its construction were the introduction by Petrovskii in the thirties of high-order hyperbolic equations (systems) generalizing the wave equation, Petrovskii's and Leray's investigations of the Cauchy problem for hyperbolic equations of arbitrary order with variable coefficients, and, finally, the study of mixed problem for hyperbolic equations of arbitrary order (first-order systems), which took several decades and was completed by Agmon, Kreiss, and Sakamoto. In our opinion, the most adequate approach to the proof of solvability of hyperbolic problems is the method using the "separating" operator and the related definite quadratic form. This method was suggested in the early fifties by Leray, who introduced energy quadratic forms for high-order hyperbolic equations with variable coefficients proceeding from the law of conservation of energy for a membrane (its motion is described by the wave equation). Gårding constructed an orderly theory of solvability for the Cauchy problem by combining energy estimates and methods of functional analysis. Sakamoto substantially developed the Leray-Gårding approach in application to the mixed problem. The fundamental idea of Leray is that in the case of a strictly hyperbolic operator (with simple real characteristics) the corresponding energy form is (in modern terminology) an elliptic form with parameters, and this makes it possible to derive estimates in the case of variable coefficients. The presentation of the energy method for hyperbolic equations constitutes a major part of this book.

This main investigation line is complemented with two other topics. First, it turns out that the energy method elaborated for hyperbolic equations can be extended almost automatically to parabolic equations; more precisely, to $2b$-parabolic differential equations with variable coefficients. This class of equations was also introduced by Petrovskii as a generalization of the heat equation to the case of high-order equations (systems). For these equations, the mixed problem as well as the Cauchy problem is traditionally investigated with the aid of some adequate estimates and some form of Levi's parametrix. Second, it turns out that the energy method makes it possible to investigate completely the solvability of the Cauchy problem and mixed problem for parabolic equations as well. The only natural change that should be made in all the constructions is to replace the ordinary principal homogeneous parts of operators by principal quasihomogeneous parts that are obtained if the time derivatives are taken with weight $2b$. This is a general situation in the theory of parabolic equations (for the heat equation we have $b = 1$). What has been said accounts for the term "quasihomogeneous principal part" in the title of the book. Moreover, within the framework of the energy estimate method the theories of hyperbolic and parabolic operators can be combined as a unified theory of operators with dominant principal quasihomogeneous part. The latter operators are characterized by the property that the correctness (well-posedness) conditions of the Cauchy problem and mixed problem are stated for them only in terms of principal quasihomogeneous parts of equations and boundary operators, and therefore they are stable with respect to the perturbations of the problem by arbitrary lower degree terms.

However, there is another possible extension of the class of operators in question, and the main innovation of the suggested presentation is related to it. Namely, we study a new class of the so-called $(2b+1)$-hyperbolic operators. In this case the weight $2b+1$ is assigned to the operator of time differentiation. In contrast to parabolic and hyperbolic operators, the $(2b+1)$-hyperbolic operators have no classical prototypes. Nevertheless, a simple and natural example of an equation corresponding to $b = 1$ can be given:

$$\pm\frac{\partial u}{\partial t} + \frac{\partial^3 u}{\partial x_1^3} + \cdots + \frac{\partial^3 u}{\partial x_n^3} = 0.$$

In the case of one spatial variable ($n = 1$) this equation coincides with the linear part of the Korteweg-de Vries (KdV) equation, and the linear parts of some multi-dimensional KdV analogs are also 3-hyperbolic operators. For $(2b+1)$-hyperbolic equations with variable coefficients it is also possible to construct an equally complete theory for the Cauchy problem and the mixed problem using the method of energy estimates. Finally, all three classes of operators under consideration can be combined to form the class of operators with dominant principal quasihomogeneous part. From the onset, the presentation could have been given for this general class of operators, which, however, would have been very cumbersome, and therefore we preferred to work with special classes and to present the unified result only in the end.

The book consists of four chapters, each starting with a detailed introduction. The first chapter compiles all the necessary material related to polynomials that in the subsequent chapters serve as symbols for all the three above-mentioned classes of operators. Properties of these polynomials (including their description which makes it possible to combine them into a unified class) that are necessary for constructing energy estimates are studied in detail. The second chapter is devoted to the Cauchy problem which is a constituent of the mixed problem. However, we consider the Cauchy problem separately because the theory of solvability for this problem is of interest by itself, and this facilitates the presentation in the last two most difficult chapters devoted to the mixed problem. It should be noted that we do not touch upon all aspects of the Cauchy problem and consider only those related to the mixed problem. For instance, we do not practically discuss the possibility of extending the correctness theory for the Cauchy problem to larger classes of operators with both variable and constant coefficients and do not investigate the exact exponential classes of correctness for the Cauchy problem either. For detailed presentation of these questions see the books and survey by Gindikin and Volevich **[GV1, 2, 8]** and **[Gi3]**.

In Chapter II we first reformulate the results of Chapter I related to polynomials in the form of estimates for the corresponding differential operators with constant coefficients and then describe the general scheme of studying the Cauchy problem for equations with variable coefficients using energy estimates and specialize this scheme in application to hyperbolic, $2b$-parabolic, and $(2b + 1)$-hyperbolic equations. After that a unified theorem is presented. Chapter II is completed by presenting the theory of the Cauchy problem for three classes of systems generalizing the classes of scalar operators considered in the book. Unfortunately, neither a satisfactory theory of the mixed problem for high-order hyperbolic systems nor a similar theory for $(2b + 1)$-hyperbolic systems has yet been created, and therefore the questions of solvability of the mixed problem for systems of equations are beyond the scope of the present book.

Chapters III and IV are devoted to the mixed problem proper, and the presentation of the material in them is in many respects "in parallel" with that in Chapter II. Chapter III considers systematically a more difficult case of hyperbolic equations, and the last comparatively small chapter compiles some additional facts making it possible to consider the mixed problem for $2b$-parabolic and $(2b+1)$-hyperbolic equations and to derive the general result unifying all the three above-mentioned theories. We note that the combined presentation of the three independent theories for mixed problem and the Cauchy problem makes it necessary to state a large number of "similar" assertions. As a rule, a thorough proof is carried out only once, say, for hyperbolic equations, and in the other cases we dwell only on some specific details or (if these details are not of fundamental importance) leave the proofs to the reader.

Sections in each chapter and formulas in each section are numbered autonomously; the subsections have double numbering (for example, 2.1 means 1 of §2). Within the subsections autonomous numbering of theorems, propositions, lemmas, etc. is used. If in a subsection there is only one theorem, proposition, lemma, etc., it has no number. When referring to the material in another subsection of the given section we write only the number of the subsection: for example, Proposition 3.1. If the given subsection contains several propositions, we write, for instance, Proposition 1, 2, etc. of 3.1. When mentioning formulas belonging to other sections we first indicate the number of the corresponding section: for example, (3.20) means formula (20) in §3. When referring to the material in other chapters we write the number of the chapter before the numbers of the sections, subsections, and formulas. For instance, §IV.2, IV.2.1, and (IV.2.7) mean, respectively, §2 of Chapter IV, subsection 1 in §2 of Chapter IV, and formula (7) in §2 of Chapter IV.

The authors express their deep gratitude to Professor V. M. Volosov for translating the book and for valuable advice and various refinements that he has made in the translation.

CHAPTER I

Basic Classes of Polynomials

Introduction

The present book is devoted to the Cauchy problem and mixed problem for differential operators, and the prime objective of the investigation is the case of operators with variable coefficients. This specifies the choice of the material in this chapter whose aim is detailed description of classes of polynomials that in the subsequent chapters will serve as symbols of differential operators for which the Cauchy problem and mixed problem are studied.

For symbols of differential operators with constant coefficients Petrovskii's classical condition is known (see **[P1]**), which is equivalent to Hadamard's correctness condition for the Cauchy problem in various function spaces including the spaces of functions continuous and bounded together with a finite number of derivatives.

Let $P(\tau, \xi)$ be the symbol of a differential operator and let the variables τ and $\xi_1, \dots, \xi_n$ be dual to time t and spatial coordinates $x_1, \dots, x_n$, respectively; it is assumed that P is resolved with respect to the highest power of τ. In this case Petrovskii's correctness condition requires that

$$(1) \qquad P(\xi, \tau) \neq 0, \qquad \tau \in \mathbb{C}^1, \quad \operatorname{Im} \tau \leqslant \gamma_0, \quad \xi \in \mathbb{R}^n,$$

for some $\gamma_0 \in \mathbb{R}^n$. Here it is important that one has to pass to the complex domain for the variable τ. We stress that in the case of equations with variable coefficients the sufficient conditions for solvability of the Cauchy problem substantially differ from a condition of the type (1) on the symbol. In all the known results on the solvability of the Cauchy problem for equations with variable coefficients (and the more so in the case of mixed problem) a condition of the type (1) holds for the classes of symbols in question under an arbitrary variation of lower monomials (which results in a change of the constant γ_0), and the monomials can be classified into leading and lower ones in different ways. When one uses the ordinary degree of monomials, the stability of Petrovskii's condition (1) with respect to the lower terms leads to the class of strictly hyperbolic (hyperbolic in Petrovskii's sense; see **[P2]**) polynomials (operators). The classical example of this kind is the wave operator

$$\frac{\partial^2}{\partial t^2} - \sum_{j=1}^{n} \frac{\partial^2}{\partial x_j^2}.$$

However, the simple example of the heat operator

$$\frac{\partial}{\partial t} - \sum_{j=1}^{n} \frac{\partial^2}{\partial x_j^2}$$

shows that in some cases it is natural to introduce a"weighted" degree by assigning the degree qk to the monomial τ^k ($q = 2$ in the case of the heat operator), whereas the degrees of the monomials $\xi_1^{\alpha_1} \cdots \xi_n^{\alpha_n}$ are defined in the ordinary way. Under this definition of the weighted degree and for even $q = 2b$, the stability of condition (1) with respect to the lower monomials means that P is a $2b$-parabolic polynomial (parabolic in Petrovskii's sense; see **[P1]**). The strictly hyperbolic and $2b$-parabolic operators are two classical types of operators with variable coefficients for which it is possible to construct a theory for the Cauchy problem and mixed problem. Chapters II and III are mainly devoted to this question.

In the present book an important addition to these classical operators is a new class of the so-called q-hyperbolic operators, where q is an odd number greater than 1. The simplest example is the operator

$$\frac{\partial u}{\partial t} - \sum a_j(t, x) \frac{\partial^3}{\partial x_j^3},$$

where $a_j(t, x)$ are real functions. In the q-hyperbolic case the symbols satisfy a version of stability condition with respect to the lower terms, which is related to the transition into the complex domain along a certain direction in the space of variables $\xi_1, \ldots, \xi_n$. The main result in this chapter is that for the given method of defining the leading and lower terms and the given (quite natural) definition of stability of Petrovskii's condition with respect to the lower terms we obtain one of the above-mentioned classes of polynomials.

In conclusion we make several remarks. Note that there are some more general constructions for separating out leading and lower monomials in a polynomial. Retaining the definition with the weighted degree and abandoning the "equivalence" of the variables $\xi_1, \ldots, \xi_n$ we arrive at the so-called pluriparabolic polynomials described in Appendix to §2. Abandoning the separation of leading terms by means of the weighted degree and using Newton's polygons or polyhedra for their definition we can obtain quite general classes of operators with variable coefficients for which the Cauchy problem is correct (see **[GV2]**).

It is also possible to consider some more general classes of symbols for which Petrovskii's condition is stable relative to various transitions into the complex domain with respect to $\xi_1, \ldots, \xi_n$. These are the so-called exponentially correct polynomials (see **[GV1]**). Under some additional conditions the Cauchy problem can be considered for them in the case of variable coefficients (see Chapter II, §2). However, it is unlikely that the mixed problem can be considered at this level of generality. For this reason such general classes of operators lie on the periphery of this book.

Finally, all classes of polynomials $P(\xi, \tau)$ considered in this chapter are characterized by means of real polynomials in the variables $(\operatorname{Re}\tau, \operatorname{Im}\tau, \xi)$:

$$H_P(\xi, \sigma, \gamma) = -\operatorname{Im}\left[P(\xi, \sigma + i\gamma)\overline{\frac{\partial P}{\partial \tau}(\xi, \sigma + i\gamma)}\right]. \tag{2}$$

These polynomials are symbols of quadratic forms that were for the first time considered by Leray **[1]** in connection with the derivation of energy estimates for hyperbolic operators. These forms are the central object in estimates presented in Chapters II and III. The thing is that because the symbols have real zeros, the hyperbolic and q-hyperbolic operators with variable coefficients do not admit adequate estimates.

However, if the multiplicities of the real zeros of P are no higher than 1, then polynomial (2) has no real zeros, and the corresponding form with variable coefficients admits upper and lower estimates in terms of forms with constant ("frozen") coefficients.

The presentation of the material in this chapter is organized in the following way. Section 1 compiles some facts that relate to polynomials correct in Petrovskii's sense and are necessary for the sequel. In particular, Borok's **[Bo]** description of polynomials in two variables, which are correct in Petrovskii's sense, in terms of Puiseux's series for their roots is presented and an equivalent definition of Petrovskii's condition in terms of polynomials (2) is given.

In §2 strictly hyperbolic and $2b$-parabolic polynomials are described. The class of q-stable correct polynomials is introduced, and it is proved that it is exhausted by strictly hyperbolic and q-parabolic polynomials (where q are even numbers). In §3 we describe q-hyperbolic polynomials. In §4 the notion of weak q-stability is introduced. Here the main result is that the class of weakly q-stable polynomials is exhausted by q-stable and q-hyperbolic polynomials.

Notation. We deal with the $(n+1)$-dimensional real space $\mathbb{R}_y^{n+1}$, $y = (y_1, \dots, y_{n+1})$. Set $D = (D_1, \dots, D_{n+1})$, $D_k = \partial/i\partial y_k$, $k = 1, \dots, n+1$, $i = \sqrt{-1}$.

We denote by $\mathbb{R}_\eta^{n+1}$ the dual space of $\mathbb{R}_y^{n+1}$. Let the coordinates $\eta = (\eta_1, \dots, \eta_{n+1})$ be chosen so that the duality is determined by means of the bilinear form

$$\langle y, \eta\rangle = y_1\eta_1 + \cdots + y_{n+1}\eta_{n+1}. \tag{3}$$

If $\alpha = (\alpha_1, \dots, \alpha_{n+1})$ is a set of nonnegative integers (a multiindex), then we denote $|\alpha| = \alpha_1 + \cdots + \alpha_{n+1}$, $\eta^\alpha = \eta_1^{\alpha_1} \cdots \eta_{n+1}^{\alpha_{n+1}}$, and $D^\alpha = D_1^{\alpha_1} \cdots D_{n+1}^{\alpha_{n+1}}$.

Let us split the variables $y \in \mathbb{R}^{n+1}$ into the time variable $t = y_{n+1} \in \mathbb{R}$ and the spatial variables $x = (x_1, \dots, x_n) \in \mathbb{R}^n$, $x_j = y_j$, $j = 1, \dots, n$. Denote by σ and $\xi = (\xi_1, \dots, \xi_n)$ the dual variables $\sigma = \eta_{n+1}$ and $\xi_j = \eta_j$, $j = 1, \dots, n$, respectively.

We consider the complexification of the space $\mathbb{R}_\eta^{n+1}$ in the form of the space $\mathbb{C}_Z^{n+1}$, $Z = (\zeta, \tau)$, $\tau = \sigma + i\gamma \in \mathbb{C}^1$, $\zeta = (\zeta_1, \dots, \zeta_n) \in \mathbb{C}_\zeta^n$, $\zeta_j = \xi_j + i\omega_j$, $j = 1, \dots, n$. Setting $\omega = (\omega_1, \dots, \omega_n)$ we write $\zeta = \xi + i\omega$. The complex coordinates (ζ, τ) and the coordinates (x, t) can be regarded as dual variables relative to the holomorphic extension of the linear form (3).

Denote $\partial_\zeta = (\partial_{\zeta_1}, \dots, \partial_{\zeta_n})$, $\partial_{\zeta_k} = \partial/\partial\zeta_k$, $\partial_\tau = \partial/\partial\tau$, $D_t = \partial/i\partial t$, $D_x = (D_1, \dots, D_n)$.

§1. Polynomials correct in Petrovskii's sense

1.1. Basic definitions. We will deal with the polynomials

$$P(\xi, \tau) = \sum_{j=0}^{m} P_j(\xi)\tau^{m-j}, \tag{1}$$

where $P_j(\xi)$ are polynomials in the variables $\xi_1, \dots, \xi_n$. If $P_0(\xi) \equiv \text{const}$, then polynomial (2) is said to be resolved with respect to the highest power of τ. Assuming that this constant is equal to 1, we write such a polynomial in the form

$$P(\xi, \tau) = \tau^m + \sum_{j=1}^{m} P_j(\xi)\tau^{m-j}. \tag{2}$$

DEFINITION. A polynomial P is said to be correct in Petrovskii's sense if
(i) P is resolved with respect to the highest power of τ, i.e., has the form (2);
(ii) there is $\gamma_0 \in \mathbb{R}$ such that

$$P(\xi,\tau) \neq 0, \qquad \operatorname{Im}\tau < \gamma_0, \quad (\xi, \operatorname{Re}\tau) \in \mathbb{R}^{n+1}. \tag{3}$$

REMARK. As was already said in the introduction to this chapter, Petrovskii introduced condition (3) when describing the class of differential operators $P(D_x, D_t)$ with constant coefficients for which the Cauchy problem

$$\begin{aligned} P(D_x, D_t)u(x,t) &= f(x,t), \qquad 0 \leqslant t \leqslant T, \quad x \in \mathbb{R}^n, \\ D^j u(x,0) &= \varphi_j(x), \qquad j = 0,\dots,m, \quad x \in \mathbb{R}^n, \end{aligned}$$

is well-posed.

For detailed discussion of this question see **[P1]**, **[GV1]**, and also §2 in Chapter II of the present book.

Condition (3) can be restated in terms of the roots of polynomial (2). To this end polynomial (2) is factorized as a polynomial in τ (depending on the parameter ξ):

$$P(\xi,\tau) = \prod_{j=1}^{m}(\tau - \lambda_j(\xi)), \quad \xi \in \mathbb{R}^n, \tag{4}$$

where $\lambda_j(\xi)$ is a multi-valued algebraic function. For each fixed ξ we formally index the branches $\lambda_1(\xi),\dots,\lambda_m(\xi)$ of the function, which however does not mean that its single-valued branches are separated out globally.

Condition (3) can be restated in terms of the roots in the following way: there is γ_0 such that

$$\operatorname{Im}\lambda_j(\xi) > \gamma_0, \qquad \xi \in \mathbb{R}^n, \quad j = 1,\dots,m. \tag{3'}$$

1.2. Quasi-homogeneous polynomials correct in Petrovskii's sense. Let $Q = (Q_1, \dots, Q_{n+1})$ be a vector. A polynomial $P_0(\xi,\tau)$ is said to be Q-homogeneous if

$$P_0(\lambda^{Q_1}\xi_1,\dots,\lambda^{Q_{n+1}}\tau) = \lambda^{m(Q)}P_0(\xi_1,\dots,\tau) \quad \forall\lambda > 0. \tag{5}$$

We will be interested in Q-homogeneous polynomials corresponding to $Q_1 = \cdots = Q_n = 1$, $Q_{n+1} = q \geqslant 1$. For brevity, they will be called q-homogeneous polynomials. For q-homogeneous polynomials $P_0(\xi,\tau)$ Definition 1.1 is equivalent to the condition

$$P_0(\xi,\tau) \neq 0, \qquad \operatorname{Im}\tau < 0, \quad \xi \in \mathbb{R}^n. \tag{6}$$

Indeed, if condition (3) holds for $P = P_0$ with some γ_0, then, replacing in this condition ξ and σ by $\rho\xi$ and $\rho^q\sigma$, respectively, we obtain, by virtue of (5),

$$P_0(\xi, \sigma + i\gamma\rho^{-q}) \neq 0, \qquad \gamma \leqslant \gamma_0, \quad (\sigma,\xi) \in \mathbb{R}^{n+1}.$$

Since $\rho > 0$ is arbitrary, on replacing $\gamma\rho^{-q}$ by γ we derive condition (6).

On the other hand, if (6) holds, then condition (ii) in Definition 1.1 is sure to hold, and (i) also holds. Indeed, from q-homogeneity it follows that the polynomials $P_j(\xi)$ in the representation of $P_0(\xi,\tau)$ in the form (1) are homogeneous polynomials of degree $l + jq$, $l \geqslant 0$. If $l \geqslant 1$, then $P_0(0,\tau) \equiv 0$ for all $\tau \in \mathbb{C}$, and condition (6) is violated. Therefore $P_0(\xi)$ is a homogeneous polynomial of degree zero, whence $P_0(\xi) \equiv \text{const}$, i.e., condition (i) holds.

A q-homogeneous polynomial $P_0(\xi, \tau)$, which is resolved with respect to the highest power of τ, can be factorized as follows:

$$P_0(\xi, \tau) = \prod_{j=1}^{m} (\tau - \lambda_{0j}(\xi)), \quad \xi \in \mathbb{R}^n. \tag{7}$$

The roots $\lambda_{0j}(\xi)$ can be chosen so that they are homogeneous functions of ξ of degree q:

$$\lambda_{0j}(\rho\xi) = \rho^q \lambda_{0j}(\xi) \quad \forall \rho > 0, \ \forall \xi \in \mathbb{R}^n, \quad j = 1, \dots, m. \tag{8}$$

Condition (6) can be rewritten in terms of the roots in the form

$$\operatorname{Im} \lambda_{0j}(\xi) \geqslant 0, \qquad j = 1, \dots, m, \quad \xi \in \mathbb{R}^n. \tag{9}$$

Petrovskii's condition (6) imposes strong constraints on the number q and on the roots $\lambda_{0j}(\xi)$. In this case we have the following

PROPOSITION. *A q-homogeneous polynomial $P_0(\xi, \tau)$ is correct in Petrovskii's sense if and only if the conditions below hold*:
(a) *q is a positive integer;*
(b) *if q is odd, then* $\operatorname{Im} \lambda_{0j}(\xi) \equiv 0$, $j = 1, \dots, m$;
(c) *if q is even, then*

$$\operatorname{Im} \lambda_{0j}(\xi) \geqslant 0 \quad \text{for} \quad \xi \in \mathbb{R}^n, \quad j = 1, \dots, m.$$

The proof is based on the following lemma.

LEMMA. *Let k and p be coprime natural numbers and let $z \in \mathbb{C}$. Then the inequality*

$$\operatorname{Im}\bigl(z(\rho^{1/p})^k\bigr) \geqslant \infty, \quad -\infty < \rho < \infty,$$

holds for all branches of $\rho^{1/p}$ if and only if $p = 1$ and
(α) $\operatorname{Im} z \geqslant 0$ *for even k,*
(β) $\operatorname{Im} z = 0$ *for odd k.*

PROOF. For definiteness, let $\rho > 0$. Then

$$\arg\bigl(z(\rho^{1/p})^k\bigr) = \arg z + 2\pi lk/p, \quad l = 0, \dots, p-1. \tag{10}$$

If $p \geqslant 2$, then there is a natural number l_0 lying between 0 and $p-1$ such that the fractional part of the ratio $l_0 k/p$ exceeds $1/2$. Therefore inequality (10) cannot hold simultaneously for the two branches of $\rho^{1/p}$ corresponding to $l = 0$ and $l = l_0$ (because their arguments differ by more than π). Now let $p = 1$ and let k be even. Then the sign of the left-hand side of (10) coincides with that of $\operatorname{Im} z$, which implies inequality (10). In case k is odd, the left-hand side of (10) has opposite signs for ρ and $-\rho$, which implies condition (β).

PROOF OF THE PROPOSITION. 1) We first consider the case of a polynomial $P_0(\rho, \tau)$ in two variables $\rho \in \mathbb{R}$ and $\tau \in \mathbb{C}$. According to (8), the roots $\lambda_{0j}(\rho)$ have the form

$$\lambda_{0j}(\rho) = a_j \rho^q,$$

and, by condition (9), they satisfy the inequality

$$\operatorname{Im}(a_j \rho^q) > 0, \quad \rho \in \mathbb{R}.$$

In view of (2) and the homogeneity condition, the polynomial P_0 has the form

$$P_0(\xi,\tau) = \tau^m + \sum_{(m-j)q+|\alpha|=mq} a_{\alpha j}\xi^\alpha \tau^{m-j},$$

and if at least one of the coefficients $a_{\alpha j}$ is nonzero, then $q = |\alpha|/j$ is a rational number. Applying the lemma we prove conditions (a), (b), and (c) for $n = 2$.

2) In the case $n > 1$ we associate with $P_0(\xi_1,\dots,\xi_n,\tau)$ the polynomial $Q(\rho,\tau) = P_0(\rho\xi_1,\dots,\rho\xi_n,\tau)$ depending on the parameters $\xi_1,\dots,\xi_n$. Applying to this polynomial the proposition, which has already been proved for $n = 1$, and setting $\rho = 1$ we establish the proposition for $n \geqslant 1$.

1.3. The real form of Petrovskii's correctness condition. We now present an equivalent description of polynomials correct in Petrovskii's sense. It plays a key role in the method of energy estimates for differential operators, which is developed in the next chapter.

Let $P(\xi,\tau)$ be a polynomial of degree m with respect to τ. We introduce the following notation:

$$\begin{aligned} &P^{(l)}(\xi,\tau) = \partial^l P(\xi,\tau)/\partial\tau^l, \quad l = 1,\dots,m \quad (P^{(0)} \equiv P), \\ &H_P(\eta,\gamma) = -\operatorname{Im}\big[P(\xi,\tau+i\gamma)\overline{P^{(1)}(\xi,\sigma+i\gamma)}\big]. \end{aligned} \tag{11}$$

Polynomial (11) can be effectively described in terms of the roots of polynomial (1). Let the polynomial P be resolved with respect to the highest power of τ, i. e., have the form (2). We write down factorization (4) for it. Differentiating this factorization with respect to τ we obtain

$$P^{(1)}(\xi,\tau) = \sum_{k=1}^m \prod_{j\neq k} (\tau - \lambda_j(\xi)). \tag{12}$$

The substitution of (4) and (12) into (11) results in

$$H_P(\eta,\gamma) = \sum_{k=1}^m (-\gamma + \operatorname{Im}\lambda_k(\xi)) \prod_{j\neq k} |\sigma + i\gamma - \lambda_j(\xi)|^2. \tag{13}$$

PROPOSITION. *For a polynomial $P(\xi,\tau)$ of degree m with respect to τ, which is resolved with respect to the highest power of τ, the following conditions are equivalent.*

(i) *The polynomial P is correct in Petrovskii's sense, i. e., condition* (ii) *of Definition* 1.1 *holds.*

(ii) *There are γ_0 and $c > 0$ such that*

$$c\sum_{l=1}^m (\gamma_0-\gamma)^{2l-1}\big|P^{(l)}(\xi,\sigma+i\gamma)\big|^2 \leqslant H_P(\xi,\sigma,\gamma), \tag{14}$$
$$\gamma \leqslant \gamma_0, \quad \eta = (\xi,\sigma) \in \mathbb{R}^{n+1},$$

where the constant c on the right-hand side of (14) *depends on m and does not depend on the coefficients of the polynomial P.*

(iii) *There is γ_0 such that*

$$H_P(\eta,\gamma) > 0 \quad \textit{for} \quad \gamma < \gamma_0, \quad \eta \in \mathbb{R}^{n+1}.$$

PROOF. (i)⇒(ii). Let $\lambda_j(\xi)$, $j = 1, \dots, m$, be the roots of the polynomial P. If P is correct in Petrovskii's sense, then condition $(3')$ holds for some real γ_0. Therefore, by (12) and (13), we have

$$(\gamma_0 - \gamma)\left|P^{(1)}(\xi, \sigma, \gamma)\right|^2 \leqslant m(\gamma_0 - \gamma)\sum_{k=1}^{m}\prod_{j\neq k} |\sigma + i\gamma - \lambda_j(\xi)|^2 \leqslant mH_P(\xi, \sigma, \gamma).$$

Using the relations

$$P^{(l)}(\xi, \tau) = \sum_{1\leqslant i_1 \leqslant \cdots \leqslant i_{m-l} \leqslant m} (\tau - \lambda_{i_1}(\xi)) \cdots (\tau - \lambda_{i_{m-l}}(\xi))$$

for $l = 2, \dots, m$ we obtain inequality (14).

(ii)⇒(iii) is obvious because $P^{(m)}(\xi, \tau) \equiv m!$.

(iii)⇒(i) is obvious because we have $H_P(\eta, \gamma) = 0$ when $P(\xi, \tau) = 0$.

1.4. Puiseux's expansion for the roots of polynomials correct in Petrovskii's sense. According to the classical theory of algebraic functions, in the case $n = 1$ every root $\lambda_j(\xi)$ of polynomial (2) for an integer p is an analytic function of $\xi^{1/p}$ in the neighborhood of $|\xi| = \infty$, and this function has no essential singularities at infinity, i. e.,

$$\lambda_j(\xi) = \sum_{-\infty}^{k_j} \lambda_{kj}(\xi^{1/p})^k_j, \tag{15}$$

where the number k_j can be positive or negative or zero. The different values of the root $\xi^{1/p}$ determine p different values of (15), i. e., p different roots of the polynomial which have the form (15) at infinity (see **[H1,** Appendix A]). Expansions (15) are called Puiseux's series for the roots of a polynomial in two variables.

Petrovskii's condition makes it possible to extract additional information about the character of Puiseux's expansions for the roots of a polynomial. We have the following

PROPOSITION (**[Bo]**). *Polynomial* (2) *is correct in Petrovskii's sense if and only if Puiseux's series for the roots* $\lambda_j(\xi)$, $j = 1, \dots, m$, *in the neighborhood of the point* $|\xi| = \infty$ *have one of the forms below*:

(a) $\lambda_j(\xi) = c_{j1}\xi^{\kappa_{j1}} + \cdots + c_{jl}\xi^{\kappa_{jl}} + o\left(|\xi|^{\kappa_{jl}}\right)$, $|\xi| \to \infty$, *where* $\kappa_{j1} > \cdots > \kappa_{jl}$ *are integers,* κ_{jl} *is an even number, and* $c_{j1}, \dots, c_{j,l-1}$ *are real numbers,* $\operatorname{Im} c_{jl} > 0$;

(b) $\lambda_j(\xi) = c_{j1}\xi^{\kappa_{j1}} + \cdots + c_{jl}\xi^{\kappa_{jl}} + O(1)$, $|\xi| \to \infty$, *where* $\kappa_{j1} > \cdots > \kappa_{jl}$ *are integers and the coefficients* $c_{j1}, \dots, c_{jl}$ *are real*;

(c) $\lambda_j(\xi) = O(1)$, $|\xi| \to \infty$.

PROOF. Because condition $(3')$ obviously holds for all the enumerated types of the roots, it remains to prove the *necessity* of the above expansions. As was noted earlier, in the general case Puiseux's series for the roots $\lambda_j(\xi)$ of polynomial (2) have the form (15). Let $\kappa_{j1} = k_j/p$ be the greatest exponent in this expansion. If $\kappa_{j1} \leqslant 0$, then $\lambda_j(\xi)$ is a root of the type (c), and the proposition is proved. In the case $\kappa_{j1} > 0$ we have

$$\lambda_j(\xi) = c_{j1}\xi^{\kappa_{j1}} + o\left(\xi^{\kappa_{j1}}\right) \quad \text{for} \quad |\xi| \to \infty.$$

If condition $(3')$ holds for some γ_0, then

$$\operatorname{Im}\left(c_j\xi^{\kappa_{j1}}\right) + o\left(\xi^{\kappa_{j1}}\right) \geqslant \gamma_0, \qquad -\infty < \xi < \infty, \quad \kappa_{j1} = k + j/p.$$

If we replace ξ by $\rho\xi$, $\rho > 0$, in this relation and divide both sides by $\rho^{\kappa_{j1}}$, then for $\rho \to +\infty$ we obtain

$$\operatorname{Im}(c_{j1}\xi^{\kappa_{j1}}) \geqslant 0, \qquad -\infty < \xi < \infty, \quad \kappa_{j1} = k_j/p.$$

Applying Lemma 1.2 we conclude that κ_{j1} is an integer and that one of the following two conditions holds:

$\operatorname{Im} c_{j1} = 0$,

$\operatorname{Im} c_{j1} > 0$, where κ_{j1} is an even number.

If the second of these conditions holds, then the root $\lambda_j(\xi)$ belongs to the type (a), i. e., the proposition is proved. Let c_{j1} be a real number and let $c_{j2}\xi^{\kappa_{j2}}$ be the next expansion term. If $\kappa_{j2} > 0$, then we have

$$\operatorname{Im} \lambda_j(\xi) = \operatorname{Im}(c_{j2}\xi^{\kappa_{j2}}) + o(\xi^{\kappa_{j2}}),$$

and the above argument can be used to prove that κ_{j2} is an integer and that either c_{j2} is real or we have $\operatorname{Im} c_{j2} > 0$ and κ_{j2} is an even number. Continuing this process we complete the proof of the proposition.

§2. Stable correct polynomials

In this section we consider a class of polynomials correct in Petrovskii's sense whose correctness condition is preserved under arbitrary variations of the coefficients in lower monomials. Complete description of this class of polynomials will be given.

2.1. q-stable correct polynomials. For a given $q \geqslant 1$ we define the q-degree of a monomial $\xi^\alpha\tau^j$ by setting

$$\deg_q(\xi^\alpha\tau^j) = |\alpha| + jq.$$

If $Q(\xi, \tau)$ is an arbitrary polynomial in the variables $\xi \in \mathbb{R}^n$ and $\tau \in \mathbb{C}$, i. e.,

$$Q(\xi, \tau) = \sum q_{\alpha j}\xi^\alpha\tau^j,$$

then we denote by $\deg_q(Q)$ the maximum value of $\deg_q(\xi^\alpha\tau^j)$ for all monomials $\xi^\alpha\tau^j$ entering the polynomial with nonzero coefficients $q_{\alpha j}$. The monomials for which

$$\deg_q(\xi^\alpha\tau^j) = \deg_q Q$$

are called q-leading monomials (q-leading terms) of the polynomial Q; accordingly, the other monomials (terms) are q-lower monomials (terms). The sum of q-leading terms of a polynomial Q will be called the principal q-homogeneous part and will be denoted $Q_q(\xi, \tau)$:

$$Q_q(\xi, \tau) = \sum_{|\alpha|+qj} q_{\alpha j}\xi^\alpha\tau^j.$$

It is clear that $Q_q(\xi, \tau)$ is a q-homogeneous polynomial.

REMARK. In the case $q = 1$ we will speak of leading and lower monomials; for $q = 1$ the principal q-homogeneous part of a polynomial will be called the principal (homogeneous) part.

DEFINITION. Let $q \geqslant 1$. A polynomial $P(\xi, \tau)$ is said to be q-stable correct if for any polynomial $Q(\xi, \tau)$, $\deg_q Q < \deg_q P$, there is $\gamma_0(Q)$ such that

$$(P+Q)(\xi, \tau) \neq 0, \qquad \operatorname{Im} \tau \leqslant \gamma_0(Q), \quad (\xi, \operatorname{Re} \tau) \in \mathbb{R}^{n+1}. \tag{1}$$

We state some immediate consequences of this definition.

(a) A polynomial P is q-stable correct if and only if its principal q-homogeneous part P_q possesses this property, i. e., there is γ_0 such that

$$P_q(\xi, \tau) \neq 0, \qquad \operatorname{Im} \tau \leqslant \gamma_0, \quad (\xi, \operatorname{Re} \tau) \in \mathbb{R}^{n+1}. \tag{1$'$}$$

(b) According to the remark in 1.2, condition (1$'$) implies that $\gamma_0 = 0$ and that the polynomial $P_q(\xi, \tau)$ is resolved with respect to the highest power of τ, i. e., has the form

$$P(\xi, \tau) = \tau^m + \sum_{\alpha, j < m} a_{ij} \xi^\alpha \tau^j. \tag{2}$$

(c) In view of (b), the definition can be restated in the following way: a polynomial P is q-stable correct if all polynomials $P+Q$, where $\deg_q Q < \deg_q P$, are correct in Petrovskii's sense.

Below we will formulate definitions of strictly hyperbolic and q-parabolic polynomials and present their description. The description will imply that strictly hyperbolic and q-parabolic polynomials are q-stable correct. The main result of this section is that the class of q-stable correct polynomials is exhausted by the above classes of polynomials.

2.2. Strictly hyperbolic polynomials.

DEFINITION 1. A homogeneous polynomial $P_0(\xi, \tau)$ of degree m is said to be strictly hyperbolic if

(i) the polynomial $P_0(\xi, \tau)$ is resolved with respect to the highest power of τ, i. e., has the form (2);

(ii) the roots $\lambda_{0j}(\xi)$, $j = 1, \ldots, m$, are real for each $\xi \in \mathbb{R}^n$;

(iii) for $\xi \neq 0$ the roots $\lambda_{0j}(\xi)$ are distinct, i. e., there is $\delta > 0$ such that

$$|\lambda_{0j}(\xi) - \lambda_{0k}(\xi)| > \delta |\xi|, \qquad j \neq k, \quad j, k = 1, \ldots, m. \tag{3}$$

We note that (ii) and the normalization condition for the coefficient in τ^m (it is equal to 1) imply that the coefficients of the polynomial P_0 are real.

PROPOSITION. *For a homogeneous polynomial of degree m the following conditions are equivalent.*

(I) *P_0 is a strictly hyperbolic polynomial* (*in the sense of Definition* 1) *of degree m.*

(II) *There is a constant $C_0 > 0$ such that*

$$\begin{aligned} C_0^{-1} |\operatorname{Im} \tau| (|\tau| + |\xi|)^{2m-2} &\leqslant H_{P_0}(\xi, \operatorname{Re} \tau, \operatorname{Im} \tau) \\ &\leqslant C_0 |\operatorname{Im} \tau| (|\tau| + |\xi|)^{2m-2}, \\ \operatorname{Im} \tau \leqslant 0, \quad &(\xi, \operatorname{Re} \tau) \in \mathbb{R}^{n+1}. \end{aligned} \tag{4}$$

(III) *There is a constant $C_0' > 0$ such that*

$$C_0' |\operatorname{Im} \tau| (|\tau| + |\xi|)^{m-1} \leqslant |P(\xi, \tau)|, \qquad \operatorname{Im} \tau \leqslant 0, \quad (\xi, \operatorname{Re} \tau) \in \mathbb{R}^{n+1}. \tag{5}$$

PROOF. (I)⇒(II). According to (1.13) and the condition that the roots are real, we have

$$H_{P_0}(\xi, \operatorname{Re}\tau, \operatorname{Im}\tau) = -\operatorname{Im}\tau \sum_{k=1}^{m} \prod_{j\neq k} \Big[\big(\operatorname{Re}\tau - \lambda_{0j}(\xi)\big)^2 + (\operatorname{Im}\tau)^2 \Big]$$

$$\overset{\text{def}}{=} -\operatorname{Im}\tau\, H(\xi, \operatorname{Re}\tau, \operatorname{Im}\tau).$$

It is clear that $H(\xi, \sigma, \gamma)$ is a nonnegative homogeneous function of degree $2m-2$. To prove the desired assertion it suffices to show that

$$(6) \qquad H(\xi, \sigma, \gamma) \neq 0, \qquad |\xi|^2 + \sigma^2 + \gamma^2 = 1, \quad \gamma \leqslant 0.$$

Indeed, since the polynomial P_0 is resolved with respect to the highest power of τ, its roots λ_{0j} are continuous functions of ξ, and, consequently, the function $H(\xi, \sigma, \gamma)$ is continuous in all its arguments. Therefore it follows from (6) that the infimum on the hemisphere in this condition is greater than a positive constant. Hence, $C_0 > 0$ can be chosen so that

$$C_0^{-1} \leqslant H(\xi, \sigma, \gamma) \leqslant C_0, \qquad |\xi|^2 + \sigma^2 + \gamma^2 = 1, \quad \gamma \leqslant 0.$$

By homogeneity, it follows that

$$C_0^{-1}(|\xi| + |\sigma| + |\gamma|)^{2m-2} \leqslant H(\xi, \sigma, \gamma) \leqslant C_0(|\xi| + |\sigma| + |\gamma|)^{2m-2},$$

which proves the desired assertion.

As to condition (6), it is obvious for $\gamma < 0$. Now let $\gamma = 0$. If $|\xi| = 0$, then we have $|\sigma| = 1$ and $H(0, \sigma, 0) = m$. Furthermore, if $|\xi| \neq 0$, then, according to condition (iii) of Definition 1, at most one of the numbers $\sigma - \lambda_{0j}(\xi)$, $j = 1, \dots, m$, can vanish, which proves (6).

(II)⇒(III). Since $\partial P_0/\partial\tau$ is a homogeneous polynomial of degree $m-1$, we have

$$H_{P_0}(\xi, \operatorname{Re}\tau, \operatorname{Im}\tau) \leqslant |P_0(\xi, \tau)| \cdot |P_0^{(1)}(\xi, \tau)| \leqslant \operatorname{const}(|\xi| + |\tau|)^{m-1} |P_0(\xi, \tau)|.$$

Substituting this estimate into the left-hand inequality (4) we obtain (5).

(III)⇒(I). By (5), we have

$$(5') \qquad |P_0(\xi, \tau)| \neq 0, \qquad \operatorname{Im}\tau < 0, \quad (\xi, \operatorname{Re}\tau) \in \mathbb{R}^{n+1}.$$

The argument in §1.2 implies that the polynomial P_0 is resolved with respect to the highest power of τ.

By homogeneity, we have $P_0(\xi, -\tau) = (-1)^m P_0(-\xi, \tau)$, and therefore, by virtue of (5′),

$$(5'') \qquad |P_0(\xi, \tau)| \neq 0, \qquad \operatorname{Im}\tau > 0, \quad (\xi, \operatorname{Re}\tau) \in \mathbb{R}^{n+1}.$$

According to (5′) and (5″), the polynomial P_0 has only real roots, and hence it remains to prove condition (iii) in Definition 1. Assume the contrary, i. e., let for some $\xi^0 \neq 0$ the polynomial P_0 have two coincident roots, say $\lambda_{01}(\xi^0) = \lambda_{02}(\xi^0)$. We consider inequality (5) along the ray

$$\xi(t) = t\xi^0, \qquad \tau(t) = \lambda_{01}(\xi(t)) - i\gamma, \quad \gamma > 0.$$

On this ray the left-hand side of (5) has a lower bound

$$\operatorname{const} \gamma(\gamma + t)^{m-1}.$$

Noting that $\lambda_{0j}(\xi(t)) = t\lambda_{0j}(\xi^0)$, we obtain an upper bound for the right-hand side of (5):

$$|P_0(\xi(t), \tau(t))| = \left|\prod_{j=1}^{m}(\xi(t) - \lambda_{0j}(\xi(t)) - i\gamma)\right|$$
$$\leqslant \gamma^2 \prod_{j=3}^{m} |t(\lambda_{01}(\xi^0) - \lambda_{0j}(\xi^0)) - i\gamma| \leqslant \text{const}\, \gamma^2 (t+\gamma)^{m-2}.$$

Finally, passing to the limit for $t \to +\infty$ we arrive at a contradiction.

DEFINITION 2. A polynomial $P(\xi, \tau)$ is said to be strictly hyperbolic (hyperbolic in Petrovskii's sense) if its principal homogeneous part possesses this property.

THEOREM. *For a polynomial of degree m the following conditions are equivalent.*
(I) *The polynomial P is strictly hyperbolic.*
(II) *There are $C > 0$ and γ_0 such that*

$$\begin{aligned} C^{-1}|\operatorname{Im}\tau|(|\tau| + |\xi|)^{2m-2} &\leqslant H_P(\xi, \operatorname{Re}\tau, \operatorname{Im}\tau) \\ &\leqslant C|\operatorname{Im}\tau|(|\tau| + |\xi|)^{m-1}, \\ \operatorname{Im}\tau \leqslant \gamma_0, \quad (\xi, \operatorname{Re}\tau) &\in \mathbb{R}^{n+1}. \end{aligned} \tag{7}$$

(III) *There are $C' > 0$ and γ_0' such that*

$$C'|\operatorname{Im}\tau|(|\tau| + |\xi|)^{m-1} \leqslant |P(\xi, \tau)|, \qquad \operatorname{Im}\tau \leqslant \gamma_0', \quad (\xi, \operatorname{Re}\tau) \in \mathbb{R}^{n+1}. \tag{8}$$

PROOF. (I)$\Rightarrow$(II). Let $P = P_0 + P'$, where P_0 is a homogeneous polynomial of degree m, and let the degree of the polynomial P' be less than m. By Definition 2, the polynomial P_0 is strictly hyperbolic, and the equivalent assertions of the above proposition hold for it. According to the definition of the polynomial H_P, we have

$$H_P = H_{P_0} + H', \quad H' = -\operatorname{Im}(P'\overline{P}^{(1)}),$$

where H' is a polynomial of degree at most $2m-2$ in the variables ξ, $\operatorname{Re}\tau$, and $\operatorname{Im}\tau$. If $\operatorname{Im}\tau < \gamma_0$ and, for definiteness, $\gamma_0 < -1$, then

$$\begin{aligned} |H'(\xi, \operatorname{Re}\tau, \operatorname{Im}\tau)| &\leqslant K\sum_{l=0}^{2m-2}(|\tau| + |\xi|)^l \leqslant K\sum_{l=0}^{2m-2}|\gamma_0|^{l-2m-2}|\operatorname{Im}\tau|^{2m-2-l}(|\tau| + |\xi|)^l \\ &\leqslant K(1 - |\gamma_0|^{-1})(|\tau| + |\xi|)^{2m-2} \leqslant (2C_0)^{-1}|\operatorname{Im}\tau|(|\tau| + |\xi|)^{2m-2}, \end{aligned}$$

where the right-hand inequality holds for sufficiently large $-\gamma_0$. In view of this inequality and (4), we obtain inequality (7) with the constant $C = 2C_0$.

(II)$\Rightarrow$(III). Because the degree of the polynomial $P^{(1)}$ does not exceed $m-1$, the repetition of the above estimation for the polynomial H' results in

$$|P^{(1)}(\xi, \tau)| \leqslant \text{const}(1 - |\gamma_0|^{-1})(|\tau| + |\xi|)^{m-1}.$$

Substituting this inequality into the left-hand inequality (7) we derive (8).

Replacing P, ξ, and τ in (8) with $P_0 + P'$, $t\xi$, and $t\tau$, respectively, dividing both sides of this inequality by t^m, and passing to the limit as $t \to +\infty$, we get (5). Applying the proposition we prove the strict hyperbolicity of P_0 and, consequently, of P as well.

COROLLARY. *Every strictly hyperbolic polynomial is q-stable correct for $q = 1$.*

PROOF. By definition, a strictly hyperbolic polynomial is resolved with respect to the highest power of τ, and, according to (8), the polynomial is correct in Petrovskii's sense. Since the condition of strict hyperbolicity relates only to the principal homogeneous part, the addition of arbitrary lower terms preserves the strict hyperbolicity of polynomials and, consequently, the correctness in Petrovskii's sense.

2.3. q-parabolic polynomials.

DEFINITION 1. Let $q \geqslant 2$ be an even number. A q-homogeneous polynomial $P(\xi, \tau)$ is said to be q-parabolic if

(i) P_0 is resolved with respect to the highest power of τ;

(ii) the roots $\lambda_{0j}(\xi)$, $j = 1, \dots, m$, of the polynomial P_0 have strictly positive imaginary parts for $|\xi| \neq 0$, i. e., there is $\kappa > 0$ such that

$$\operatorname{Im} \lambda_{0j}(\xi) \geqslant \kappa |\xi|^q, \qquad j = 1, \dots, m, \quad \xi \in \mathbb{R}^n.$$

PROPOSITION. *For a given $q \geqslant 1$ and a q-homogeneous polynomial $P_0(\xi, \tau)$ the following conditions are equivalent.*

(I) *q is even and P_0 is a q-parabolic polynomial of degree m in the variable τ.*

(II) *There is $C_0 > 0$ such that*

$$\begin{aligned} C_0^{-1}(|\operatorname{Im}\tau| + |\xi|^q)(|\tau| + |\xi|^q)^{2m-2} &\leqslant H_{P_0}(\xi, \operatorname{Re}\tau, \operatorname{Im}\tau) \\ &\leqslant C_0(|\operatorname{Im}\tau| + |\xi|^q)(|\tau| + |\xi|^q)^{m-1}, \\ \operatorname{Im}\tau \leqslant 0, \quad (\xi, \operatorname{Re}\tau) &\in \mathbb{R}^{n+1}. \end{aligned} \tag{9}$$

(III) *There is $C_0' > 0$ such that*

$$\begin{aligned} C_0^{-1}(|\operatorname{Im}\tau| + |\xi|^q)(|\tau| + |\xi|^q)^{2m-2} &\leqslant |P_0(\xi, \tau)|, \\ \operatorname{Im}\tau \leqslant 0, \quad (\xi, \operatorname{Re}\tau) &\in \mathbb{R}^{n+1}. \end{aligned} \tag{10}$$

PROOF. According to condition (ii) of Definition 1, we have

$$-\operatorname{Im}\tau + \lambda_{0j}(\xi) \geqslant |\operatorname{Im}\tau| + \kappa|\xi|^q, \quad \operatorname{Im}\tau \leqslant 0. \tag{11}$$

Therefore, by (1.13), the function H_{P_0} has a lower bound:

$$H_{P_0}(\xi, \operatorname{Re}\tau, \operatorname{Im}\tau) \geqslant (|\operatorname{Im}\tau| + \kappa|\xi|^q) H(\xi, \operatorname{Re}\tau, \operatorname{Im}\tau), \tag{9'}$$

where

$$H(\xi, \sigma, \gamma) = \sum_{k=1}^{m} \prod_{j \neq k} [(\sigma - \operatorname{Re}\lambda_{0j}(\xi))^2 + (|\gamma + \kappa|\xi|^q)^2]. \tag{12}$$

Because this function is quasi-homogeneous, i. e.,

$$H(\rho\xi, \rho^q\sigma, \rho^q\gamma) = \rho^{(2m-2)q} H(\xi, \sigma, \gamma),$$

to prove (9) is suffices to show that

$$H(\xi, \sigma, \gamma) \neq 0, \qquad |\xi|^{2q} + \sigma^2 + \gamma^2 = 1, \quad \gamma \leqslant 0. \tag{13}$$

It immediately follows from (12) that (13) holds for $|\xi| + |\gamma| > 0$. In case $|\xi| = |\gamma| = 0$, we have $|\tau| = 1$ and $H = m$.

(II)⇒(III) is implied by the fact that $|P_0^{(1)}(\xi,\tau)| < \mathrm{const}(|\tau + |\xi|^q)^{m-1}$.

(III)⇒(I). By virtue of inequality (10), the q-homogeneous polynomial P_0 is correct in Petrovskii's sense, i. e., one of the conditions in Proposition 1.2 holds for it. Moreover, according to (10), the polynomial P_0 is nonzero for $\operatorname{Im}\tau = 0$ and $|\xi| > 0$, i. e., for $|\xi| > 0$ all its roots have positive imaginary parts. According to Proposition 1.2, the number q is even so that the q-parabolicity of the polynomial P_0 is proved.

REMARKS. 1) From (11) it readily follows that

$$K_0^{-1}(|\tau| + |\xi|^q) \leqslant |\tau - \lambda_{0j}(\xi)| \leqslant K_0(|\tau| + |\xi|^q), \quad \operatorname{Im}\tau \leqslant 0.$$

Multiplying these inequalities we find that

$$K_1(|\tau| + |\xi|^q)^m \leqslant |P_0(\xi,\tau)|, \qquad \operatorname{Im}\tau \leqslant 0, \quad (\xi, \operatorname{Re}\tau) \in \mathbb{R}^{n+1}. \tag{14}$$

This inequality trivially implies (10) so that (14) can be included in the list of equivalent assertions of the proposition.

2) By the elementary inequality

$$|\operatorname{Im}\tau|^{1/q}|\xi|^{q-1} \leqslant \frac{1}{q}|\operatorname{Im}\tau| + \frac{q-1}{q}|\xi|^q,$$

relation (10) implies a weaker condition

$$\begin{gathered} C_0''(|\operatorname{Im}\tau| + |\operatorname{Im}\tau|^{1/q}|\xi|^{q-1})(|\tau| + |\xi|^q)^{m-1} \leqslant |P_0(\xi,\tau)|, \\ \operatorname{Im}\tau \leqslant 0, \quad (\xi, \operatorname{Re}\tau) \in \mathbb{R}^{n+1}. \end{gathered} \tag{15}$$

Nevertheless, inequality (15) can be included in the list of equivalent assertions of the proposition. The necessity of the condition of q-parabolicity for the validity of this inequality will be proved in the next subsection.

DEFINITION 2. Let $q \geqslant 2$ be an even number. A polynomial $P(\xi,\tau)$ is said to be q-parabolic if its principal q-homogeneous part possesses this property.

THEOREM 2. *For a given $q \geqslant 1$ and a polynomial $P(\xi,\tau)$ the following conditions are equivalent*:

(I) *q is even and the polynomial P is q-parabolic.*

(II) *There are $C > 0$ and γ_0 such that*

$$\begin{aligned} C^{-1}(|\operatorname{Im}\tau| + |\xi|^q)(|\tau| + |\xi|^q)^{2m-2} &\leqslant H_P(\xi, \operatorname{Re}\tau, \operatorname{Im}\tau) \\ &\leqslant C(|\operatorname{Im}\tau| + |\xi|^q)(|\tau| + |\xi|^q)^{2m-2}, \\ \operatorname{Im}\tau \leqslant \gamma_0, &\quad (\xi, \operatorname{Re}\tau) \in \mathbb{R}^{n+1}. \end{aligned} \tag{16}$$

(III) *There are $C' > 0$ and γ_0' such that*

$$\begin{gathered} C'(|\operatorname{Im}\tau| + |\xi|^q)(|\tau| + |\xi|^q)^{m-1} \leqslant |P_0(\xi,\tau)|, \\ \operatorname{Im}\tau \leqslant \gamma_0', \quad (\xi, \operatorname{Re}\tau) \in \mathbb{R}^{n+1}. \end{gathered} \tag{17}$$

PROOF. As in Theorem 2.2, we set $P = P_0 + P'$, where P_0 is the principal q-homogeneous part of P and the q-degree of the polynomial P' is less than that of P_0, which is equal to mq, i. e., $\deg_q P' \leqslant mq - 1$. Therefore the polynomial

$$H'(\xi, \operatorname{Re}\tau, \operatorname{Im}\tau) = -\operatorname{Im}\big[P'(\xi,\tau)\overline{P^{(1)}(\xi,\tau)}\big]$$

can be represented as a linear combination of the monomials

$$\xi^\alpha (\operatorname{Re}\tau)^j (\operatorname{Im}\tau)^k, \quad |\alpha| + q(j+k) \leqslant 2mq - q - 1. \tag{18}$$

It follows that the degree of $\operatorname{Re}\tau$ in these monomials cannot exceed $2m-2$. Consequently, monomials (18) and hence the polynomial H' have an upper bound

$$|\operatorname{Im}\tau|^{-\kappa}(|\operatorname{Im}\tau| + |\xi|^q)(|\tau| + |\xi|^q)^{2m-2},$$

where $\kappa = 2mq - q - |\alpha| - (j+k)q \geqslant 2mq - q - (2mq - q - 1) \geqslant 1$.

These inequalities directly imply that

$$|H'(\xi, \operatorname{Re}\tau, \operatorname{Im}\tau)| < (1/2C_0)(|\operatorname{Im}\tau| + |\xi|^q)(|\operatorname{Im}\tau| + |\xi|^q)^{2m-2},$$

provided that $\operatorname{Im}\tau < -\gamma_0$ and $-\gamma_0$ is sufficiently large. Inequality (16) now follows from (9).

(II)⇒(III) is implied by $|P^{(1)}| < \operatorname{const}(|\tau| + |\xi|^q)^{m-1}$.

(III)⇒(I). Setting $P = P_0 + P'$, replacing ξ and τ by $\rho\xi$ and $\rho^q\tau$, respectively, dividing both sides of the inequality by ρ^{-mq}, and passing to the limit for $\rho \to \infty$, we derive (10), after which it remains to apply the proposition.

Repeating the argument in Corollary 2.2 we obtain

COROLLARY. *Every q-parabolic polynomial is q-stable correct.*

2.4. Description of q-stable correct polynomials. The foregoing results in the present section can be unified as the following

THEOREM. *For a given $q \geqslant 1$ and a polynomial $P(\xi,\tau)$ of degree m with respect to τ the conditions below are equivalent.*

(I) *P is a q-stable correct polynomial.*

(II) *There is γ_0 such that*

$$\begin{gathered} C(|\operatorname{Im}\tau| + |\operatorname{Im}\tau|^{1/q}|\xi|^{q-1})(|\tau| + |\xi|^q)^{m-1} \leqslant |P(\xi,\tau)|, \\ \operatorname{Im}\tau \leqslant \gamma_0, \quad (\xi, \operatorname{Re}\tau) \in \mathbb{R}^{n+1}, \end{gathered} \tag{19}$$

for some $C > 0$.

(III) *The number q is either equal to 1 or is even; for $q = 1$ the polynomial P is strictly hyperbolic, and for even q it is q-parabolic.*

PROOF. (III)⇒(II). For $q = 1$ inequality (19) becomes inequality (8), which was proved in Theorem 2.2. For even q inequality (19) is a consequence of (17) (cf. Remark 2 in 2.3).

(II)⇒(I). 1) Inequality (19) implies that the polynomial P satisfies condition (ii) of Definition 1.1.

2) As was already done many times, by means of dilatations $(\xi,\tau) \longmapsto (\rho\xi, \rho^q\tau)$ we derive from (19) the following inequality for the principal q-homogeneous part P_q of the polynomial P:

$$C(|\operatorname{Im}\tau| + |\operatorname{Im}\tau|^{1/q}|\xi|^{q-1})(|\tau| + |\xi|^q)^{m-1} \leqslant |P_q(\xi,\tau)|, \quad \operatorname{Im}\tau \leqslant 0. \tag{20}$$

It follows from (20) that the polynomial P_q is correct in Petrovskii's sense. Therefore, by virtue of 1.2, the polynomial P_q (together with P) is resolved with respect to the highest power of τ, and q is an integer.

3) We have thus proved that, in view of (19), the polynomial P is correct in Petrovskii's sense. It now remains to show that the correctness condition is preserved under arbitrary perturbations of lower terms. By virtue of (19), to this end it suffices to prove that the q-lower monomials of P have a lower bound in the form of the left-hand side of (19) with a constant that can be made arbitrarily small by choosing a sufficiently large $-\gamma_0$.

4) By what has been said, the desired assertion is implied by the following elementary

LEMMA. *Let q and m be positive integers, let α be a multiindex, and let*

$$|\alpha| + jq < mq.$$

Then

$$|\xi^\alpha \tau^j| \leqslant |\operatorname{Im}\tau|^{-\kappa}(|\operatorname{Im}\tau| + |\operatorname{Im}\tau|^{1/q}|\xi|^{q-1})(|\tau| + |\xi|^q)^{m-1}, \quad \kappa > 0. \tag{21}$$

PROOF. For the given integers we have

$$|\alpha| \leqslant mq - 1, \quad j \leqslant m - 1.$$

Set $|\alpha| = (m-1-j)q + k$. Then $k \leqslant q - 1$. Assume (for definiteness) that $k \geqslant 0$. It follows that

$$|\xi^\alpha \tau^j| \leqslant |\xi|^k [|\xi|^{(m-1-j)q}|\tau|^j].$$

The second factor on the right-hand side does not exceed

$$(|\tau| + |\xi|^q)^{m-1}.$$

As to the first factor, it does not exceed

$$\begin{aligned}(|\operatorname{Im}\tau|^{(q-1)/q} + |\xi|^{q-1})^{k/(q-1)} &= (|\operatorname{Im}\tau|^{(q-1)/q} + |\xi|^{q-1})^{(q-1-k)/(q-1)} \\ &\quad \times (|\operatorname{Im}\tau|^{q-1/q} + |\xi|^{q-1}) \\ &\leqslant |\operatorname{Im}\tau|^{-(q-k)/q}(|\operatorname{Im}\tau| + |\operatorname{Im}\tau|^{1/q}|\xi|^{q-1}).\end{aligned}$$

Taking the product of these inequalities we obtain (20) with the constant $\kappa = m - j - |\alpha|/q$.

REMARK. The implication (III)$\Rightarrow$(II) also follows from Corollaries 2.2 and 2.3.

(I)$\Rightarrow$(III). As was already mentioned, the q-stable correctness of P implies the correctness of the q-principal part P_q, whence it follows that q is an integer and the polynomial P is resolved with respect to the highest power of τ. Moreover, the roots $\lambda_{0j}(\xi)$ are real for odd q and have nonnegative imaginary parts for even q.

We first prove the desired assertion for the case $n = 1$ when there is only one variable ξ. It will be convenient to change the notation and replace the variable ξ by ρ. We can assume, without loss of generality, that $P = P_q$, i. e., the polynomial P is q-homogeneous. The roots of the polynomial $P(\rho, \tau)$ have the form

$$\lambda_{0j}(\rho) = c_j \rho^q, \quad j = 1, \dots, m.$$

We will prove that if $q > 1$, then

$$\operatorname{Im} c_j > 0, \quad j = 1, \dots, m. \tag{22}$$

This will imply that q cannot assume odd values exceeding 1 (in this case the roots must be real), and for even q the polynomial q is parabolic.

Condition (22) is proved by contradiction. Assume, for instance, that $\operatorname{Im} c_1 = 0$. Consider the polynomial $\widetilde{P}(\rho, \tau)$ that is obtained from the polynomial P if in the factorization of the latter the root $\lambda_{01}(\rho)$ is replaced by $c_1\rho^q - i\varepsilon\rho^{q-1}$, $\varepsilon > 0$, while the other roots remain unchanged. Then the polynomial

$$\widetilde{P}(\rho, \tau) = P(\rho, \tau) + r(\rho, \tau), \quad \deg_q r < \deg_q P,$$

is not correct in Petrovskii's sense because one of its roots does not satisfy the conditions of Proposition 1.4. On the other hand, the polynomials P and $\widetilde{P}$ differ only in q-lower terms, and, by virtue of the q-stable correctness of P, the polynomial $\widetilde{P}$ must be correct in Petrovskii's sense. We arrive at a contradiction, which proves (22).

It now remains to show that for $q = 1$ the (real) roots are distinct. Assume the contrary, i. e., let $\lambda_{01}(\rho) = \lambda_{02}(\rho) = a\rho$. We consider the polynomial $\widetilde{P}(\rho, \tau)$ that is obtained from $P(\rho, \tau)$ by replacing the roots λ_{01} and λ_{02} by $a\rho + \sqrt{\rho}$ and $a\rho - \sqrt{\rho}$, respectively, while the other roots remain unchanged; here $\pm\sqrt{\rho}$ are the two values of the function $\sqrt{\rho}$. Noting that

$$(\tau - a\rho + i\sqrt{\rho})(\tau - a\rho - i\sqrt{\rho}) = (\tau - a\rho)^2 + \rho,$$

we conclude that the polynomial $\widetilde{P}(\rho, \tau)$ is not correct in Petrovskii's sense although the degree of r is less than that of P. Hence, in the case $n = 1$ the desired assertion is proved.

We now pass to the case $n > 1$. Assume that there is a point ξ^0, $\xi^0 \neq 0$, at which one of the following two conditions holds:

$$\operatorname{Im} \lambda_{01}(\xi^0) = 0, \quad q > 1, \tag{23}$$

$$\lambda_{01}(\xi^0) = \lambda_{02}(\xi^0), \quad q = 1. \tag{24}$$

In these two cases we consider the polynomial $P^0(\rho, \tau) = P(\rho\xi^0, \tau)$. As was already proved, there is a polynomial $r(\rho, \tau)$, consisting of q-lower monomials of P_0 such that the polynomial $P_0(\rho, \tau) + r(\rho, \tau)$ is not correct in Petrovskii's sense. However, in this case the polynomial

$$P(\xi, \tau) + r(\langle \xi, \xi^0 \rangle |\xi^0|^{-2}, \tau)$$

is not correct either, because the correctness condition does not hold along the ray $\xi = \rho\xi^0$. We arrive at a contradiction, which proves that conditions (23) and (24) cannot hold. Therefore $q > 1$ can assume only even values, and in this case the roots have strictly positive imaginary parts. In the case $q = 1$ condition (14) does not hold, and the roots are real and distinct. The proof of the theorem is completed.

Appendix to §2. Pluriparabolic polynomials

As was shown above, if the degrees of the monomials $\xi^\alpha \tau^j$ are defined using the weight $Q = (1, \dots, 1, q)$, then only parabolic and strictly hyperbolic polynomials remain correct in Petrovskii's sense under arbitraryperturbations of q-lower terms. If we change the weight Q by assigning the same degree as that of τ to some of the variables $\xi_1, \dots, \xi_n$ and consider the corresponding class of Q-stable correct polynomials, then we obtain the class of pluriparabolic polynomials (see **[Gi1]** and **[GV3]**). These polynomials are hyperbolic with respect to one group of variables and parabolic with respect to the other.

1. We somewhat change the notation. Represent the space $\mathbb{R}^{n+1}$ as the direct sum of the subspace $\mathbb{R}^k$ of the variables $\sigma = (\sigma_1, \dots, \sigma_k)$ and the subspace $\mathbb{R}^l$ of the variables $\zeta = (\zeta_1, \dots, \zeta_l)$, $l + k = n + 1$. We separate out the variable σ_1, and let $\sigma' = (\sigma_2, \dots, \sigma_l)$ and $\tau = \sigma_1 + i\sigma_1$.

Let $q > 1$ be an even number. In what follows we will assign the weight q to the variables σ and the weight 1 to the variables ζ. We set $Q = (q, \dots, q, 1, \dots, 1)$.

DEFINITION 1. A q-homogeneous polynomial is said to be strictly pluriparabolic[1] if

(a) the polynomial $P_0(\sigma, 0)$ is strictly hyperbolic;

(b) there is $\kappa > 0$ such that

$$\operatorname{Im} \lambda_{0j}(\sigma', \zeta) \geqslant \kappa |\zeta|^q, \quad j = 1, \dots, m,$$

where $\lambda_{0j}(\sigma', \zeta)$ are the roots of the polynomial P_0 with respect to σ_1.

DEFINITION 2. A polynomial $P(\sigma, \zeta)$ is said to be strictly pluriparabolic if for some $Q = (q, \dots, q, 1, \dots, 1)$ its q-principal part possesses this property.

The strong hyperbolicity of the polynomial $P(\sigma, 0)$ implies that, like $P(\xi, \zeta)$, it is resolved with respect to σ_1^m (the coefficient of σ_1^m is assumed to be equal to 1):

$$P(\sigma, \zeta) = \sigma_1^m + \sum_{j \geqslant 1} P_j(\sigma', \zeta)\sigma_1^{m-j}. \tag{1}$$

Replacing the vector q in Definition 2.2 by Q, we obtain Q-stable Petrovskii's condition. We can now state a generalization of Propositions 2.2 and 2.3. For simplicity, we will assume in advance that the polynomial P has the form (1).

PROPOSITION. *For a given positive and even q and a q-homogeneous polynomial P_0 of type* (1) *the following conditions are equivalent.*

(i) *$P_0(\sigma, \zeta)$ satisfies Q-stable Petrovskii's condition for $Q = (q, \dots, q, 1, \dots, 1)$.*

(ii) *For some $C > 0$ we have*

$$C(|\operatorname{Im} \tau| + |\zeta|^q)(|\tau| + |\sigma'| + |\zeta|^q)^{m-1} \leqslant |P_0(\tau, \sigma', \zeta)|, \quad \operatorname{Im} \tau \leqslant 0, \quad \eta = (\sigma, \zeta) \in \mathbb{R}^{n+1}. \tag{2}$$

(iii) *The polynomial P_0 is strictly pluriparabolic.*

PROOF. The implication (ii)⇒(i) is obvious, and (i)⇒(ii) follows from the general considerations presented in the next section (see Lemma 3.3). Thus, we confine ourselves to proving the equivalence of (ii) and (iii).

(ii)⇒(iii). Setting $\zeta = 0$ in (2) we obtain

$$C|\operatorname{Im} \tau|(|\tau| + |\sigma'|)^{m-1} \leqslant |P_0(\tau, \sigma', 0)|, \quad \operatorname{Im} \tau \leqslant 0.$$

[1] More precisely, it should be called a strictly q-pluriparabolic polynomial.

Therefore Proposition 2.3 implies that the polynomial $P_0(\sigma, 0)$ is strictly hyperbolic.

It now remains to prove condition (b) in Definition 1. By quasihomogeneity, it suffices to show that there is $\kappa > 0$ such that

$$P_0(\sigma_1 + i\gamma, \sigma', \zeta) \neq 0 \quad \text{for} \quad \gamma \leqslant \kappa,\ |\zeta| = 1,\ \sigma \in \mathbb{R}^k. \tag{3}$$

In view of (2), we should consider only the case $\gamma \geqslant 0$. Setting $\operatorname{Im} \tau = 0$ in (2) and assuming that $|\zeta| = 1$ we find

$$|P_0(\sigma, \zeta)| > C_1(1 + |\sigma|)^{m-1},$$

whence

$$\begin{aligned} |P_0(\sigma_1 + i\gamma, \sigma', \zeta)| &> |P_0(\sigma, \zeta)| - |P_0(\sigma_1 + i\gamma, \sigma', \zeta) - P_0(\sigma_1, \sigma', \zeta)| \\ &> C_1(1 + |\sigma|)^{m-1} - C_2\gamma\left[(1 + |\sigma|)^{m-1} + \gamma^{m-1}\right]. \end{aligned}$$

Choosing $\kappa < C_1/4C_2$ and $\gamma \leqslant \kappa \leqslant 1$ we obtain (3).

2. As in 2.3 and 2.4, to prove the implication (iii)$\Rightarrow$(ii) we consider the real polynomial

$$H_{P_0}(\sigma, \zeta, \gamma) = -\operatorname{Im}\left[P_0(\sigma_1 + i\gamma, \sigma', \zeta)\overline{P_0^{(1)}(\sigma_1 + i\gamma, \sigma', \zeta)}\right].$$

We will prove the following

LEMMA. *If $P_0(\sigma, \zeta)$ is a q-homogeneous strictly pluriparabolic polynomial for some $Q = (q, \dots, q, 1, \dots, 1)$, then there exists $C > 0$ such that*

$$\begin{gathered} C^{-1} \leqslant (|\gamma| + |\zeta|^q)^{-1}(|\gamma| + |\sigma| + |\zeta|^q)^{-2m+2} H_P(\sigma, \zeta, \gamma) \leqslant C, \\ \gamma \leqslant 0, \quad (\sigma, \zeta) \in \mathbb{R}^{n+1}. \end{gathered} \tag{4}$$

REMARK. Inequality (2) is a trivial consequence of (4), and hence (4) can be included in the list of equivalent conditions of the proposition.

PROOF OF THE LEMMA. According to (1.13) and condition (b) of Definition 1, for $\gamma < 0$ we have

$$H_{P_0}(\sigma, \zeta, \gamma) \leqslant (|\gamma| + \kappa|\zeta|^q) H(\sigma, \zeta, \gamma),$$

where

$$H(\sigma, \zeta, \gamma) = \sum_{k=1}^{m} \prod_{k \neq j} |\sigma_1 + i\gamma - \lambda_{0j}(\sigma', \zeta)|^2$$

is a q-homogeneous polynomial. Thus, to prove (4) we should show that

$$H(\sigma, \zeta, \gamma) \neq 0 \quad \text{for} \quad \gamma \leqslant 0, \quad \sigma^2 + |\zeta|^{2q} + \gamma^2 = 1. \tag{5}$$

Since $-\gamma + \operatorname{Im} \lambda_{0j}(\sigma', \zeta) \geqslant |\gamma| + \kappa|\zeta|^q$, it suffices to check condition (5) only for $\gamma = |\zeta| = 0$, i. e., for the strictly hyperbolic polynomial $P_0(\sigma, 0)$. We now refer the reader to the proof of inequality (2.3) in Proposition 2.3.

3. We state a generalization of Theorems 2.3 and 2.4.

THEOREM. *For a given* $q \geqslant 0$ *and a polynomial* (1) *the following conditions are equivalent.*

(i) $P(\sigma, \zeta)$ *satisfies* Q*-stable Petrovskii's condition for* $Q = (q, \dots, q, 1, \dots, 1)$.

(ii) *There are* $\gamma_0 < 0$ *and* $C_1 > 0$ *such that for* $\gamma \leqslant \gamma_0$ *we have*

$$C^{-1} \leqslant H_P(\sigma, \zeta, \gamma)/(|\gamma| + |\zeta|^q)(|\gamma| + |\sigma| + |\zeta|^q)^{2m-2} \leqslant C.$$

(iia) *There are* $\gamma_0 < 0$ *and* $C_1 > 0$ *such that for* $\gamma \leqslant \gamma_0$ *the inequality*

$$C_1(|\operatorname{Im} \tau| + |\zeta|^q)(|\tau| + |\sigma'| + |\zeta|^q)^{m-1} \leqslant |P(\tau, \sigma', \zeta)|$$

holds.

(iii) *The polynomial* P *is strictly pluriparabolic.*

PROOF. (i)$\Rightarrow$(iii). If P satisfies Q-stable Petrovskii's condition, then its Q-principal Q-homogeneous part possesses the same property. Therefore, according to the proposition, (iii) holds.

(iii)$\Rightarrow$(ii). As in Theorem 2.5, we set $P(\sigma, \gamma) = P_0(\sigma, \gamma) + P'(\sigma, \gamma)$, where P_0 is the Q-principal part of P and P' is a polynomial whose Q-degree is less than that of P_0. It follows that

$$H_P(\zeta, \sigma, \gamma) > H_{P_0}(\zeta, \sigma, \gamma) - |(\operatorname{Im} P' \overline{P^{(1)}})(\zeta, \sigma, \gamma)|.$$

By virtue of (4), it suffices to prove that for each $\varepsilon > 0$ there is $\gamma(\varepsilon)$ such that

$$(6) \qquad |(\operatorname{Im}(P' \overline{P^{(1)}}))(\zeta, \sigma, \gamma)| \leqslant \varepsilon(|\gamma + |\zeta|^q)(|\gamma| + |\sigma| + |\zeta|^q)^{2m-2}$$

for $\operatorname{Im} \tau \leqslant \gamma(\varepsilon)$ and $(\sigma, \zeta) \in \mathbb{R}^{n+1}$.

The polynomial on the left-hand side of (6) is a linear combination of the monomials

$$(\gamma, \sigma)^\alpha \zeta^\beta, \qquad |\alpha| \leqslant 2m - 2, \quad q|\alpha| + |\beta| \leqslant 2mq - q - 1.$$

These monomials can be represented as

$$(\gamma, \sigma)^\alpha \zeta^{\beta'} \zeta^{\beta''}, \qquad |\alpha| q + |\beta'| \leqslant 2m - 2, \quad |\beta''| \leqslant q - 1.$$

Such expressions can be estimated using the left-hand side of (6) with the constant $\varepsilon = \operatorname{const} |\gamma|^{1/q}$.

(ii)$\Rightarrow$(iia)$\Rightarrow$(iii). The first implication is obvious. By means of dilatations $(\tau, \sigma', \gamma) \longmapsto (t^q \tau, t^q \sigma', t\zeta)$ the inequality in (iia) is reduced to an inequality for the principal Q-homogeneous part of P_0, after which it remains to apply the proposition.

As in the proposition, the implication (iia)$\Rightarrow$(i) follows from a general result that will be presented in the next section.

§3. q-hyperbolic polynomials (q odd)

In §2 we described q-stable correct polynomials whose properties were determined by the principal q-homogeneous parts. For $q = 1$ and even q these were strictly hyperbolic and parabolic polynomials, respectively. The case of polynomials corresponding to odd q is beyond the scope of the theory. In the present section we will describe a class of polynomials whose properties are determined by the principal q-homogeneous part, where q is an odd natural number. These polynomials can also be characterized by means of a stability condition with respect to the lower terms, but this will be considered in the next section.

3.1. Definition of q-hyperbolic polynomials.

DEFINITION 1. Let $q \geqslant 1$ be odd. A q-homogeneous polynomial $P_0(\xi, \tau)$ is said to be q-hyperbolic if

(i) the polynomial $P_0(\xi, \tau)$ is resolved with respect to the highest power of τ; below we will assume, without special stipulation, that it has the form (2.2);

(ii) the roots $\lambda_{0j}(\xi)$ are real and distinct for $\xi \neq 0$ so that there is $\delta > 0$ such that

$$|\lambda_{0j}(\xi) - \lambda_{0k}(\xi)| > \delta|\xi|^q, \qquad j \neq k, \quad j, k = 1, \dots, m, \quad \xi \in \mathbb{R}^n \setminus \{0\}. \tag{1}$$

We note that, by virtue of (i) and the condition that the roots $\lambda_{0j}(\xi)$ are simple, they are differentiable function of ξ outside the origin;

(iii) if $q > 1$, then there is a direction $N \in R$ such that

$$\langle N, \operatorname{grad} \lambda_{0j}(\xi) \rangle > 0, \qquad \xi \neq 0, \quad j = 1, \dots, m.$$

By homogeneity, this condition can be restated in the following way: there is a unit vector $N \in \mathbb{R}^n$ such that for some $\kappa > 0$ the inequality

$$\langle N, \operatorname{grad} \lambda_{0j}(\xi) \rangle \geqslant \kappa |\xi|^{q-1}, \qquad j = 1, \dots, m, \quad \xi \neq 0, \tag{2}$$

holds.

REMARK. For $q = 1$ this definition turns into the definition of strictly hyperbolic polynomials. It would be more appropriate to call the polynomials in question strictly q-hyperbolic polynomials, but we will not do this because the q-hyperbolic polynomials that are not "strict" will not be studied here.

We consider some classes of polynomials that satisfy the above definition. In the case $m = 1$, i. e., for polynomials having the form

$$P_0(\xi, \sigma) = \tau - Q_0(\xi), \tag{3}$$

where Q_0 is a homogeneous polynomial of degree q, Definition 1 reduces to the following conditions:

(ii′) $Q_0(\xi)$ is a real polynomial;

(iii′) there is $N \in \mathbb{R}^n$ such that $\partial Q_0/\partial N$ is a positive elliptic polynomial.

Assume that Q_0 has the form

$$Q_0(\xi) = \sum_{j=1}^{n} a_j \xi_j^q, \tag{4}$$

where the coefficients $a_1, \dots, a_n$ are real and nonzero. In this case (iii′) holds automatically, namely it suffices to take the vector $(\operatorname{sgn} a_1, \dots, \operatorname{sgn} a_n)$ as N. In the case of positive a_j this is the vector $(1, \dots, 1)$.

We now consider a q-hyperbolic polynomial in the case of one variable $\xi \in \mathbb{R}^n$. Let us factorize the polynomial $P_0(\xi, \tau)$:

$$P_0(\xi, \tau) = \prod_{j=1}^{m} (\tau - a_j \xi^q).$$

Condition (ii) means that the coefficients a_j are real, and (iii) is equivalent to the condition that all the numbers $a_1, \dots, a_n$ are of the same sign.

Let P_0 satisfy the conditions of Definition 1. It is clear that if condition (iii) holds for some $N \in \mathbb{R}^n$, then, by continuity, it holds for all close directions as well. Denote by $V(P_0)$ the set of all $N \in \mathbb{R}^n$ such that condition (iii) holds, i. e.,

$$(5) \qquad N \in V(P_0) \Longleftrightarrow \langle N, \operatorname{grad} \lambda_{0j}(\xi) \rangle > 0, \qquad |\xi| = 1, \quad j = 1, \dots, m.$$

It is obvious that the set $V(P_0)$ possesses the following properties:

(1) $V(P_0)$ is an open set;

(2) $V(P_0)$ is a conic set, i. e., if $N \in V(P_0)$, then $\lambda N \in V(P_0)$ for an arbitrary $\lambda > 0$;

(iii) $V(P_0)$ is a convex set, i. e., if $N_1, N_2 \in V(P_0)$, then $tN_1 + (1-t)N_2 \in V(P_0)$ for each $t \in [0,1]$.

In view of these conditions, $V(P_0)$ is an open convex cone.

DEFINITION 2. The open convex cone of directions $N \in \mathbb{R}^n$ satisfying condition (5) is called the cone of q-hyperbolicity of P_0.

We note that in the case $n = 1$ the cone $V(P_0)$ consists of either the ray $\{v > 0\}$ or the ray $\{v < 0\}$.

In the case of polynomial (3) we have

$$V(P_0) = \{(v_1, \dots, v_n),\ v_j \operatorname{sgn} a_j > 0,\ j = 1, \dots, n\}.$$

DEFINITION 3. Let $q \geqslant 1$ be odd. A polynomial $P(\xi, \tau)$ is said to be q-hyperbolic if its principal q-homogeneous part P_0 possesses this property. The cone of q-hyperbolicity of the polynomial P_0 is called the cone of q-hyperbolicity of the original polynomial P and is denoted $V(P)$.

3.2. Description of q-hyperbolic polynomials. The main result of this section is the following

PROPOSITION. *For an odd $q > 1$ and a q-homogeneous polynomial $P_0(\xi, \tau)$ the conditions below are equivalent.*

(I) *P_0 is a q-hyperbolic polynomial of degree m with respect to τ.*

(II) *There are positive constants R and C_0 such that the following inequality holds:*

(6)
$$\begin{aligned} C_0^{-1}(|\operatorname{Im} \tau| + v|\zeta|^{q-1})(|\tau| + |\zeta|^q)^{2m-2} &\leqslant -\operatorname{Im}\bigl[P_0(\zeta, \tau)\overline{P_0^{(1)}(\zeta, \tau)}\bigr] \\ &\leqslant C_0\bigl(|\operatorname{Im} \tau| + v|\zeta|^{q-1})(|\tau| + |\zeta|^q\bigr)^{2m-2}, \end{aligned}$$

$$(6') \qquad \zeta = \xi + ivN, \quad v \geqslant 0, \quad \operatorname{Im} \tau < -Rv^q, \quad (\xi, \operatorname{Re} \tau) \in \mathbb{R}^{n+1}.$$

(III) *There are positive constants R and C such that under conditions* (6′) *the inequality*

$$(7) \qquad C_1(|\operatorname{Im} \tau| + v|\zeta|^{q-1})(|\tau| + |\zeta|^q)^{m-1} \leqslant |P_0(\tau, \zeta)|$$

holds.

In what follows we will use the following notation for the middle term in (6): if P is a polynomial and we have $N \in \mathbb{R}^n$, then

$$H_{P,N}(\xi, v, \sigma, \gamma) = -\operatorname{Im}\bigl[P(\xi + ivN, \sigma + i\gamma)\overline{P^{(1)}(\xi + ivN, \sigma + i\gamma)}\bigr].$$

The proposition readily implies a similar assertion for nonhomogeneus polynomials. We have the following

THEOREM. *For an odd $q > 1$ and a polynomial $P(\xi, \tau)$ of degree m with respect to τ the conditions below are equivalent.*

(I) *P is a q-hyperbolic polynomial.*

(II) *There are positive constants ν_0, R, and C such that*

$$C_1(|\operatorname{Im}\tau| + \nu|\zeta|^{q-1})(|\tau| + |\zeta|^q)^{2m-2} \leqslant H_{P,N}(\xi, \nu, \operatorname{Re}\tau, \operatorname{Im}\tau) \leqslant C(|\operatorname{Im}\tau| + \nu|\zeta|^{q-1})(|\tau| + |\zeta|^q)^{2m-2}, \tag{8}$$

$$\zeta = \xi + i\nu N, \quad \nu \geqslant \nu_0, \quad \operatorname{Im}\tau < -R\nu^q, \quad (\xi, \operatorname{Re}\tau) \in \mathbb{R}^{n+1}. \tag{8'}$$

(III) *There are positive constants ν_0, R, and C_2 such that under conditions (8′) the inequality*

$$C_1(|\operatorname{Im}\tau| + \nu|\zeta|^{q-1})(|\tau| + |\zeta|^q)^{m-1} \leqslant |P(\tau, \zeta)| \tag{9}$$

holds.

Assuming that the proposition is proved we carry out

PROOF OF THE THEOREM. (II)⇒(III) follows from the fact that for $\nu \geqslant \nu_0$ the polynomial $P^{(1)}(\tau, \zeta)$ has an upper bound $\operatorname{const}(|\tau + |\zeta|)^{m-1}$.

(III)⇒(I). If ξ, ν, τ in (9) are replaced by $\rho\xi$, $\rho\nu$, $\rho^q\tau$ respectively, and the passage to the limit for $\rho \to +\infty$ is performed, then we obtain (7). Applying the proposition we derive (I).

(I)⇒(II). Set $P = P_0 + P'$, where P_0 is the principal q-homogeneous part. The polynomial P_0 satisfies inequality (6). We now show that if $\nu \geqslant \nu_0$, then

$$\left|H_{P,N}(\xi, \nu, \operatorname{Re}\tau, \operatorname{Im}\tau) - H_{P_0,N}(\xi, \nu, \operatorname{Re}\tau, \operatorname{Im}\tau)\right| \leqslant \varepsilon(\nu_0)\nu|\zeta|^{q-1}(|\tau| + |\zeta|^q)^{2m-2}, \tag{10}$$

where $\varepsilon(\nu_0) \to 0$, $\nu_0 \to +\infty$. This inequality together with (6) implies (8).

To prove (10) we note that its left-hand side is the absolute value of the polynomial

$$\operatorname{Im}\left[P'(\xi + i\nu N, \tau)\overline{P^{(1)}(\xi + i\nu N, \tau)} + P_0(\xi + i\nu N, \tau)\overline{P'^{(1)}(\xi + i\nu N, \tau)}\right],$$

which is a linear combination of the monomials

$$\xi^\alpha \nu^k (\operatorname{Im}\tau)^{j_1}(\operatorname{Re}\tau)^{j_2}, \qquad |\alpha| + k + q(j_1 + j_2) = 2mq - q - 1 - \kappa, \quad \kappa \geqslant 0.$$

Therefore, for $\nu > \nu_0$ we have the following chain of obvious inequalities (recall that $j_1 + j_2 \leqslant 2m - 2$):

$$|\xi^\alpha \nu^\kappa (\operatorname{Im}\tau)^{j_1}(\operatorname{Re}\tau)^{j_2}| = \nu^{-\kappa-1}|\nu\xi^\alpha \nu^{k+\kappa}(\operatorname{Re}\tau)^{j_1}(\operatorname{Re}\tau)^{j_2}| \leqslant \nu^{-\kappa-1}\nu|\zeta|^{|\alpha|+k+\kappa}|\tau|^{j_1+j_2} \leqslant \nu_0^{-\kappa-1}\nu|\zeta|^{q-1}(|\tau| + |\zeta|^q)^{2m-2}.$$

3.3. Remarks on the roots of q-hyperbolic polynomials. The proof of Proposition 3.2 requires an accurate analysis of the behavior of the roots of a q-hyperbolic polynomial in the complex domain. According to the definition of a q-hyperbolic q-homogeneous polynomial P_0, the roots $\lambda_0(\xi + i\nu N)$ are real and distinct for $\nu = 0$ and are smooth functions of ξ for $\xi \neq 0$. Although for $\nu \neq 0$ the roots may become multiple, we can always separate out m single-valued continuous branches $\lambda_{0j}(\xi + i\nu N)$ that are homogeneous functions of ξ and ν of degree q. In view of what has been said, when studying the roots we can confine ourselves to the unit hemisphere

$$S = \{(\xi, \nu) \in \mathbb{R}^{n+1},\ |\xi|^2 + \nu^2 = 1,\ \nu \geqslant 0\}.$$

For a given $\varepsilon > 0$ we separate out the subregions

$$S'_\varepsilon = \{(\nu, \xi) \in S,\ 0 \leqslant \nu \leqslant \varepsilon\}, \quad S''_\varepsilon = \{(\nu, \xi) \in S,\ \nu \geqslant \varepsilon\}.$$

If ε is sufficiently small, then in the region S'_ε the roots $\lambda_{0j}(\xi + i\nu N)$ are distinct and, hence, smooth functions. Since the imaginary parts of these roots vanish for $\nu = 0$, by virtue of their smoothness, the roots can be represented in the form

$$\lambda_{0j}(\xi + i\nu N) = p_j(\xi, \nu) + i\nu q_j(\xi, \nu), \tag{11}$$

where p_j and q_j are smooth real homogeneous functions in ξ and ν of degrees q and $q - 1$, respectively, and we have

$$p_j(\xi, 0) = \lambda_{0j}(\xi), \quad q_j(\xi, 0) = \langle N, \operatorname{grad} \lambda_{0j}(\xi)\rangle > 0. \tag{12}$$

By continuity, the functions $p_j(\xi, \nu)$ are distinct for small ν and the functions $q_j(\xi, \nu)$ are positive. Hence, there are $\delta, \kappa > 0$ (generally, depending on ε) such that

$$|p_j(\xi, \nu) - p_k(\xi, \nu)| \geqslant \kappa|\zeta|^q, \quad j \neq k, \quad j, k = 1, \dots, m, \tag{13}$$
$$\zeta = \xi + i\nu N, \quad (\xi, \nu) \in S'_\varepsilon,$$

$$q_j(\xi, \nu) > \kappa|\zeta|^{q-1}, \quad j = 1, \dots, m, \quad (\xi, \nu) \in S'_\varepsilon. \tag{14}$$

In the subregion S''_ε we will use a trivial inequality based on the continuity:

$$|\operatorname{Im} \lambda_{0j}(\xi + i\nu N)| \geqslant C \geqslant C\varepsilon^{-q}\nu^q = C(\varepsilon)\nu^q. \tag{15}$$

By homogeneity, what has been said can be restated as the following

LEMMA. *Let $P_0(\xi, \tau)$ be a q-homogeneous q-hyperbolic polynomial. Let $\lambda_{0j}(\xi + i\nu N)$, $j = 1, \dots, m$, be continuous single-valued branches of the roots of the polynomial $P_0(\xi + i\nu N)$ which are chosen so that they are homogeneous functions of ξ and ν of degree q. Let $N \in \mathbb{R}^n$ be a vector satisfying condition* (iii) *of Definition* 1 *in* 3.1 *(i. e., $N \in V(P_0)$). Then there is $\varepsilon > 0$ such that*

(i) *if $0 \leqslant \nu \leqslant \varepsilon|\xi|$, then the roots $\lambda_{0j}(\xi + i\nu N)$ can be represented in the form* (11), *and in this case the functions p_j and q_j satisfy* (13) *and* (14);

(ii) *if $0 \geqslant \nu \geqslant \varepsilon|\xi|$, then inequality* (15) *holds.*

3.4. Proof of Proposition 3.2 ((I)⇒(II)). The estimates for the function $H_{P,N}$ are based on the consequence of Lemma 3.3 presented below.

LEMMA. *Let $\lambda_{0j}(\xi+i\nu N)$ be the roots of the q-hyperbolic polynomial that were studied in Lemma* 3.3. *Then there is $R>0$ such that for $\operatorname{Im}\tau<-R\nu^q$ the following two-sided inequality holds*:

$$K^{-1}(|\operatorname{Im}\tau|+\nu|\zeta|^{q-1})\leqslant-\operatorname{Im}\tau+\operatorname{Im}\lambda_{0j}(\xi+i\nu N)\leqslant K(|\operatorname{Im}\tau|+\nu|\zeta|^{q-1}). \tag{16}$$

PROOF. Choose a sufficiently small ε so that for $\nu\leqslant\varepsilon|\xi|$ assertion (i) of Lemma 3.3 holds. Under these conditions, by virtue of (13) and (14), we have

$$\kappa\nu|\zeta|^{q-1}\leqslant\operatorname{Im}\lambda_{0j}(\xi+i\nu N)\leqslant\kappa_1\nu|\zeta|^{q-1}.$$

Because $-\operatorname{Im}\tau>0$, we obtain (16) for $\nu<\varepsilon|\xi|$. Now let $\nu>\varepsilon|\xi|$. Then we have $\nu>\varepsilon'|\zeta|$, and, according to (15),

$$\begin{aligned}-\operatorname{Im}\tau+\operatorname{Im}\lambda_{0j}(\xi+i\nu N)&>\frac{1}{2}|\operatorname{Im}\tau|+\left(\frac{R}{2}-C(\varepsilon)\right)\nu^q\\&\geqslant\frac{1}{2}|\operatorname{Im}\tau|+\frac{R}{4}\nu^q>\frac{1}{2}|\operatorname{Im}\tau|+\frac{R\varepsilon'^{q-1}}{4}\nu|\zeta|^{q-1},\end{aligned}$$

provided that $R\geqslant4C(\varepsilon)$. The right-hand inequality (16) is obvious for $\nu>\varepsilon|\xi|$.

Set

$$H(\xi,\nu,\sigma,\gamma)=\sum_{k=1}^{m}\prod_{j\neq k}\left[(\sigma-\operatorname{Re}\lambda_{0j}(\xi+i\nu N))^2+(|\gamma|+\nu|\zeta|^{q-1})^2\right]. \tag{17}$$

According to the lemma, the function $H_{P,N}(\xi,\nu,\sigma,\gamma)$ has upper and lower bounds expressed in terms of

$$(|\operatorname{Im}\tau|+\nu|\zeta|^{q-1})H(\xi,\nu,\sigma,\gamma).$$

Thus, we must find upper and lower estimates for the function $H(\xi,\nu,\sigma,\gamma)$ via $(|\xi|^q+\nu^q+|\sigma|+|\gamma|)^{2m-2}$. By q-homogeneity, is suffices to show that

$$H(\xi,\nu,\sigma,\gamma)\neq0,\quad(|\xi|^2+|\nu|^2)^q+\sigma^2+\gamma^2=1,\quad\nu\geqslant0,\quad-\gamma>R\nu^q. \tag{18}$$

Definition (17) implies that (18) holds for $|\gamma|\neq0$. In case $\gamma=0$ the condition $\nu=0$ is sure to hold and we have

$$H(\xi,0,\sigma,0)=\sum_{k=1}^{m}\prod_{j\neq k}(\sigma-\lambda_{0j}(\xi))^2.$$

Because (13) holds, it remains to repeat the argument in Proposition 2.2.

3.5. The end of the proof of Proposition 3.2 ((III)⇒(I)).

1) According to (7) (for $\nu=0$), the polynomial $P_0(\xi,\tau)$ is q-homogeneous and satisfies the condition

$$P_0(\xi,\tau)\neq0,\qquad\operatorname{Im}\tau<0,\quad(\xi,\operatorname{Re}\tau)\in\mathbb{R}^{n+1}.$$

It follows that the polynomial P_0 is correct in Petrovskii's sense; in particular, it is resolved with respect to the highest power of τ (see 1.2). Since q is odd, Proposition 1.2 implies that the roots $\lambda_{0j}(\xi)$ are real.

We will consecutively check the validity of conditions (ii) and (iii) of Definition 1 in 3.2.

2) We first show that the roots are simple. Assume the contrary, namely let there be $\xi^0 \neq 0$ such that $\lambda_{01}(\xi^0) = \lambda_{02}(\xi^0)$. Now we show that inequality (7) does not hold along the curve $\{\xi = \rho\xi^0, \operatorname{Re}\tau = \lambda_{01}(\xi^0)\rho^q\}$.

Along this curve the left-hand side of (7) has the following lower bound:

$$\nu|\zeta|^{q-1}(|\tau| + |\zeta|^q)^{m-1} \geqslant \text{const}\,\rho^{mq-1}.$$

To find an upper bound for the right-hand side we consider in detail the roots $\lambda_j(\rho)$, $j = 1, \dots, m$, of the polynomial

$$B(\rho, \tau) = P_0(\rho\xi^0 + i\nu N, \tau). \tag{18'}$$

By what was proved, this polynomial is correct in Petrovskii's sense, and its roots can be expanded as Puiseux's series with respect to ρ. A priori, it follows from Proposition 1.4 that

$$\lambda_j(\rho) = a_j\rho^{q_j} + o(\rho^{q_j}). \tag{19}$$

We now refine this expansion. The q-homogeneity of the polynomial P_0 implies that $\rho^{-q}\lambda_j(\rho)$ are the roots of the polynomial $P_0(\xi^0 + i\nu\rho^{-1}N, \tau)$. Because this polynomial is resolved with respect to the highest power of τ, its roots depend continuously on the parameter $i\nu\rho^{-1}N$, and for $\rho \to +\infty$ they tend to the roots $\lambda_{0j}(\xi^0)$ of the polynomial P_0. In other words, (19) takes the form

$$\lambda_j(\rho) = \lambda_{0j}(\xi^0)\rho^q + o(\rho^q). \tag{19'}$$

Since the roots $\lambda_{0j}(\xi^0)$ are real, according to Proposition 1.4, the remainder in (19′) is of the order of ρ^{q-1}, i. e.,

$$\lambda_j(\rho) = \lambda_{0j}(\xi^0)\rho^q + O(\rho^{q-1}). \tag{19''}$$

We now have

$$\begin{aligned}|P_0(\rho\xi^0 + i\nu N, \lambda_{01}(\xi^0)\rho^q + i\gamma)|\\ = \prod_{j=1}^{m}\left[(\lambda_{01}(\xi^0)\rho^q - \operatorname{Re}\lambda_j(\rho))^2 + (\gamma - \operatorname{Im}\lambda_j(\rho))^2\right]^{1/2}.\end{aligned}$$

By (19″), the factors corresponding to $j > 2$ are $o(\rho^q)$. As to the factors corresponding to $j = 1, 2$, the assumption concerning the multiplicity of the root λ_{01}, imply that they are $o(\rho^{q-1})$, whence

$$\left|P_0(\rho\xi^0 + i\nu N, \lambda_{01}(\xi^0)\rho^q + i\gamma)\right| < \text{const}\,\rho^{mq-2}.$$

Comparing this inequality with (18′) we arrive at a contradiction, which is what proves that the roots of the polynomial P_0 are simple.

3) The fact that the roots of P_0 are simple for $\xi \neq 0$ implies that they are smooth. We now check condition (iii) of Definition 1 in 3.1. Assume that it does not hold, i. e., there is $\xi^0 \neq 0$ such that

$$\langle N, \operatorname{grad}\lambda_{01}(\xi^0)\rangle = 0. \tag{20}$$

In this case expansion (19″) for $j = 1$ takes the form

$$\lambda_1(\rho) = \lambda_{0j}(\xi^0)\rho^q + O(\rho^{q-2}). \tag{21}$$

Therefore, for the above polynomial $B(\rho,\tau)$ inequality (7) does not hold for $\operatorname{Re}\tau = \lambda_{01}(\xi^0)\rho^q$. Indeed, in this case $B(\rho,\tau)$ has an upper bound const ρ^{mq-2}, whereas the left-hand side of (7) has a lower bound const ρ^{mq-1}.

4) To prove (21) we note that if $\xi^0 \neq 0$, then the roots $\lambda_{0j}(\zeta)$ are smooth functions at close points ζ, whence

$$\lambda_{0j}(\xi^0 + i\nu\rho^{-1}N) = \lambda_{0j}(\xi^0) + i\nu\rho^{-1}\langle N, \operatorname{grad}\lambda_{0j}(\xi^0)\rangle + O(\rho^{-2}), \quad \rho \to +\infty.$$

Consequently, we have

$$\lambda_j(\rho) = \lambda_{0j}(\xi^0)\rho^q + i\nu\langle N, \operatorname{grad}\lambda_{0j}(\xi^0)\rangle\rho^{q-1} + O(\rho^{q-2}). \tag{22}$$

If condition (20) holds for $j = 1$, then (22) implies (21).

§4. Weakly q-stable correct polynomials

In §2 we studied q-stable correct polynomials, i. e., polynomials whose correctness is retained under arbitrary perturbations of q-lower terms. As was shown, these polynomials correspond to even q (parabolic polynomials) or $q = 1$ (strictly hyperbolic polynomials). In §3 the class of q-hyperbolic polynomials was associated with odd q. As in the case of parabolic and strictly hyperbolic polynomials, the property of q-hyperbolicity is determined by the principal q-homogeneous part. Theorem 3.2 suggests a weakened version of the condition of q-stable correctness that holds for q-hyperbolic polynomials. It will be shown that this condition holds for q-parabolic and strictly hyperbolic polynomials as well. The main result of the present section is that these polynomials exhaust the class of weakly q-stable correct polynomials.

4.1. Definition of weakly q-stable correct polynomials. Let $q \geqslant 1$. A polynomial $P(\zeta,\tau)$ is said to be weakly q-stable correct if there is a direction $N \in \mathbb{R}^{n+1}$ such that for any polynomial $Q(\zeta,\tau)$, $\deg_q Q < \deg_q P$, there is $\nu_0(Q)$ such that for $\nu \geqslant \nu_0(Q)$ the following condition holds:

$$P(\zeta,\tau) + Q(\zeta,\tau) \neq 0, \tag{1}$$

$$\zeta = \xi + i\nu N, \quad \nu \geqslant \nu_0, \quad \operatorname{Im}\tau \leqslant \gamma_0(\nu, Q), \quad (\xi, \operatorname{Re}\tau) \in \mathbb{R}^{n+1}. \tag{1'}$$

Denote by $\mathcal{B}(P)$ the conic set of those directions $N \in \mathbb{R}^{n+1}$ for which (1) and (1′) hold.

We consider some immediate consequences of the above definition.

(a) A polynomial P is weakly q-stable correct if and only if its principal q-homogeneous part possesses this property.

(b) A q-homogeneous weakly q-stable correct polynomial is correct in Petrovskii's sense. This property follows from condition (1) with $Q \equiv 0$ if ξ and τ are replaced by $\rho\xi$ and $\rho^q\tau$, respectively, and passage to the limit as $\rho \longrightarrow \infty$ is carried out.

(c) It follows from (a) and (b) that a weakly q-stable correct polynomial is resolved with respect to the highest power of τ, and the number q can assume only integral values.

THEOREM. *Every q-hyperbolic polynomial is weakly q-stable correct. In this case the set $\mathcal{B}(P)$ coincides with the cone of q-hyperbolicity.*

PROOF. If a polynomial P is q-hyperbolic and if a direction $N \in \mathbb{R}^{n+1}$ belongs to the cone of q-hyperbolicity $V(P_0)$, then, by Theorem 3.2, inequality (3.9) holds. Since any polynomial Q whose q-degree is less than $\deg_q P = mq$, i. e., $\deg_q Q \leqslant mq - 1$ (because we deal with integers), has an upper bound $\mathrm{const}(1+|\zeta|)^{q-1}(|\tau|+|\zeta|^q)^{m-1}$, we conclude that for $\nu > \nu_0$, $\operatorname{Im}\tau < -R\nu^q$, where ν_0 is sufficiently large, the polynomial $P+Q$ satisfies inequality (3.9), and hence conditions (1) and (1′) hold, and we have $V(P) \subset \mathcal{B}(P)$. On the other hand, when proving the proposition and theorem in 3.2 we showed that (3.9) implies that N belongs to the cone of q-hyperbolicity. Below we will prove that the condition of weak q-stable correctness implies (3.9) (i. e., the condition of weak q-stable correctness can be included in the list of equivalent conditions of Theorem 3.2), which will thus prove that $V(P)$ and $\mathcal{B}(P)$ coincide.

4.2. Weak q-stable correctness of q-stable polynomials. In this subsection we will prove the following result.

THEOREM. *Every q-stable correct polynomial is weakly q-stable correct, and we have $\mathcal{B}(P) = \mathbb{R}^n$.*

Because the class of q-stable correct polynomials is exhausted by strictly hyperbolic and q-parabolic polynomials, we will show that conditions (1) and (1′) hold for these two classes for any $N \in \mathbb{R}^{n+1}$. The proof is based on a generalization of the inequalities in Theorems 2.2 and 2.3 to the expression

$$H_{P,N}(\xi,\nu,\operatorname{Re}\tau,\operatorname{Im}\tau) = -\operatorname{Im}\big[P(\xi+i\nu N,\tau)\overline{P^{(1)}(\xi+i\nu N,\tau)}\big]. \tag{2}$$

We first consider the case $q=1$.

PROPOSITION 1. *Let P be a strictly hyperbolic polynomial. Then there are $R>0$ and $\nu_0>0$ such that for any direction $N \in \mathbb{R}^{n+1}$ the following two-sided inequality holds with a positive constant C:*

$$C^{-1}|\operatorname{Im}\tau|(|\tau|+|\zeta|)^{2m-2} \leqslant H_{P,N}(\xi,\nu,\operatorname{Re}\tau,\operatorname{Im}\tau) \leqslant C|\operatorname{Im}\tau|(|\tau|+|\zeta|)^{2m-2}, \tag{3}$$

$$\zeta = \xi' + i\nu N, \quad \nu \geqslant \nu_0, \quad \operatorname{Im}\tau < -R\nu^q, \quad (\xi,\operatorname{Re}\tau) \in \mathbb{R}^{n+1}. \tag{3′}$$

PROOF. Consider the right-hand inequality in (3). Since the coefficients of the polynomial P are real (we assume that the coefficients of P are normalized in some way, say P has the form (2.2)), we have

$$H_{P,N}(\xi,0,\operatorname{Re}\tau,0) = -\operatorname{Im}\big[P(\xi,\sigma)P^{(1)}(\xi,\sigma)\big] \equiv 0.$$

Hence, polynomial (2) is represented in the form

$$H_{P,N}(\xi,\nu,\operatorname{Re}\tau,\operatorname{Im}\tau) = \nu H^1(\xi,\nu,\operatorname{Re}\tau,\operatorname{Im}\tau) + \operatorname{Im}\tau H^2(\xi,\nu,\operatorname{Re}\tau,\operatorname{Im}\tau),$$

where H^1 and H^2 are polynomials of degree at most $2m-2$. Therefore

$$\begin{aligned}|H^1(\xi,\nu,\operatorname{Re}\tau,\operatorname{Im}\tau)| &\leqslant K_0\sum_{l=0}^{2m-2}(|\xi|+|\nu|+|\operatorname{Re}\tau|+|\operatorname{Im}\tau|)^l \leqslant K_1\sum_{l=0}^{2m-2}(|\tau|+|\zeta|)^l\\ &\leqslant K_1(1-\nu_0^{-1})(|\tau|+|\zeta|)^{2m-2} \leqslant K_2(|\tau|+|\zeta|)^{2m-2},\end{aligned}$$

where $K_2 = 2K_1$ if, say, $\nu_0 \geqslant 2$. Because $\nu \leqslant R^{-1}|\operatorname{Im}\tau|$, we arrive at the right-hand inequality (3).

To prove the first inequality (3) we use (2.7).

If $\nu < R^{-1}|\operatorname{Im}\tau|$ and, say, $R \geqslant 2$, then the left-hand side of (2.7) has a lower bound

$$2^{-2m+2}C^{-1}|\operatorname{Im}\tau|(|\tau|+|\zeta|)^{2m-2} = C_1|\operatorname{Im}\tau|(|\tau|+|\zeta|)^{2m-2}. \tag{4}$$

The second term in (2.7) is equal to

$$H_{P,N}(\xi,\nu,\operatorname{Re}\tau,\operatorname{Im}\tau) + \big[H_P(\xi,\operatorname{Re}\tau,\operatorname{Im}\tau) - (H_{P,N}(\xi,\operatorname{Re}\tau,\operatorname{Im}\tau))\big].$$

The expression in square brackets is identically equal to zero for $\nu = 0$, i. e., it is a polynomial divisible by ν. It follows that this expression has an upper bound

$$C_2\nu(|\tau|+|\zeta|)^{2m-2} \leqslant C_2R^{-1}|\operatorname{Im}\tau|(|\tau|+|\zeta|)^{2m-2}.$$

For a sufficiently large R the factor C_2R^{-1} is less than half the constant C_1 in (4), which implies the left-hand inequality (3).

COROLLARY. *A strictly hyperbolic polynomial P is weakly 1-stable correct, and we have $\mathcal{B}(P) = \mathbb{R}^n$.*

PROOF. The left-hand inequality (3) and the condition $\nu \leqslant R^{-1}|\operatorname{Im}\tau|$ imply that if (3′) holds, then

$$C_3\nu(|\tau|+|\zeta|)^{m-1} < |P(\tau,\zeta)|. \tag{5}$$

Therefore, for any polynomial $Q(\tau,\zeta)$ of degree no higher than $m-1$ there is $\nu_0 > 0$ such that for $\nu \geqslant \nu_0$ inequality (5) remains valid when P and C_3 are replaced by $P+Q$ and, say, $C_3/2$, respectively.

PROPOSITION 2. *Let $q \geqslant 2$ be even and let $P(\zeta,\tau)$ be a q-parabolic polynomial of degree m with respect to τ. Then there are R and ν_0 such that for any (unit) vector $N \in \mathbb{R}^{n+1}$ the two-sided inequality*

$$\begin{aligned} C^{-1}(|\operatorname{Im}\tau|+|\zeta|^q)(|\tau|+|\zeta|^q)^{2m-2} &\leqslant H_{P,N}(\xi,\nu,\operatorname{Re}\tau,\operatorname{Im}\tau) \\ &\leqslant C(|\operatorname{Im}\tau|+|\zeta|^q)(|\tau|+|\zeta|^q)^{2m-2}, \end{aligned} \tag{6}$$

$$\zeta = \xi + i\nu N, \quad \nu \geqslant \nu_0, \quad \operatorname{Im}\tau < -R\nu^q, \quad (\xi,\operatorname{Re}\tau) \in \mathbb{R}^{n+1}, \tag{6′}$$

holds with some constant $C > 0$.

PROOF. To check the right-hand inequality (6), it should be noted that the polynomial $H_{P,N}$ is a linear combination of the monomials

$$\xi^\alpha\nu^i(\operatorname{Re}\tau)^j(\operatorname{Im}\tau)^k, \quad |\alpha|+i+(j+k)q \leqslant 2mq-q, \tag{7}$$

where, since $P(0,\sigma)$ is a real polynomial, we have $j \leqslant 2m-2$. It follows that the monomials (7) have a lower bound

$$(|\operatorname{Im}\tau|+|\zeta|^q)^{l_1}(|\tau|+|\zeta|^q)^{l_2}, \qquad l_1+l_2 \leqslant 2m-1, \quad l_2 \leqslant 2m-2. \tag{7′}$$

If $\nu \geqslant \nu_0$, $|\operatorname{Im}\tau| > R\nu_0^q$, and, say, the numbers ν_0 and $R\nu_0^q$ exceed 1, then expression (7′) can readily be estimated using the right-hand side of (6).

The proof of the left-hand inequality (6) proceeds from the left-hand inequality (2.10). We first of all find a bound for the left-hand side of (2.10) under conditions (6′). We have

$$|\xi|^q = |\xi + i\nu N - i\nu N|^q \geqslant (|\zeta| - \nu)^q \geqslant |\zeta|^q - \sum_{j=1}^{q} \binom{q}{j} |\zeta|^{q-j} \nu^j.$$

Applying the inequality

$$|\zeta|^{q-j} \nu^j \leqslant \frac{q-j}{q} \varepsilon |\zeta|^q + \frac{j}{q} \varepsilon^{-(q-j)/j} \nu^q$$

with a sufficiently small ε we find that under conditions (6′) the following relation holds:

$$|\xi|^q > \frac{1}{2} |\zeta|^q - K(q)\nu^q.$$

If $R > 2K(q)$, then

$$|\operatorname{Im}\tau| + |\xi|^q > (|\operatorname{Im}\tau| + |\zeta|^q)/2.$$

Similarly,

$$(|\tau| + |\xi|^q)^{2m-2} \geqslant 2^{-2m+2} (|\tau| + |\zeta|^q)^{2m-2}.$$

Multiplying these inequalities we find a lower estimate for the left-hand side of (2.10) in the form

$$C_1 (|\operatorname{Im}\tau| + |\zeta|^q)(|\tau| + |\zeta|^q)^{2m-2}. \tag{8}$$

As to the second term in (2.10), it is equal to

$$H_{P,N}(\xi, \nu, \operatorname{Re}\tau, \operatorname{Im}\tau) + \big[H_P(\xi, \nu, \operatorname{Re}\tau, \operatorname{Im}\tau) - H_{P,N}(\xi, \nu, \operatorname{Re}\tau, \operatorname{Im}\tau)\big].$$

The polynomial in square brackets vanishes for $\nu = 0$, i. e., it is divisible by ν and is equal to $\nu H'$, where H' is a linear combination of the monomials

$$\xi^\alpha \nu^i (\operatorname{Re}\tau)^j (\operatorname{Im}\tau)^k, \qquad |\alpha| + i + q(j+k) \leqslant 2mq - q - 1, \quad j \leqslant 2m - 2.$$

These monomials have the upper bound

$$\nu (|\operatorname{Im}\tau| + |\zeta|^q)^{(q-1)/q} (|\tau| + |\zeta|^q)^{2m-2}.$$

Noting that $\nu \leqslant R^{-1/q} |\operatorname{Im}\tau|$, we obtain an upper bound for $\nu H'$:

$$\operatorname{const} R^{-1/q} (|\operatorname{Im}\tau| + |\zeta|^q)(|\tau + |\zeta|^q)^{2m-2}.$$

For a sufficiently large R the expression const $R^{-1/q}$ is less than half the constant C_1 in (8), which implies the left-hand inequality.

COROLLARY. *Every q-parabolic polynomial P is weakly q-stable correct, and we have* $\mathcal{B}(P) = \mathbb{R}^n$.

PROOF. Note that (6) implies that under conditions (6′) we have

$$C_1(|\operatorname{Im}\tau| + |\zeta|^q)(|\tau| + |\zeta|^q)^{m-1} \leqslant |P(\tau,\zeta)|. \tag{9}$$

Furthermore, every q-lower monomial has the form

$$\zeta^\alpha \tau^j, \qquad |\alpha| + jq \leqslant mq - 1, \quad j \leqslant m - 1.$$

Any monomial of this type has the upper bound

$$\nu^{-\kappa}(|\operatorname{Im}\tau| + |\zeta|^q)(|\tau| + |\zeta|^q)^{m-1}, \qquad \kappa = mq - |\alpha| - jq \geqslant 1.$$

Thus, if we have $\nu \geqslant \nu_0$, and ν_0 is sufficiently large, then inequality (9) is retained when P is replaced by $P + Q$, where $\deg_q Q < \deg_q P$, i. e., the condition of weak q-stable correctness holds.

4.3. Description of weakly q-stable correct polynomials. Summing up the above results on strictly hyperbolic, q-parabolic, and q-hyperbolic polynomials we arrive at the following

THEOREM. *For a given $q \geqslant 1$, $N \in \mathbb{R}^{n+1}$, and a polynomial $P(\zeta,\tau)$ of degree m with respect to τ the following conditions are equivalent.*

(I) *The polynomial P is q-stable correct, and we have $N \in \mathcal{B}(P)$, i. e., for any polynomial Q, $\deg_q Q < \deg_q P$, conditions* (1) *and* (1′) *hold.*

(II) *There are $C, R > 0$, and ν_0 such that under conditions* (1′) *the inequality*

$$C(|\operatorname{Im}\tau| + \nu|\zeta|^{q-1})(|\tau| + |\zeta|^q)^{m-1} \leqslant |P(\zeta,\tau)| \tag{10}$$

holds.

(IIa) *The polynomial P is resolved with respect to the highest power of τ^m, and there is $R > 0$ such that*

$$(1 + |\zeta|)^{q-1}(|\tau| + |\zeta|^q)^{m-1} \leqslant \varepsilon(\nu)|P(\zeta,\tau)|, \tag{11}$$

$$\zeta = \xi + i\nu N, \quad \operatorname{Im}\tau < -R\nu^q, \quad \varepsilon(\nu) \longrightarrow 0, \quad \nu \longrightarrow +\infty. \tag{11′}$$

(III) *q is a positive integer, and one of the conditions below holds:*

(a) *$q = 1$, the polynomial P is strictly hyperbolic, and $N \in \mathbb{R}^n$ is an arbitrary vector;*

(b) *q is even, the polynomial P is q-parabolic, and $N \in \mathbb{R}^n$ is an arbitrary vector;*

(c) *$q > 1$, q is odd, the polynomial P is q-hyperbolic, and the vector N belongs to the cone of q-hyperbolicity $V(P)$.*

PROOF. First of all note that the equivalence of conditions (II), (IIa), and (III) is in fact proved. Indeed, we have the following implications.

(II)⇒(IIa). As was many times noted, inequality (10) implies that the polynomial P is resolved with respect to τ^m. Inequality (11) is a direct consequence of (10).

(IIa)⇒(III). Note that (IIa) implies (I), and therefore, according to the remarks in 4.1, q is an integer, and the principal q-homogeneous part P_0 of P is a polynomial correct in Petrovskii's sense.

We first consider the case of an odd q. Careful examination of the proof of the implication (III)⇒(I) in Proposition 3.2 shows that for $q = 1$ the polynomial P is strictly hyperbolic, and for $q > 1$ it is q-hyperbolic. By applying the argument in Proposition 3.2 it is easy to consider the case of an even q as well, which we leave to the reader as an exercise.

(III)$\Rightarrow$(II) follows from the results of 4.1 and 4.2. Hence, the proof of the theorem reduces to proving the equivalence of (I) and (IIa). To this end we need an auxiliary assertion (see **[GV2,** Lemma 1 in § 7.3.2]).

LEMMA. *Let $M(\eta, \nu)$ and $R(\eta, \nu)$ be some functions of the variables $\nu \in [\nu_0, \infty]$ and η, where η runs over a set Ω (common for all the functions). Assume that*

$$M(\eta, \nu) \neq 0, \quad \forall \nu \geqslant \nu_0 \quad \forall \eta \in \Omega.$$

Then the following conditions are equivalent:

(a) *$R(\eta, \nu)M^{-1}(\eta, \nu) \longrightarrow 0$, $\nu \longrightarrow +\infty$, uniformly with respect to Ω;*

(b) *for each $a \in \mathbb{C}$ there is $\nu_0(a)$ such that*

$$M(\eta, \nu) + aR(\eta, \nu) \neq 0, \qquad \nu \in \nu_0(a), \quad \eta \in \Omega,$$

where the function $\nu_0(a)$ can be chosen so that it has a lower bound on any compact subset of $\mathbb{C}$.

PROOF. (a)$\Rightarrow$(b) is obvious.

(b)$\Rightarrow$(a). Assume that condition (a) is not satisfied. Then for some $\varepsilon > 0$ there is a sequence (η_j, ν_j), $\nu_j \longrightarrow \infty$, such that

$$|R(\eta_j, \nu_j)M^{-1}(\eta_j, \nu_j)| \geqslant \varepsilon.$$

Set $a_j = -M(\eta_j, \nu_j)/R(\eta_j, \nu_j)$. Then we have $|a_j| < 1/\varepsilon$, and, by virtue of condition (b), the sequence $\nu_0(a_j)$ has a lower bound $\overline{\nu}_0$. Since $\nu_j \longrightarrow \infty$, we have $\nu_j > \overline{\nu}_0$ for sufficiently large j. Consequently,

$$M(\eta_j, \nu_j) + a_j R(\eta_j, \nu_j) = 0, \quad \nu_j > \overline{\nu}_0.$$

It is the resulting contradiction that proves the lemma.

We now show that (I)$\Rightarrow$(IIa). To this end we take the monomial

$$Q(\zeta, \tau) = \zeta^\alpha \tau^j, \quad qj + |\alpha| \leqslant mq - 1,$$

as the polynomial Q. Then for an arbitrary $a \in \mathbb{C}$ there are $\nu_0(a)$, $\gamma_0(a, \nu)$ such that

$$P(\xi + i\nu N, \tau) + aQ(\xi + i\nu N, \tau) \neq 0, \qquad \nu \geqslant \nu_0(a), \quad \operatorname{Im} \tau < \gamma_0(a, \nu).$$

We now set $\tau = \sigma + i\gamma + i\gamma_0(a, \nu)$, $\eta = (\xi, \sigma, \gamma)$,

$$\Omega = \{\eta \in \mathbb{R}^{n+2},\ \gamma \geqslant 0\},$$
$$M(\eta, \nu) = -P(\xi + i\nu N, \tau), \quad R(\eta, \nu) = -Q(\xi + i\nu N, \tau).$$

Applying the lemma to M and R we conclude that there exists a function $\varepsilon(\nu) \to 0$, $\nu \to +\infty$, such that

$$|\zeta^\alpha \tau^j| < \varepsilon(\nu)|P(\zeta, \tau)|, \quad |\alpha| + qj \leqslant mq - 1,$$
$$\nu \geqslant \nu_0 = \nu_0(1), \quad \operatorname{Im} \tau \leqslant \gamma_0(1, \nu), \quad \zeta = \xi + i\nu N.$$

These inequalities imply (11). The theorem is proved.

CHAPTER II

The Cauchy Problem

Introduction

This chapter is devoted to the Cauchy problem for partial differential operators (generally, with variable coefficients) whose symbols are polynomials introduced in Chapter I. We will deal with operators $P(x, t, D_x, D_t)$ resolved with respect to the highest power of D_t, i. e., having the form

$$P(x, t, D_x, D_t) = D_t^m + \sum_{j=1}^{m} P_j(x, t, D_x) D_t^{m-j}, \tag{1}$$

where $P_j(x, t, D_x)$ are differential operators in $\mathbb{R}_x^n$ depending on the parameter t. By the Cauchy problem for equation (1) we mean the problem of determining a (smooth) function $u(x, t)$ satisfying the equation

$$P(x, t; D_x, D_t)u(x, t) = f(x, t), \qquad x \in \mathbb{R}^n, \quad t > 0, \tag{2}$$

and the initial conditions

$$D_t^j u(x, t)|_{t=0} = \varphi_j(x), \qquad j = 0, \dots, m-1, \quad x \in \mathbb{R}^n. \tag{3}$$

In the case when $\varphi_0 \equiv \dots \equiv \varphi_{m-1} \equiv 0$ we obtain the so-called homogeneous Cauchy problem. By means of the transformation

$$u = v + \sum_{j=0}^{m-1} t^j \varphi_j(x)/j!,$$

the problem (2), (3) can be reduced to a similar problem for v with zero initial conditions. If the right-hand side of (3) is identically equal to zero, then equation (2) remains valid for $t \in \mathbb{R}$ if the functions $u(x, t)$ and $f(x, t)$ are extended as zero to $t < 0$, i. e., we obtain an equation in $\mathbb{R}^{n+1}$:

$$P(x, t; D_x, D_t)u(x, t) = f(x, t), \quad (x, t) \in \mathbb{R}^{n+1}, \tag{4}$$

$$u(x, t) = f(x, t) = 0 \quad \text{for} \quad t < 0. \tag{5}$$

In what follows, by the Cauchy problem we mean problem (4), (5).

Sections 1 and 2 are mainly of introductory character, and they differ in style from the rest of the chapter, where all the proofs are carried out thoroughly and with all technical details. Section 1 presents some facts of the theory of distributions and function spaces that are used in the sequel. The reader is assumed to be familiar with the majority of these facts, and they are given here to fix the notation. All the questions touched upon in §1 are presented in the required form and with detailed proofs in the initial sections of Chapters 1 and 2 of the authors' book **[GV1]**.

In §2 the technique of Fourier-Laplace transforms is applied to formulate the theorems of Chapter I in terms of estimates and solvability conditions for partial differential operators with constat coefficients in the spaces introduced in §1. The Fourier and Laplace transforms make it possible to represent the solution as a result of the application of a pseudodifferential operator to the right-hand side of the equation (or, which is the same, as a result of convolution of the fundamental solution with the right-hand side). For some classes of differential operators with variable coefficients the Cauchy problem can be solved by means of the so-called parametrix method, which is a specialized version of the method of successive approximations in which at the first step an equation with constant (frozen) coefficients is solved. This method, which is presented in the concluding part of §2, imposes very strong constraints on the symbols of the operators in question, which makes it inapplicable to hyperbolic and q-hyperbolic operators with variable coefficients of the principal (q-principal) part.

An alternative to the parametrix type methods (that represent the solution explicitly as a Neumann series) is implicit functional methods based on the proof of the absence of the kernel of the adjoint operator. In other words, let there be given a closed unbounded operator A from a Banach space E into a Banach space F and let the domain of A contain a dense subdomain $E_0 \subset E$. Let $A^* : F' \to E'$ be the adjoint operator and let F_0' be a dense subset contained in F'. Then to prove the unique solvability of the equation

$$Ae = f, \qquad e \in E, \quad f \in F, \tag{6}$$

it suffices to establish the following pair of a priori estimates:

$$|e, E| \leqslant C|Ae, F| \quad \forall e \in E_0, \tag{7}$$

$$|h, F'| \leqslant C|A^*h, E'| \quad \forall h \in F_0'. \tag{7'}$$

This approach makes it easy to prove the solvability theorems for differential equations defined throughout the space. However, in studying the Cauchy problem (4), (5) there arise difficulties relating to the fact that we deal with spaces of functions that are equal to 0 for $t < 0$. The corresponding conjugate spaces are some quotient spaces. In §3 a special technique is developed (see **[V1]**), which in this case as well makes it possible to confine oneself to a pair of a priori estimates for functions defined throughout the space. The main subject of the remaining part of this chapter is the derivation of these estimates for strictly hyperbolic, q-parabolic, and q-hyperbolic operators. They are derived using the method of Leray's separating operator **[Le]** based on estimates for the quadratic forms

$$\int w^2(x,t)P(x,t;D_x,D_t)u\overline{P^{(1)}(x,t,D_x,D_t)u}\,dx\,dt \tag{8}$$

and some more general forms. Here we have $P^{(1)}(x,t,\xi,\tau) = \partial P(x,t,\xi,\tau)/\partial\tau$, and $w(x,t)$ is equal to either $\exp(\gamma t)$ or $\exp(\gamma t + \nu\langle N, x\rangle)$. In the case of constant coefficients the desired estimates follow from those presented in Chapter I for the polynomials $H_P(\xi,\sigma,\gamma)$. In §§4–6 it is proved that if some natural uniformity conditions (with respect to the parameters x and t) are imposed on the symbols of strictly hyperbolic, q-parabolic, and q-hyperbolic operators, then the upper and lower bounds (with constants not depending on γ and ν) for the forms (8) can be expressed in terms of similar forms for the operator $P(x^0, t^0, D_x, D_t)$ with "frozen" coefficients. The proofs of this result for all the three classes of operators are based on the same ideas but differ in some details. In this connection, the proof for the hyperbolic case will be carried out more thoroughly. In the case of q-parabolic and q-hyperbolic operators the proofs will be presented more superficially, and attention will be drawn only to the new technical details.

The a priori estimates make it possible to deduce the unique solvability of the Cauchy problem from algebraic conditions on the symbol of the operator. We also prove the necessity of these conditions for the validity of the a priori estimates and, hence, the validity of the theorem on the unique solvability for any q-lower terms. We thus obtain exact analogs of the theorem on equations with constant coefficients for equations with variable coefficients.

In §7 some broad classes of strictly hyperbolic, q-parabolic, and q-hyperbolic systems are considered. Based on the results obtained in the scalar case, we prove the solvability of the (generalized) Cauchy problem for these systems.

In Appendixes 1 and 2 we consider some classes of hyperbolic and q-hyperbolic systems for which energy estimates cannot be obtained by reduction to the scalar case.

§1. Some auxiliary notions. Basic spaces of functions and distributions

In this section we give a survey (without proofs) of some facts relating to distributions, Fourier transforms, pseudodifferential operators, and Sobolev's spaces with exponential weights. For detailed presentation of all questions mentioned in §1 see **[GV1]**.

1.1. Following L. Schwartz, we denote by $\mathcal{D} = \mathcal{D}(\mathbb{R}^n)$ and $\mathcal{D}(\Omega)$ the space of infinitely differentiable functions with compact support and the set of those functions whose supports belong to a region $\Omega \in \mathbb{R}^n$.

Let $\mathcal{D}'(\Omega)$ and $\mathcal{D}' = \mathcal{D}'(\mathbb{R}^n)$ denote the spaces of continuous linear functionals on $\mathcal{D}(\Omega)$ and $\mathcal{D}$, respectively, endowed with the strong topology of conjugate space.

The elements of $\mathcal{D}'$ are called distributions. By (f, φ) we will denote the value of the functional $f \in \mathcal{D}'$ on the element $\varphi \in \mathcal{D}$. If f is a locally absolutely summable measurable function ($f \in L_1^{\text{loc}}$), then the functional

$$(f, \varphi) = \int_{\mathbb{R}^{n+1}} f(y)\varphi(y)\, dy \tag{1}$$

on $\mathcal{D}$ is associated with it. In other words, the canonical embedding $L_1^{\text{loc}} \longrightarrow \mathcal{D}'$ is defined (for a given element of volume dy). It will be convenient to interpret the various function spaces defined below as subsets in $\mathcal{D}'$.

Following L. Schwartz, we denote by $\mathcal{S}$ the space of infinitely differentiable functions such that for any $m \in \mathbb{Z}_+$ and $l \in \mathbb{R}$ their norms

$$|\varphi|_{(l)}^{(m)} = \sup_{y \in \mathbb{R}^{n+1}, |\alpha| \leqslant m} (1 + |y|^2)^{l/2} |D^\alpha u(y)|$$

are bounded. We denote by $\mathcal{S}'$ the conjugate space endowed with the topology of strongly conjugate space. The notation (f, φ) will be retained for the values of the functional $f \in \mathcal{S}'$ on the test functions $\varphi \in \mathcal{S}$. The elements $f \in \mathcal{S}'$ are usually called tempered distributions.

If f and φ are test functions (i. e., elements of $\mathcal{D}$ and $\mathcal{S}$, respectively), then the integration by parts in (1) results in

$$(D^\alpha f, \varphi) = (f, (-D)^\alpha \varphi). \tag{2}$$

Based on this relation, we define the derivative $D^\alpha f$ of an element $f \in \mathcal{D}'$ as a functional satisfying (2). We thus define in $\mathcal{D}'$ the action of differential operators with constant coefficients:

$$(P(D)f, \varphi) = (f, P(-D)\varphi). \tag{3}$$

1.2. The Fourier transform of a function $u(y) \in \mathcal{D}, \mathcal{S}$ is the function

$$\widehat{u}(\eta) = (2\pi)^{-(n+1)/2} \int \exp(-i\langle y, \eta\rangle) u(y)\, dy, \tag{4}$$

and we have

$$\widehat{D^\alpha u}(\eta) = \eta^\alpha \widehat{u}(\eta). \tag{5}$$

Thus, if $P(D)$ is a differential operator with constant coefficients, then

$$(\widehat{P(D)u})(\eta) = P(\eta)\widehat{u}(\eta). \tag{5'}$$

For the Fourier transform we have the following inversion formula:

$$u(y) = (2\pi)^{-(n+1)/2} \int \exp(i\langle y, \eta\rangle) \widehat{u}(\eta)\, d\eta. \tag{4'}$$

Under the transition to Fourier transforms the bilinear form (1) goes into

$$(f, \varphi) = (\widehat{f}, I\widehat{\varphi}), \quad (I\widehat{\varphi})(\eta) \stackrel{\text{def}}{=} \widehat{\varphi}(-\eta). \tag{6}$$

Along with the form (1), we will consider the Hermitian form

$$[f, \varphi] = (f, \overline{\varphi}) = \int f(y)\overline{\varphi(y)}\, dy. \tag{1'}$$

For this form formula (6) implies

$$[f, \varphi] = [\widehat{f}, \widehat{\varphi}]. \tag{6'}$$

Setting $\varphi = f$ we obtain the classical Parseval relation:

$$\int |f(y)|^2\, dy = \int |\widehat{f}(\eta)|^2\, d\eta. \tag{7}$$

The Fourier operator determined by integral (4) transforms the space $\mathcal{S}$ into itself, and, by conjugacy, it also transforms the space $\mathcal{S}'$ into itself, and for an element $f \in \mathcal{S}'$ the generalized Fourier transform is defined by means of (6) (here $\varphi \in \mathcal{S}$ should be taken).

1.3. By a pseudodifferential operator we mean an operator having the form

$$(a(y, D)\varphi)(y) = (2\pi)^{-(n+1)/2} \int \exp(-i\langle y, \eta\rangle) a(y; \eta)\widehat{\varphi}(\eta)\, d\eta. \tag{8}$$

The function $a(y,\eta)$ is called the symbol of the operator (8). If the symbol is a polynomial with respect to the variable η, i. e., $a(y,\eta) = \sum a_\alpha(y)\eta^\alpha$, then (8) can be rewritten as

$$(a(y;D)\varphi)(y) = \sum_\alpha a_\alpha(y)D^\alpha\varphi.$$

If a symbol does not depend on y, then the corresponding operator (8) is called a pseudodifferential operator with constant coefficients:

$$(a(D)\varphi)(y) = (2\pi)^{-(n+1)/2}\int \exp(-i\langle y,\eta\rangle)a(\eta)\widehat{\varphi}(\eta)\,d\eta. \tag{9}$$

If a symbol $a(\eta)$ belongs to the space $\mathcal{M}$ of infinitely differentiable functions such that the function $a(\eta)$ and all its derivatives $\partial^\alpha a(\eta)$ increase at infinity no faster than a power of η (depending on α), then the operator of multiplication by $a(\eta)$ transforms $\mathcal{S}$ into itself. In this case the operator $a(D)$, which is the composition of the Fourier operator and the operator of multiplication by $a(\eta)$, also transforms $\mathcal{S}$ into itself.

By conjugacy, the operator $a(D)$ can be defined on $\mathcal{S}'$:

$$(a(D)\varphi) = (f, a(-D)\varphi), \qquad f\in\mathcal{S}', \quad \varphi\in\mathcal{S}, \quad a\in\mathcal{M}. \tag{9'}$$

This operator transforms $\mathcal{S}'$ into itself.

1.4. We now define some spaces that will constantly be used in the sequel.

Denote by $H^{(s)}$ the space of elements $f\in\mathcal{S}'$ such that $\widehat{f}$ is a locally square summable function and the norm

$$\|f\|_{(s)} = \left(\int (1+|\eta|^2)^{s/2}|\widehat{f}(\eta)|^2\,d\eta\right)^{1/2} \tag{10}$$

is finite. By the Parseval relation (7), for $s=0$ the norm (10) goes into the ordinary L_2 norm. For natural numbers s the norm (10) is equivalent to Sobolev's norm

$$\left(\sum_{|\alpha|\leqslant s}\|D^\alpha f\|^2\right)^{1/2}. \tag{10'}$$

Recalling the definition of the pseudodifferential operator with symbol $(1+|\eta|^s)^{1/2}$ (see (9)) we can rewrite the norm (10) in the form

$$\|f\|_{(s)} = \|(1+|D|^2)^{s/2}f\|.$$

In other words, we can define $H^{(s)}$ as the space of functions $f\in\mathcal{S}'$ such that $(1+|D|^2)^{s/2}f\in L_2$. Each element $f\in H^{(s)}$ can be represented as

$$f = (1+|D|^2)^{-s/2}g, \qquad g\in L_2, \quad \|g\| = \|f\|_{(s)}. \tag{11}$$

According to the definition of the norm (10), we have $\|f\|_{(s')}\leqslant\|f\|_{(s)}$ for $s'<s$ so that the natural embeddings

$$i_s^{s'}: H^{(s)}\longrightarrow H^{(s')}, \quad s'<s,$$

are defined, and for $s>s'$ the space $H^{(s)}$ can be regarded as a subset of $H^{(s')}$. We can consider the projective and inductive limits of the scale of the spaces $H^{(s)}$:

$$H^{(\infty)} = \bigcap_s H^{(s)}, \quad H^{(-\infty)} = \bigcup_s H^{(s)}.$$

The space $H^{(\infty)}$ consists of infinitely differentiable functions f (Sobolev's embedding theorem) which, together with all their derivatives, belong to $D^\alpha f$. The space $H^{(\infty)}$ is dense in any space $H^{(s)}$, $s > -\infty$, so that $H^{(s)}$ can be regarded as the completion of $H^{(\infty)}$ with respect to the norm (10).

By (6) and the Schwarz inequality, we have

$$|(f,\varphi|) = |(1+|D|^2)^{-s/2} f, ((1+|D|^2)^{s/2}\varphi)| \leqslant \|f\|_{(-s)}\|\varphi\|_{(s)} \tag{12}$$

for any $f, \varphi \in H^{(\infty)}$. Since $H^{(\infty)}$ is dense in an arbitrary space $H^{(s)}$, the form (1) can be continued by continuity from $H^{(\infty)} \times H^{(\infty)}$ to $H^{(-s)} \times H^{(s)}$. The embedding $H^{(-s)} \longrightarrow (H^{(s)})'$ induced by the form (9′) is the canonical isomorphism of $H^{(-s)}$ and the Banach conjugate space of $H^{(s)}$:

$$(H^{(s)})' = H^{(-s)}. \tag{13}$$

These isomorphisms remain valid in the case $s = +\infty$ as well.

1.5. If Φ is a space of functions or distributions, and we have $\omega \in \mathbb{R}^{n+1}$, then by $\Phi_{[\omega]}$ we denote the space of elements f such that $f \exp\langle y, \omega\rangle \in \Phi$. If Φ is a Banach space, then $\Phi_{[\omega]}$ can also be turned into a Banach space by endowing it with the norm

$$\|f, \Phi_{[\omega]}\| = \|\exp\langle y, \omega\rangle f, \Phi\|.$$

Let Φ' be the conjugate space of Φ. Then the space $(\Phi')_{[-\omega]}$ can be identified with the conjugate space of $\Phi_{[\omega]}$:

$$(\Phi_{[\omega]})' = (\Phi')_{[-\omega]}. \tag{14}$$

If the Fourier transform is defined in the space Φ, then in the space $\Phi_{[\omega]}$ we can define the complex Fourier (Laplace) transform

$$\begin{aligned} \widehat{\varphi}(\xi + i\omega) &= (2\pi)^{-(n+1)/2} \int \exp(-i\langle y, \eta\rangle)(\exp\langle \omega, y\rangle \varphi(y))\, dy \\ &= (2\pi)^{-(n+1)/2} \int \exp(-i\langle y, \eta + i\omega\rangle)\varphi(y)\, dy. \end{aligned} \tag{15}$$

The inversion formula for this transform takes the form

$$\varphi(y) = (2\pi)^{-(n+1)/2} \int \exp(i\langle y, \eta + i\omega\rangle)\widehat{\varphi}(\eta + i\omega)\, d\eta. \tag{15′}$$

Proceeding from (15) and (15′), we can define pseudodifferential operators on $\Phi_{[\omega]}$:

$$\begin{aligned} {}_{[\omega]}a(D)\varphi &= (2\pi)^{-(n+1)/2} \int_{\mathbb{R}^{n+1}} \exp(i\langle y, \eta + i\omega\rangle) a(\eta + i\omega)\widehat{\varphi}(\eta + i\omega)\, d\eta \\ &= \exp\langle -y, \omega\rangle a(D + i\omega)\exp\langle y, \omega\rangle \varphi. \end{aligned} \tag{16}$$

It should be stressed that the symbol $a(\zeta)$ in (16) is assumed to be defined only for $\operatorname{Im}\zeta = \omega$. We note that in the case of differential operators we have the identity

$$\exp\langle -y, \omega\rangle P(D + i\omega)\exp\langle y, \omega\rangle = P(D). \tag{17}$$

This identity also remains valid for pseudodifferential operators having holomorphic symbols and acting, for instance, in the space $\mathcal{D}$. More precisely, if we have $\Phi = \mathcal{D}$, and the holomorphic function $a(\zeta)$ increases at infinity no faster than a power of $|\operatorname{Re}\zeta|$, then, according to Cauchy's theorem, the right-hand side of (16) does not depend on ω, and we arrive at the original definition of (9).

As a rule, we will deal with the spaces $\Phi_{[\omega]}$ corresponding to $\omega = (0, \dots, 0, \gamma)$, and in this case, instead of $\Phi_{[0,\dots,\gamma]}$, we will write $\Phi_{[\gamma]}$:

$$\Phi_{[\gamma]} = \{f, \exp(\gamma t) f \in \Phi\}.$$

According to the above definitions, $H^{(s)}_{[\gamma]}$ is the space of distributions such that their Fourier transforms $\widehat{f}(\xi, \sigma + i\gamma)$ are locally square integrable and the integral

$$\iint (1 + |\xi|^2 + \sigma^2)^s |\widehat{f}(\xi, \sigma + i\gamma)|^2 \, d\xi \, d\sigma$$

is convergent. As to the norm in $H^{(s)}_{[\gamma]}$, it is convenient to take it in the form of the expression

$$\Big(\iint\limits_{\operatorname{Im} \tau = \gamma} (1 + |\xi|^2 + |\tau|^2)^s |\widehat{f}(\xi, \tau)|^2 \, d\xi \, d\sigma \Big)^{1/2}, \tag{18}$$

which is equivalent to the norm $\| \exp(\gamma t) f \|_{(s)}$ with equivalence constants depending on γ. In the case of positive integers s, applying Newton's binomial formula to the integrand:

$$(1 + |\xi|^2 + |\tau|^2)^s = \sum_{k+\beta \leqslant s} \frac{s!}{\beta! k! (s - k - \beta)!} |\tau|^{2\beta} |\xi|^{2k}$$

we conclude that the norm (18) is equivalent to the norm

$$\Big(\sum_{|\alpha| + \beta \leqslant s} \| \exp(\gamma t) D_x^\alpha D_t^\beta f \|^2 \Big)^{1/2} \tag{18$'$}$$

with equivalence constants depending on γ.

By (13) and (14), we have

$$\left(H^{(s)}_{[\gamma]} \right)' = H^{(-s)}_{[-\gamma]}. \tag{19}$$

1.6. If Φ is a space of functions or distributions in $\mathbb{R}^{n+1}$, then by Φ_+ we will denote the subspace of elements $f \in \Phi$ such that their supports belong to the half-space $\mathbb{R}^{n+1}_+ = \{(x, t) \in \mathbb{R}^{n+1}, t \geqslant 0\}$.

Below we describe the spaces $H^{(s)}_{[\gamma]+}$; for detailed proofs see **[GV1,** Chapter 2, §1]. We first consider the case $s = 0$.

PROPOSITION 1. *The inclusion $f \in H_{[\gamma]+}$ holds if and only if $f \in H_{[\rho]}$ for $\rho < \gamma$ and the norms $\| \exp(\gamma t) f \|$ are uniformly bounded with respect to $\rho < \gamma$. Moreover,*

$$\sup_{\rho < \gamma} \| \exp(\rho t) f \| = \| \exp(\gamma t) f \|, \quad f \in H_{[\gamma]+}. \tag{20}$$

According to (15), relation (20) can be rewritten in the form

$$\sup_{\rho < \gamma} \Big(\iint\limits_{\operatorname{Im} \tau = \rho} |\widehat{f}(\xi, \tau)|^2 \, d\xi \, d\xi \Big)^{1/2} = \| \exp(\gamma t) f \|, \quad f \in H_{[\gamma]+}. \tag{21}$$

PROPOSITION 2 (Paley-Wiener). *A function $f(y)$ belongs to $H_{[\gamma]+}$ if and only if the Fourier-Laplace transform $\widehat{f}(\xi,\tau)$ is a holomorphic function of τ for* $\operatorname{Im}\tau < \gamma$ *and almost all $\xi \in \mathbb{R}^n$, the integrals on the right-hand side of* (21) *exist, and relation* (21) *holds.*

A generalization of Proposition 2 to the spaces $H^{(s)}_{[\gamma]+}$ is

PROPOSITION 2′. *The inclusion $f \in H^{(s)}_{[\gamma]+}$ holds if and only if the Fourier-Laplace transform $\widehat{f}(\xi,\tau)$ is a holomorphic function of τ for* $\operatorname{Im}\tau < \gamma$ *and almost all $\xi \in \mathbb{R}^{n+1}$ and the integral*

$$\iint_{\operatorname{Im}\tau\leqslant\rho} (1+|\xi|^2+|\tau|^2)^s |f(\xi,\tau)|^2\, d\xi\, d\sigma \tag{22}$$

is uniformly bounded for $\rho < \gamma$.

COROLLARY. *The space $H^{(\infty)}_{[\gamma]+}$ ($H^{(-\infty)}_{[\gamma]+}$) consists of functions (distributions) f such that their Fourier-Laplace transforms $\widehat{f}(\xi,\tau)$ are holomorphic functions of τ for* $\operatorname{Im}\tau<\gamma$ *and almost all $\xi \in \mathbb{R}^{n+1}$, and for $\rho<\gamma$ integrals* (22) *are uniformly bounded for all $s\in\mathbb{R}$ (some $s \in \mathbb{R}$).*

Let a function $a(\xi,\tau)$ be continuous in all directions and holomorphic with respect to the variable τ, and let it increase at infinity no faster than a power of $|\xi|+|\operatorname{Re}\tau|$. Then (cf. 1.5) the values of the pseudodifferential operators ${}_{[\rho]}a(D)f$ do not depend on ρ so that the operator

$$a(D) \overset{\text{def}}{=} {}_{[\rho]}a(D)f : H^{(-\infty)}_{[\gamma]+} \longrightarrow H^{(-\infty)}_{[\gamma]+}$$

is well defined.

The space $H^{(s)}$ was defined above as the image of $H^{(0)}=H$ under the action of the pseudodifferential operator $(1+|D|^2)^{-s/2}$. This operator can be replaced by any other operator such that the upper and lower bounds of its symbol can be estimated via $(1+|\eta|)^{-s}$. In the case of the spaces $H^{(s)}_{[\gamma]+}$ the pseudodifferential operator with symbol

$$\delta_s^+(\xi,\tau) = \left(i\tau+\gamma_0+\sqrt{1+|\xi|^2}\right)^s \tag{23}$$

can be taken as the "graduating operator". This symbol is holomorphic with respect to τ for $\operatorname{Im}\tau<\gamma_0$, and the upper and lower bounds of its modulus can be expressed in terms of $(|\xi|+|\tau|^2)^{s/2}$ with constants depending on γ_0 only. Hence, the pseudodifferential operator

$$\delta_s^+(D) = {}_{[\gamma]}\delta_s^+(D) \quad (\gamma\leqslant\gamma_0) \tag{23′}$$

transforms the space $H^{(\infty)}_{[\gamma]+}$ into itself, and we can define the norm in $H^{(s)}_{[\gamma]+}$ using this operator:

(24)

$$\|f\|_{(s),[\gamma]} \overset{\text{def}}{=} \|\exp(\gamma t)\delta_s^+(D)f\| = \|\delta_s^+(D_x, D_t+i\gamma)\exp(\gamma t)f\|$$
$$= \left(\iint \left(\sigma^2+\left(\sqrt{1+|\xi|^2}+\gamma_0-\gamma\right)^2\right)^s |\widehat{f}(\xi,\sigma+i\gamma)|^2\, d\xi\, d\sigma\right)^{1/2}.$$

For $\gamma \leqslant \gamma_0$ this norm is equivalent to the norm (18) with equivalence constants depending only on γ_0.

Proposition 1 can be generalized in the following way.

PROPOSITION 1′. (i) $f \in H^{(s)}_{[\gamma]}$ *if an only if the distribution* f *can be represented in the form*

$$f = \delta^{+}_{-s}(D)g, \qquad g \in H_{[\gamma]+}, \quad \|\exp(\gamma t)g\| = \|f\|_{(s),[\gamma]}.$$

(ii) $f \in H^{(s)}_{[\gamma]}$ *if and only if* $f \in H^{(s)}_{[\rho]}$ *for any* $\rho < \gamma$ *and the norms* $\|f\|_{(s),[\rho]}$ *are uniformly bounded for* $\rho < \gamma$. *Moreover,*

$$\sup_{\rho<\gamma} \|f\|_{(s),[\rho]} = \|f\|_{(s),[\gamma]}. \tag{25}$$

1.7. We will also consider the intersection and union of the spaces $H^{(s)}_{[\gamma]}$. For $\gamma' \leqslant \gamma''$ we denote by $H^{(s)}_{[\gamma',\gamma'']}$ the intersection

$$H^{(s)}_{[\gamma',\gamma'']} = H^{(s)}_{[\gamma']} \cap H^{(s)}_{[\gamma'']}, \quad \gamma' \leqslant \gamma'',$$

endowed with the natural norm of intersection of two Banach spaces:

$$\|u\|_{(s),[\gamma_1]} + \|u\|_{(s),[\gamma_2]}. \tag{26}$$

PROPOSITION 1. $f \in H^{(s)}_{[\gamma',\gamma'']}$, $\gamma' \leqslant \gamma''$, *if and only if* $f \in \bigcap_{\gamma'<\rho<\gamma''} H^{(s)}_{[\rho]}$ *and the norm*

$$\|f\|_{(s)[\gamma',\gamma'']} \overset{\text{def}}{=} \sup_{\gamma'<\gamma<\gamma''} \|f\|_{(s)[\rho]} \tag{27}$$

is finite. Moreover, the norms (26) *and* (27) *are equivalent.*

For $s = 0$ the space $H^{(s)}_{[\gamma',\gamma'']}$ consists of functions square summable with weight $\exp(\gamma' t) + \exp(\gamma'' t)$. For $s \neq 0$ it consists of functions having the form

$$_{[\rho]}\delta^{+}_{s}(D)g, \qquad g \in H^{(s)}_{[\gamma',\gamma'']}, \quad \gamma' \leqslant \rho \leqslant \gamma''. \tag{28}$$

REMARK. The right-hand side of (27) makes sense for the infinite value $\gamma' = -\infty$ as well. Indeed, according to Proposition 1′ in 1.7, for $\gamma' = -\infty$ the norm (27) is finite if and only if $f = 0$ for $t < 0$; in this case (27) coincides with $\|f\|_{(s),[\gamma'']}$. Thus, in Proposition 1 only γ'' should be assumed to be bounded, whereas γ' can take both finite and infinite values.

For $\gamma' > \gamma''$ we denote by $H^{(s)}_{[\gamma',\gamma'']}$ the linear span of the union $H^{(s)}_{[\gamma']} + H^{(s)}_{[\gamma'']}$, and the norm can be taken in the form

$$\|u\|_{(s)[\gamma',\gamma'']} = \|\chi u\|_{(s),[\gamma'']} + \|(1-\chi)u\|_{(s),[\gamma']}, \quad \gamma' > \gamma'', \tag{29}$$

where

$$\chi(t) \in C^{\infty}(\mathbb{R}), \quad \chi(t) = 1 \text{ for } t \geqslant 0, \quad \chi(t) = 0 \text{ for } t \leqslant -1.$$

If $s = 0$, then we can take

$$\left(\int (\exp(-\gamma' t) + \exp(-\gamma'' t))^{-2} |u(y)|^2 \, dy\right)^{1/2}$$

as a norm in $H^{(s)}_{[\gamma',\gamma'']}$, $\gamma' > \gamma''$. For $s = 0$ there is an analog of representation (28).

The main fact that will be used in the sequel is the duality of the above spaces corresponding to the cases $\gamma' \leqslant \gamma''$ and $\gamma' > \gamma''$.

PROPOSITION 2. *The form* (1) *can be defined on*

$$H^{(\infty)}_{[\gamma_1,\gamma_2]} \times H^{(\infty)}_{[-\gamma_1,-\gamma_2]}$$

and extended by continuity to $H^{(s)}_{[\gamma_1,\gamma_2]} \times H^{(-s)}_{[-\gamma_1,-\gamma_2]}$. *The embedding*

$$H^{(-s)}_{[-\gamma_1,-\gamma_2]} \longrightarrow \left(H^{(s)}_{[\gamma_1,\gamma_2]}\right)'$$

induced by this form is the canonical isomorphism of $H^{(-s)}_{[-\gamma_1,-\gamma_2]}$ *and the Banach conjugate space of* $H^{(s)}_{[\gamma_1,\gamma_2]}$:

$$\left(H^{(s)}_{[\gamma_1,\gamma_2]}\right)' = H^{(-s)}_{[-\gamma_1,-\gamma_2]}. \tag{30}$$

In the case when $-\gamma_1 = \gamma'$, $-\gamma_2 = \gamma''$, *and* $\gamma' > \gamma''$ *we conclude that the norm* (29) *is equivalent to that of the Banach conjugate space of* $H^{(-s)}_{[-\gamma',-\gamma'']}$.

REMARK. The space $H^{(s)}_{[\gamma',\gamma'']}$, where $\gamma' > \gamma''$, can be considered for $\gamma' = \infty$ as well. In this case it is realized as the quotient space with respect to the subspace of elements that vanish for $t > 0$ (see the next subsection), and the duality (30) also remains valid.

For detailed proofs of the above results see **[GV1,** 2.5].

1.8. In conclusion we make some remarks about the spaces of functions (distributions) defined for t belonging to an interval on the real line.

Let Φ be a space of functions (distributions). Denote by $\Phi[a,\infty)$ the subspace of elements $f \in \Phi$ whose support belongs to the half-space $\{t \geqslant a,\ x \in \mathbb{R}^n\}$. In particular, we have $\Phi_+ = \Phi[0,\infty)$. For $a < b$ we denote by $\Phi[a,b)$ the quotient space

$$\Phi[a,b) = \Phi[a,\infty)/\Phi[b,\infty). \tag{31}$$

If Φ is a Banach space with norm $\|f,\Phi\|$, then $\Phi[a,b)$ can be endowed with the quotient norm

$$\|f,\Phi[a,b)\| = \inf_{f_1\in\Phi[b,\infty)} \|f_0+f_1,\Phi\|, \tag{32}$$

where f_0 is a representative of the residue class f.

We can similarly define the subspace $\Phi(-\infty,a]$ of elements whose supports belong to the subspace $\{t \leqslant a, x \in \mathbb{R}^n\}$, and for $a < b$ we denote by $\Phi(a,b]$ the quotient space

$$\Phi(a,b] = \Phi(-\infty,b]/\Phi(-\infty,a]. \tag{31$'$}$$

For the above spaces, say for $\Phi = H^{(s)}$, the following duality relation holds:

$$(\Phi[a,b))' = (\Phi)'(a,b], \quad (\Phi(a,b])' = (\Phi)'[a,b). \tag{33}$$

As to the spaces $H^{(s)}[a,b]$, here we need the following

PROPOSITION. *For any $\rho \in \mathbb{R}$ the space $H^{(s)}[a,b]$ coincides with $(H^{(s)}_{[\rho]})[a,b)$, and the quotient norm in $H^{(s)}[a,\infty)/H^{(s)}[b,\infty)$ is equivalent to that of the space $H^{(s)}_{[\rho]}[a,\infty)/H^{(s)}_{[\rho]}[b,\infty)$.*

For the proof see [**GV1,** 2.2.3, Lemma 1].

§2. The Cauchy problem for differential operators with constant coefficients

As was mentioned in the introduction to the present chapter, in this section the basic results on polynomials that were established in Chapter I will be interpreted in terms of the Cauchy problem for differential operators with constant coefficients. First we formulate Petrovskii's correctness condition for a polynomial $P(\xi,\tau)$ as a solvability condition for the Cauchy problem in the case of the equation

$$P(D_x, D_t)u = f. \tag{1}$$

It should be noted that here and henceforth this equation will be understood in the sense of the theory of distributions, i. e.,

$$(u, P(-D_x, -D_t)\varphi) = (f,\varphi) \quad \forall\varphi \in D. \tag{1'}$$

Equation (1) in spaces of the type $H^{(s)}_{[\gamma]+}$ will be related to the Cauchy problem (1.4), (1.5). This equation will also be considered in quotient spaces corresponding to finite time intervals.

Furthermore, the results of 1.2 will be formulated in terms of solvability conditions and estimates for the Cauchy problem. The results of 1.3 can also be formulated in a similar way in the language of differential operators. To this end it is necessary to consider spaces with exponential weight $\exp(\gamma t + \nu\langle N, x\rangle)$. However, we will put off these questions until §6.

The concluding part of the present section is unrelated to the further presentation. As will be seen from the consideration below, in the case of constant coefficients the question of existence of a solution to the Cauchy problem (of course, when Petrovskii's condition holds) is solved automatically because the solution is constructed "explicitly" using the Fourier transform. In the case of variable coefficients this question is by far not so simple. At the end of this section we present a special method of successive approximations (the so-called parametrix method) that makes it possible to construct solutions for equations with variable coefficients. However, this method is inapplicable to hyperbolic and q-hyperbolic equations with variable coefficient of the principal q-homogeneous part. An alternative approach to the solvability problem will be suggested in §3.

2.1. The Cauchy problem for partial differential operators correct in Petrovskii's sense. We first prove the following

PROPOSITION. *For a differential operator $P(D_x, D_t)$ with constant coefficients the conditions below are equivalent.*

(I) *There are γ_0 and μ such that for all $s \in \mathbb{R}$ and $\gamma \leqslant \gamma_0$ for any right-hand side $f \in H^{(s)}_{[\gamma]}$ there exists a unique solution $u \in H^{(s-\mu)}_{[\gamma]}$ to equation (1).*

(I_+) *There are γ_0 and μ such that for all $s \in \mathbb{R}$ and $\gamma \leqslant \gamma_0$ for any right-hand side $f \in H^{(s)}_{[\gamma]+}$ there exists a unique solution $u \in H^{(s-\mu)}_{[\gamma]+}$ to equation (1).*

(II) *There are* γ_0, μ, *and* $C > 0$ *such that for all* $s \in \mathbb{R}$ *the inequality*

$$\|u\|_{(s-\mu),\gamma} \leqslant C\|P(D)u\|_{(s),\gamma} \tag{2}$$

holds for any function $u \in H^{(\infty)}_{[\gamma]}$.

(II$_+$) *Inequality* (2) *holds for any function* $u \in H^{(\infty)}_{[\gamma]+}$.

(III) *There is* γ_0 *such that*

$$P(\xi,\tau) \neq 0, \qquad \operatorname{Im}\tau \leqslant \gamma_0, \quad \forall(\xi, \operatorname{Re}\tau) \in \mathbb{R}^{n+1}. \tag{3}$$

PROOF. (I)⇒(II). We associate with equation (1) the (in general, unbounded) operator

$$\mathcal{P} : H^{(s)}_{[\gamma]} \longrightarrow H^{(s-\mu)}_{[\gamma]} \quad (u \longmapsto P(D)u) \tag{4}$$

with domain $D_{\mathcal{P}} = H^{(s+M)}_{[\gamma]}$, where $M + \mu \geqslant \deg P$. This operator is closable. Indeed, assume that a sequence $u_j \in H^{(s+M)}_{[\gamma]}$ is such that $u_j \to 0$, $j \to \infty$, with respect to the norm of the space $H^{(s)}_{[\gamma]}$ and that $\mathcal{P}u_j \to f$ in $H^{(s-\mu)}_{[\gamma]}$. We have to show that $f \equiv 0$.

Note that the operator (4) can be regarded as the restriction of the continuous operator

$$H^{(s)}_{[\gamma]} \longrightarrow H^{(s-\mu-M)}_{[\gamma]} \quad (u \longmapsto P(D)u).$$

Consequently, we have $\mathcal{P}u_j \to 0$ in $H^{(s-\mu-M)}_{[\gamma]}$, and hence $f \equiv 0$.

According to the above conditions, the inverse operator $\mathcal{P}^{-1}$ is defined on the right space in (4) and is closed. By the Banach theorem, the operator $\mathcal{P}^{-1}$ is bounded, which implies (2).

(II)⇒(III). According to the Parseval relation, (2) is equivalent to the inequality

$$\int |\widehat{u}(\xi,\sigma+i\gamma)|^2|\delta^+_{s-\mu}(\xi,\sigma+i\gamma)|^2 \left[1 - C^2|\delta^+_{\mu}(\xi,\sigma+i\gamma)P(\xi,\sigma+i\gamma)|^2\right] d\xi\, d\sigma < 0.$$

Since an arbitrary positive function belonging to $L_1(\mathbb{R}^{n+1}_\eta)$ can be approximated with functions $|\widehat{u}(\xi,\sigma+i\gamma)\delta^+_s(\xi,\sigma+i\gamma)|^2$, this inequality implies that the expression in square brackets is nonnegative, whence

$$|P(\xi,\tau)| > C^{-1}|\delta^+_{-\mu}(\xi,\tau)|, \quad \operatorname{Im}\tau = \gamma \leqslant \gamma_0, \tag{3$'$}$$

i. e., condition (3) holds.

(III)⇒(I). If (3) holds, then, by the Tarski-Seidenberg theorem (see **[H1,** Appendix A]), there are $C > 0$ and $\mu \in \mathbb{R}$ such that (3′) holds. Using the Fourier transform we can solve equation (1) explicitly and find the solution

$$u = {}_{[\gamma]}P^{-1}(D)f \quad \forall f \in H^{(s)}_{[\gamma]}. \tag{5}$$

Multiplying (3′) by $|\widehat{u}(\xi,\sigma+i\gamma)\delta_s(\xi,\sigma+i\gamma)|$, squaring, and integrating over $\mathbb{R}^{n+1}$, we obtain (2), which implies the uniqueness.

(I$_+$)⇒(II$_+$) is proved in the same way as (I)⇒(II). To this end it is only necessary to replace $H^{(s)}_{[\gamma]}$ and $H^{(s-\mu)}_{[\gamma]}$ in (4) with $H^{(s)}_{[\gamma]+}$ and $H^{(s-\mu)}_{[\gamma]+}$, respectively.

(II$_+$)⇒(III). Inequality (2) is invariant with respect to the transformations $u(x,t) \longmapsto u(x,t+t_0)$ so that if it holds for all $u \in H^{(\infty)}_{[\gamma]+}$, then the same is true for the functions $u \in H^{(\infty)}_{[\gamma]}$ with supports bounded from the left in the t-direction. However, the set of these functions is dense in $H^{(s)}_{[\gamma]}$, which implies that (2) holds for arbitrary $u \in H^{(\infty)}_{[\gamma]}$.

(III)$\Rightarrow$(I_+). If (3) holds, then (3′) also holds, and therefore the operator (5) (see 1.6) transforms $H^{(s)}_{[\gamma]+}$ into $H^{(s-\mu)}_{[\gamma]+}$, which is what proves (I_+).

REMARK. Despite some similarity, condition (III) in the above proposition differs from Petrovskii's correctness condition in 1.1. This distinction manifests itself in that

(a) in (3) it is not assumed that the polynomial P is resolved with respect to the highest power of τ;

(b) in (3) the half-plane $\operatorname{Im}\tau \leqslant \gamma_0$ is fixed, whereas in Petrovskii's conditions it is assumed that there exists a half-plane corresponding to some γ_0.

It should be noted that the fact that the polynomial P is resolved with respect to the highest power of τ is not a necessary condition for the solvability of the Cauchy problem with zero initial conditions. If the Cauchy problem with nonzero initial data is considered for an operator with symbol

$$P(\xi,\tau) = P_0(\xi)\tau^m + \sum_{j=1}^{m} P_j(\xi)\tau^{m+j},$$

then the relation $P_0(\xi) \neq 0$ for every $\xi \in \mathbb{R}^n$ together with (3) is a necessary and sufficient condition for the unique solvability of problem (1.2), (1.3). For detailed discussion of this question see **[GV1,** Chapter 2, §2].

As for property (b), we arrive at condition (3) in an arbitrary half-plane by considering the homogeneous Cauchy problem in a finite strip $\{a \leqslant t \leqslant b,\ x \in \mathbb{R}^n\}$. We have the following result.

THEOREM. *For a polynomial $P(\xi,\tau)$ the conditions below are equivalent*:

(I) *there is γ_0 such that condition* (3) *holds*;

(II) *there is μ such that for any $a < b$ and $s \in \mathbb{R}$ and an arbitrary right-hand side $f \in H^{(s)}[a,b)$ there exists a unique solution $u \in H^{(s-\mu)}[a,b)$ to equation* (1).

REMARK. We say that $u \in H^{(s-\mu)}[a,b)$ is a solution to (1) if for any representatives $u_0 \in H^{(s-\mu)}[a,\infty)$ and $f_0 \in H^{(s)}[a,\infty)$ of the residue classes u and f we have

$$P(D)u_0 - f_0 \in H^{(s)}[b,\infty).$$

In the case of spaces of smooth functions we can say that the equation holds at each point in the strip $\{a \leqslant t \leqslant b,\ x \in \mathbb{R}^n\}$.

PROOF OF THE THEOREM. (I)$\Rightarrow$(II). 1) If (3) holds, then, by the above-mentioned Tarski-Seidenberg theorem, there are $C > 0$, $\mu \in \mathbb{R}$ such that condition (3′) holds.

2) If (3′) holds, then the the above proposition is true. Since the operator $P(D)$ commutes with the translations, the assertion of the proposition remains valid if the space $H^{(s)}_{[\gamma]+} = H^{(s)}_{[\gamma]}[0,\infty)$ is replaced by $H^{(s)}_{[\gamma]+}[a,\infty)$ with an arbitrary $a \in \mathbb{R}$.

3) To proceed to quotient spaces we need the following

LEMMA. *Let the condition* (3′) *hold, $u \in H^{(s-\mu)}_{[\gamma]+}[a,\infty)$, and $f = P(D)u \in H^{(s)}_{[\gamma]+}[b,\infty)$, where $b > a$. Then $u \in H^{(s-\mu)}_{[\gamma]+}[b,\infty)$.*

PROOF. If (3′) holds, then, according to 2), for the right-hand side $f \in H^{(s)}_{[\gamma]}[b,\infty)$, $\gamma \leqslant \gamma_0$, there is a solution $v \in H^{(s)}_{[\gamma]}[b,\infty)$. Therefore $P(D)(u-v) = 0$. Because in the space $H^{(-\infty)}_{[\gamma]}[a,\infty)$ the uniqueness takes place, we have $u \equiv v \in H^{(s-\mu)}_{[\gamma]}[b,\infty)$.

4) According to Proposition 1.8, the space $H^{(s)}[a,b)$ can be realized as the quotient space $H^{(s)}_{[\gamma]}[a,\infty)/H^{(s)}_{[\gamma]}[b,\infty)$ for any γ, say for $\gamma \leqslant \gamma_0$. Denote by $f_0 \in H^{(s)}_{[\gamma]}[a,\infty)$ a representative of the residue class f. By 2), there is $u_0 \in H^{(s-\mu)}_{[\gamma]}[a,\infty)$ such that $P(D)u_0 = f_0$. Let u be the residue class of $u_0 \mod H^{(s-\mu)}_{[\gamma]}[b,\infty)$. This class does not depend on the choice of f_0 in the residue class f. Indeed, let $f_0' \in H^{(s)}_{[\gamma]}[a,\infty)$ be another representative of this residue class and let u_0' be the corresponding solution to equation (1). Then we have $P(D)(u-u_0') = f - f_0' \in H^{(s)}_{[\gamma]}[b,\infty)$, and, according to the lemma, $u - u_0 \in H^{(s-\mu)}_{[\gamma]}[b,\infty)$. Thus, the solvability of (1) in the quotient space is proved.

5) It remains to check the uniqueness. Assume that $u \in H^{(s-\mu)}_{[\gamma]}[a,b)$ and that $P(D)u = 0$ in the sense of $H^{(s)}_{[\gamma]}[a,b)$. The latter assumption means that if $u_0 \in H^{(s)}_{[\gamma]}[a,\infty)$ is a representative of the residue class u, then $P(D)u_0 \in H^{(s)}_{[\gamma]}[b,\infty)$. For $\gamma \leqslant \gamma_0$ we can apply the lemma to conclude that $u_0 \in H^{(s)}_{[\gamma]}[b,\infty)$, i. e., $u = 0$.

(II)⇒(I). Let $a = 0$ and let $s - \mu$ be a sufficiently large negative number. Then for $f = \delta(x)$ the equation possesses a solution $G(x) \in H^{(s)}_{[\gamma]}[0,\infty)$, $\gamma \leqslant \gamma_0$, so that

$$P(D)G - \delta(x) \in H^{(s)}_{[\gamma]}[b,\infty).$$

Applying the Fourier transform and the estimates for the Fourier transforms of the elements of the right-hand space we can show (see **[GV1,** 2.4.9]) that

$$\{P(\tau,\xi) = 0, \quad \operatorname{Im}\tau \leqslant 0\} \Longrightarrow \{\operatorname{Im}\tau > C_1 \ln(1+|\tau|+|\xi|)\}.$$

By the above-mentioned Tarski-Seidenberg theorem, we conclude that the roots of the polynomial $P(\xi,\tau)$ lie in the half-plane $\operatorname{Im}\tau \geqslant \gamma_0$, which proves the desired assertion.

2.2. The Cauchy problem for q-stable correct partial differential operators with constant coefficients. With every $q \geqslant 1$ we associate the space of distributions $H^{(s|q)}$ with the norm

$$\|u\|_{(s|q)} = \left(\int (1+|\sigma|^2+|\xi|^{2q})^s |\hat{u}(\xi,\sigma)|^2\, d\xi\, d\sigma\right)^{1/2}. \tag{6}$$

For $q = 1$ this space becomes $H^{(s)}$, and the norm (6) becomes the ordinary norm $\|\cdot\|_{(s)}$. The spaces $H^{(s|q)}_{[\gamma]}$, $H^{(s|q)}_{[\gamma]+}$ are defined in a natural way. In these spaces the norm

$$\|u\|_{(s|q),\gamma} = \left(\int\limits_{\operatorname{Im}\tau=\gamma} (1+|\tau|^2+|\xi|^{2q})^s |\hat{u}(\xi,\tau)|^2\, d\xi\, d\sigma\right)^{1/2} \tag{7}$$

is defined (cf. (1.18)). This norm is equivalent to the one determined by the pseudodifferential operator $\delta^+_{(s|q)}(D)$:

$$\|u\|_{(s|q),\gamma} = \|\exp(\gamma t)\delta^+_{(s|q)}(D)u\| = \|\delta^+_{(s|q)}(D_x, D_t + i\gamma)\exp(\gamma t)u\|, \tag{7′}$$

where (cf. (1.18))

$$\delta^+_{(s|q)}(\xi,\tau) = \left(i(\tau+\gamma_0) + (1+|\xi|^2)^{q/2}\right)^s.$$

Proposition 1′ in 1.6 is transferred to the space $H^{(s)}_{[\gamma]}$ in a trivial manner: $f \in H^{(s)}_{[\gamma]+}$ if and only if $f = \delta^+_{(s|q)}(D)g$, $g \in H_{[\gamma]+}$ and $\|\exp(\gamma t)g\| = \|f\|_{(s|q)}$.

THEOREM. *For $q \geqslant 1$ and a polynomial $P(\xi,\tau)$ the following conditions are equivalent.*

(I) *For every $Q(\xi,\tau)$, $\deg_q Q < \deg_q P$, there exist $\gamma_0 = \gamma_0(Q)$ and $\lambda = \lambda(Q)$ such that the equation*

$$(P+Q)(D)u = f \tag{8}$$

possesses a unique solution $f \in H^{(s|q)}_{[\gamma]}$, $\gamma \leqslant \gamma_0$, for any $u \in H^{(s-\lambda|q)}_{[\gamma]+}$.

(II) *There is $\gamma_0 < 0$ such that for each $s \in \mathbb{R}$ the inequality*

$$\begin{aligned}|\gamma|\|u\|_{(s+m-1|q),[\gamma]} + |\gamma|^{1/q}\|(1+|D_x|^2)^{(q-1)/2}u\|_{(s+m-1|q),[\gamma]}\\ \leqslant C\|P(D)u\|_{(s|q),[\gamma]} \quad \forall u \in H^{(\infty)}_{[\gamma]}\end{aligned} \tag{9}$$

holds, where m is the highest degree of τ in the polynomial $P(\xi,\tau)$.

(III) *Condition* (iii) *of Theorem* 1.2.5 *holds.*

The equivalent conditions also include conditions (I_+) and (II_+), which are obtained from (I) and (II) by replacing the spaces $H^{(s)}_{[\gamma]}$ by $H^{(s)}_{[\gamma]+}$.

PROOF. According to Proposition 2.1, for every Q there is $\gamma_0(Q)$ such that

$$(P+Q)(\xi,\tau) \neq 0, \quad \operatorname{Im}\tau \leqslant \gamma_0(Q),$$

i. e., the polynomial $P(\xi,\tau)$ is q-stable correct. In other words, conditions (I) and (I_+) are equivalent to the condition of q-stable correctness of the polynomial P. By Theorem 1.2.5, conditions (I) and (I_+) are equivalent to condition (III) and the algebraic inequality (1.2.22). Repeating the argument of Proposition 2.1 we prove that (II) and (II_+) are equivalent to (III). The theorem is proved.

REMARK. 1) In the case of strictly hyperbolic polynomials ($q = 1$) the two terms on the right-hand side of (9) coincide, i. e., the inequality

$$\gamma\|u\|_{(s+m-1|q)} \leqslant C\|P(D)u\|_{(s),[\gamma]} \tag{9′}$$

holds. In turn, this inequality is equivalent to the condition of strict hyperbolicity.

2) In the case of q-parabolic polynomials (q is even) stronger inequalities (as compared to (9)) can be proved:

$$|\gamma|\,\|u\|_{(s+m-1|q),[\gamma]} + \sum_{|\beta|\leqslant q/2}\|D^\beta u\|_{(s+m-1|q),[\gamma]} \leqslant C\|P(D)u\|_{(s|q),[\gamma]}, \tag{9″}$$

$$\|u\|_{(s+m-1|q),[\gamma]} \leqslant C\|P(D)u\|_{(s),[\gamma]}. \tag{9‴}$$

They are equivalent to the condition of q-parabolicity.

The differential operators $P(D_x, D_y)$ whose symbols $P(\xi,\tau)$ satisfy the equivalent conditions of the theorem are called q-stable correct operators. For $q = 1$ and for even values of q they are called strictly hyperbolic and q-parabolic operators, respectively.

Condition (I) of the theorem contains no precise information about the degree of smoothness of the solutions to equation (1) in the case of q-stable correct differential operators. We now present the related condition, which is equivalent to the conditions of the theorem.

In view of inequality (9), it appears natural to consider a scale of norms involving an additional parameter that takes into account the differentiability with respect to the spatial variables. We define $H^{(s|q),\lambda}$ as the space of functions (distributions) with the norm

$$\|u\|_{(s|q),\lambda} = \|(1+D_x|^2)^{\lambda/2}u\|_{(s|q)}. \tag{10}$$

As a rule, in the case $\lambda = 0$ we will omit the corresponding index. The spaces $H^{(s|q),\lambda}_{[\gamma]}$ and $H^{(s|q),\lambda}_{[\gamma]+}$ are defined in a natural way. For the norms in these spaces we take

$$\|u\|_{(s|q),\lambda,[\gamma]} = \left(\int_{\operatorname{Im}\tau=\gamma} (1+|\tau|^2+|\xi^{2q})^s(1+|\xi|^2)^\lambda |\widehat{u}(\xi,\tau)|^2\,d\xi\,d\sigma \right)^{1/2}. \tag{11}$$

Proceeding from inequality (9), we define the norm

$$|[u]|_{(s|q),\lambda,[\gamma]} = |\gamma|\,\|u\|_{(s|q),\lambda,[\gamma]} + |\gamma|^{1/q}\|u\|_{(s|q),\lambda+(q-1)/2,[\gamma]} \tag{12}$$

in the space $H^{(s|q),\lambda+(q-1)/2}_{[\gamma]}$. This is an intermediate norm between $\|\cdot\|_{(s|q),\lambda,[\gamma]}$ and $\|\cdot\|_{(s+1|q),\lambda,[\gamma]}$ in the sense that the inequality

$$d^{-1}\|u\|_{(s|q),\lambda,[\gamma]} \leqslant |[u]|_{(s|q),\lambda,[\gamma]} \leqslant d\|u\|_{(s+1|q),\lambda,[\gamma]}$$

holds with a constant $d = d(\gamma_0)$ not depending on γ for $\gamma < \gamma_0 < 0$.

We note that in the new notation inequality (9) takes the form

$$|[u]|_{(s+m-1|q),[\gamma]} \leqslant C\|P(D)u\|_{(s|q),[\gamma]}. \tag{13}$$

The space $H^{(s+m-1|q),(q-1)/2}_{[\gamma]}$ endowed with the norm (12) will be denoted $\mathcal{H}^{(s+m-1|q)}_{[\gamma]}$.

The condition below can be included in the set of equivalent conditions of the theorem.

(IV) For any differential operator $Q(D)$, $\deg_q Q < \deg_q P$, there is $\gamma_0(Q)$ such that for any $f \in H^{(s|q)}_{[\gamma]}$ equation (8) possesses a unique solution $u \in \mathcal{H}^{(s+m-1|q)}_{[\gamma]}$ that satisfies the inequality

$$|[u]|_{(s+m-1|q),[\gamma]} \leqslant C(Q)\|P+Q(D)u\|_{(s|q)[\gamma]} \tag{14}$$

with a constant C not depending on γ.

Replacing the spaces $H^{(s|q)}_{[\gamma]}$ and $\mathcal{H}^{(s+m-1|q)}_{[\gamma]}$ in (IV) by $H^{(s|q)}_{[\gamma]+}$ and $\mathcal{H}^{(s+m-1|q)}_{[\gamma]+}$, respectively, we obtain condition (IV$_+$).

Condition (IV) can be formulated in the operator form. Indeed, repeating the argument of Proposition 2.1 we can prove that the operator

$$\mathcal{P} : \mathcal{H}^{(s+m-1|q)}_{[\gamma]} \longrightarrow H^{(s|q)}_{[\gamma]} \quad (u \longmapsto (P+Q)(D)u) \tag{15}$$

has a bounded inverse operator whose norm is bounded by a constant not depending on $\gamma \leqslant \gamma_0(Q)$.

2.3. Remarks on the Cauchy problem for a differential operator with variable coefficients (the parametrix method). Consider a differential operator with variable coefficients:

$$P(y, D) = \sum_{\alpha} a_{\alpha}(y) D^{\alpha}. \tag{16}$$

The function $P(y, \eta)$ is called the symbol of the operator. We will consider operators (16) whose coefficients $a_{\alpha}(y)$ are infinitely differentiable functions not depending on y for sufficiently large $|y|$, i. e.,

$$a_{\alpha}(y) = a_{\alpha} + a'_{\alpha}(y), \quad a'_{\alpha}(y) \in \mathcal{D}. \tag{17}$$

If f is a distribution, then the distribution $P(y, D)f$ is found from the relation

$$(P(y, D)f, \varphi) = (f, {}^{t}P(y, D)\varphi), \tag{16'}$$

where

$${}^{t}P(y, D)\varphi = \sum_{\alpha} (-D)^{\alpha}(a_{\alpha}(y)\varphi(y)).$$

A simple calculation shows that the symbol ${}^{t}P$ has the form

$${}^{t}P(y, \eta) = \sum_{\beta > 0} \frac{1}{\beta!} P^{(\beta)}_{(\beta)}(y, -\eta). \tag{18}$$

Here we use the standard notation

$$a^{(\alpha)}_{(\beta)}(y, \eta) = \partial^{\alpha}_{\eta} D^{\beta}_{y} a(y, \eta).$$

In the case $\beta = 0$ or $\alpha = 0$ we will omit the corresponding index and write $a^{(\alpha)}$ or $a_{(\beta)}$, respectively. For the operator (16) we will consider the equation

$$P(y, D)u = f, \tag{19}$$

which is understood in the sense of distribution theory, i. e.,

$$(u, {}^{t}P(y, D)\varphi) = (f, \varphi).$$

The ways of imposing the initial conditions corresponding to the homogeneous Cauchy problem will be discussed below. As was shown above, in the case of constant coefficients the solution to (19) in the spaces $H^{(s)}_{[\gamma]}$ or $H^{(s)}_{[\gamma]+}$ is determined by the pseudodifferential operator (5). Therefore it is natural to try to construct the solution in the case of variable coefficients by applying the pseudodifferential operator with symbol $P^{-1}(y, \eta)$ not to the right-hand side f, but to an unknown function $g(y)$ to be found:

$$\begin{aligned} u(y) &= \gamma P^{-1}(y, D) g(y) \\ &= (2\pi)^{-(n+1)/2} \int_{\operatorname{Im} \tau = \gamma} \exp(i\langle x, \xi\rangle + it\tau) P^{-1}(y; \xi, \tau) \widehat{g}(\xi, \tau)\, d\xi\, d\sigma. \end{aligned} \tag{20}$$

Applying the operator $P(y, D)$ to both sides of (20), inserting it under the integral sign on the right-hand side, and representing the expression

$$P(y, D)\left[P^{-1}(y, \xi, \tau) \exp(i\langle x, \xi\rangle + it\tau)\right]$$

according to the Leibniz-Hörmander formula, we derive a pseudodifferential equation for g:

$$g + {}_{[\gamma]}H(y, D_x, D_t)g = f, \tag{21}$$

where

$$H(y, \xi, \tau) = \sum_{\beta>0} \frac{1}{\beta!} P^{(\beta)}(y; \xi, \tau) D_y^\beta P^{-1}(y; \xi, \tau).$$

If we can show that

$$\left| D_y^\beta H(y; \xi, \tau) \right| < C_\beta(\operatorname{Im} \tau) \longrightarrow 0, \quad \operatorname{Im} \tau \to -\infty, \tag{22}$$

for $|\beta| \leqslant N(s)$, then some simple estimates for pseudodifferential operators will imply that for sufficiently large $-\gamma$ the operator

$${}_{[\gamma]}H(y; D_x, D_t) : H^{(s)}_{[\gamma]+} \longrightarrow H^{(s)}_{[\gamma]+}$$

is bounded, and its norm is less than 1. Therefore the equation (21) possesses a solution $g \in H^{(s)}_{[\gamma]+}$ determined by Neumann's series:

$$g = \sum_{j=0}^{\infty} \left(-H(y; D_x, D_t)\right)^j f.$$

DEFINITION 1. A polynomial $P(\xi, \tau)$ is said to be exponentially correct if for all $\omega \in \mathbb{R}^n$ the polynomial $P_\omega(\xi, \tau) = P(\xi + i\omega, \tau)$ satisfies Petrovskii's condition.

It can be shown (see **[GV1,** Proposition 3.4.1]) that a polynomial $P(\xi, \tau)$ is exponentially correct if and only if for each $\alpha > 0$ we have

$$|P^{(\alpha)}(\xi, \tau) P^{-1}(\xi, \tau)| \to 0, \quad \operatorname{Im} \tau \to -\infty,$$
$$\text{uniformly with respect to } (\xi, \operatorname{Re} \tau) \in \mathbb{R}^{n+1}.$$

DEFINITION 2. A symbol $P(y; \xi, \tau)$ is said to be an exponentially correct symbol of constant strength if

(i) for every $y^0 \in \mathbb{R}^{n+1}$ the polynomial $P(y^0; \xi, \tau)$ is exponentially correct;

(ii) there are constants $A > 0$ and γ_0 such that for each $y', y'' \in \mathbb{R}^{n+1}$ we have

$$A^{-1} \leqslant |P(y'; \xi, \tau) P^{-1}(y''; \xi, \tau)| \leqslant A, \quad \operatorname{Im} \tau \leqslant \gamma_0.$$

For symbols $P(y; \xi, \tau)$ satisfying the conditions of Definition 2 inequalities (22) hold, and, consequently, the homogeneous Cauchy problem is solvable. For detailed proof of the existence and uniqueness of the resulting solution and an investigation of its smoothness properties see **[GV1,** Chapter 5].

It should be noted that strictly hyperbolic, parabolic, and pluriparabolic operators are exponentially correct.

As to the condition of constant strength, it is very restrictive. For example, a strictly hyperbolic symbol satisfies this condition if and only if its principal homogeneous part has constant coefficients. Indeed, let

$$|P(y', \xi\tau)| \leqslant A|P(y'', \xi, \tau)|, \qquad \operatorname{Im} \tau \leqslant \gamma_0, \quad (\xi, \operatorname{Re} \tau) \in \mathbb{R}^{n+1}.$$

If we replace ξ and $\operatorname{Re}\tau$ in this inequality by $\lambda\xi$ and $\lambda\operatorname{Re}\tau$, respectively, divide the inequality by λ^m, where $m = \deg P$, and pass to the limit for $\lambda \to +\infty$, then we obtain the inequality

$$|P_0(y', \xi, \sigma)| \leqslant A|P_0(y'', \xi, \sigma)| \quad \forall(\xi, \sigma) \in \mathbb{R}^{n+1}$$

for the principal homogeneous parts. If $\sigma = \lambda(\xi)$ is a real root of the polynomial $P_0(y'', \xi, \sigma)$, then, by virtue of the inequality, it also serves as a real root of the polynomial $P(y', \xi, \sigma)$. Because the left- and right-hand sides in this inequality can be interchanged, the polynomials $P_0(y', \eta)$ and $P_0(y'', \eta)$ have the same real roots. In view of their hyperbolicity, there are no other roots, and therefore the polynomials coincide up to a constant factor.

As to the q-parabolic symbols, they satisfy the two-sided inequality

$$A^{-1}(1 + |\tau| + |\xi|^q)^m \leqslant |P(y, \xi, \tau)| \leqslant A(1 + |\tau| + |\xi|^q)^m,$$
$$\operatorname{Im}\tau \leqslant \gamma_0, \quad (\operatorname{Re}\tau, \xi) \in \mathbb{R}^{n+1},$$

which automatically implies condition (ii).

§3. Method of energy estimates

The main idea forming the basis of the results in this section is presented in the introduction to the present chapter. The functional scheme below for the solvability of the Cauchy problem is oriented to q-stable correct operators. To study the q-hyperbolic operators it is necessary to pass to norms involving the weight $\exp\langle N, x\rangle$, and this will be done in §6.

We will deal with differential operators $P(y, D)$ resolved with respect to the highest derivative with respect to t:

$$P(y, D) = D_t^m + \sum_{j=1}^{m} P_j(y, D_x)D_t^{m-j} = \sum a_{\alpha j}(y) D_x^\alpha D_t^j, \tag{1}$$

where the coefficients $a_{\alpha j}(y)$ satisfy the stabilization condition (2.17):

$$a_{\alpha j}(y) = a_{\alpha j} + a'_{\alpha j}(y), \quad a'_{\alpha j}(y) \in \mathcal{D}. \tag{2}$$

As in 1.1.4, we denote

$$P^{(l)}(y; \xi, \tau) = \partial^l P(y; \xi, \tau)/\partial\tau^l, \quad l = 1, \dots, m, \tag{3}$$

and let $P^*(y, D)$ be the formally adjoint operator relative to the Hermitian form (1.1′):

$$\left[P(y; D)f, g\right] = \left[f, P^*(y, D)g\right]. \tag{4}$$

In this case we have $P^*(y, D) = \overline{{}^tP(y, D)}$, which implies a formula for the symbol of the operator P^*:

$$P^*(y, \eta) = \sum_{\beta \geqslant 0} \overline{P}_{(\beta)}^{(\beta)}(y; \eta)/\beta!. \tag{5}$$

3.1. A priori estimates guaranteeing the solvability of the homogeneous Cauchy problem. Consider the differential equation with variable coefficients

$$P(y;D)u = f, \tag{6}$$

which is understood in the sense of distribution theory:

$$(u, {}^tP(y,D)\varphi) = (f,\varphi) \quad \forall\varphi \in \mathcal{D}. \tag{6'}$$

We prove the following

THEOREM. *Assume that for the operator* (1), *its formally adjoint operator, and an arbitrary* $M > 0$ *there is* $\gamma_0 = \gamma_0(M)$ *such that for* $|s| \leqslant M$ *and* $\gamma \leqslant \gamma_0$ *the inequalities*

$$\sum_{l=1}^{m}(\gamma_0-\gamma)^l \|P^{(l)}(y,D)u\|_{(s),[\gamma]} \leqslant C_s\|P(y,D)u\|_{(s),[\gamma]} \quad \forall u \in H^{(\infty)}_{[\gamma]}, \tag{7}$$

(7′)

$$\sum_{l=1}^{m}(\gamma_0-\gamma)^l \|P^{*(l)}(y,D)u\|_{(-s),[-\gamma]} \leqslant C^*_s\|P^*(y,D)u\|_{(-s),[-\gamma]} \quad \forall u \in H^{(\infty)}_{[-\gamma]}$$

take place. Then the following assertions hold.

Existence. *For each* $\overline{M} > 0$ *there is* $\gamma_0(\overline{M})$ *such that for* $s \in [-\overline{M}, \overline{M}]$, $\gamma \leqslant \gamma_0(\overline{M})$, *and an arbitrary right-hand side* $f \in H^{(s)}_{[\gamma]}$ *there exists a unique solution* $u \in H^{(s)}_{[\gamma]+}$ *to equation* (6), *and we have*

$$P^{(l)}(y;D)u \in H^{(s)}_{[\gamma]+}, \quad l = 1,\dots,m. \tag{8}$$

Uniqueness. *If* $u \in H^{(-\infty)}_{[\gamma_1]+}$ *for some* γ_1 *and if* $P(y;D)u = 0$ (*in the sense of* (6′)), *then* $u \equiv 0$.

3.1.1. We begin with the uniqueness, which is simpler. If $u \in H^{(-\infty)}_{[\gamma]}$, then, according to the definition of the space $H^{(-\infty)}$, there is s_1 such that $u \in H^{(s_1)}_{[\gamma_1]+}$, and, consequently, $u \in H^{(s_1)}_{[\gamma_1]+}$ for any $\gamma \leqslant \gamma_1$. Take a sequence of functions $u_j \in H^{(\infty)}_{[\gamma_1]+}$ converging to u in $H^{(s_1)}_{[\gamma_1]}$. Then the sequence $P(y;D)u_j$ converges to $P(y,D)u \equiv 0$ in $H^{(s_2)}_{[\gamma_1]}$, where $s_2 \leqslant s_1 - \deg P$. Choose M such that $s_2 \in [-M, M]$, and let $\gamma \leqslant \gamma_0(M)$. Then, by the condition of the theorem, inequality (7) holds, and we have

$$\|u_j\|_{(s_2),[\gamma]} \leqslant \text{const}\,\|P(y;D)u_j\|_{(s_2),[\gamma]} \longrightarrow 0, \quad j \to \infty,$$

whence $u = \lim u_j \equiv 0$.

3.1.2. First we consider the existence of a solution in the spaces $H^{(s)}_{[\gamma]}$, which is a consequence of the weakened inequality (7′).

PROPOSITION. *Let for some fixed* s *and* γ *the inequality*

$$\|v\|_{(-s),[\gamma]} \leqslant C\|P^*(y;D)v\|_{(-s),[-\gamma]} \quad \forall v \in H^{(\infty)}_{[-\gamma]} \tag{9'}$$

hold. Then equation (6) *for* $f \in H^{(s)}_{[\gamma]}$ *has a solution* $u \in H^{(s)}_{[\gamma]}$, *and it satisfies the inequality*

$$\|u\|_{(s),[\gamma]} \leqslant C\|P(y,D)u\|_{(s),[\gamma]}. \tag{9}$$

The proof is quite standard. Consider the linear functional $L(\psi) = (f, \varphi)$ on the lineal $\{\psi,\ \psi = {}^tP\varphi,\ \varphi \in H^{(\infty)}_{[-\gamma]}\}$. By inequality (9′),

$$|L(\psi)| < \|f\|_{(s),[\gamma]}\|\varphi\|_{-(s),[-\gamma]} \leqslant C\|f\|_{(s),[\gamma]}\|{}^tP\varphi\|_{-(s),[-\gamma]}.$$

Here we have used the fact that $\overline{{}^tPv} = P^*\overline{v}$ so that in the case of (9′) it is allowable to replace P^* by ${}^tP$. Since $\psi = {}^tP\varphi$, the norm of the functional does not exceed $c\|f\|_{(s),[\gamma]}$. Consequently, by the Hahn-Banach theorem, there is an element $u \in H^{(s)}_{[\gamma]}$, $\|u\|_{(s),[\gamma]} \leqslant C\|f\|_{(s),[\gamma]}$, such that $(u, \psi) = L(\psi)$, i. e., $(u, {}^tP\varphi) = (f, \varphi)$.

3.1.3. In view of Proposition 3.1.2, the proof of the theorem reduces to establishing the fact that if the support of the right-hand side $f \in H^{(s)}_{[\gamma]}$ belongs to the half-space $t \geqslant 0$, then the support of the corresponding solution $u \in H^{(s)}_{[\gamma]}$ also belongs to it. According to 1.7 (see the remark), we have

$$H^{(s)}_{[\gamma]+} = H^{(s)}_{[-\infty,\gamma]}.$$

The main technical detail in the proof of the solvability of equation (6) in the subspaces $H^{(s)}_{[\gamma]+}$ consists in the reduction of the solvability problem to establishing the following two simpler facts:

(a) the solvability in $H^{(s)}_{[\gamma]}$ (Proposition 3.1.2);

(b) the solvability in the spaces $H^{(s)}_{[\gamma',\gamma'']}$ corresponding to finite values of γ' and γ''.

A sufficient condition for the validity of (b) is an a priori estimate for the operator P^* in the conjugate space $H^{(-s)}_{[-\gamma',-\gamma'']}$. Literally repeating the arguments in Proposition 3.1.2 we prove

PROPOSITION. *Let the inequality*

$$\|v\|_{(-s),[-\gamma',-\gamma'']} \leqslant C^*\|P^*(y, D)v\|_{(s),[-\gamma',-\gamma'']} \quad \forall v \in H^{(\infty)}_{[-\gamma',-\gamma'']} \tag{10′}$$

hold for some fixed s, γ', and γ'', $\gamma' < \gamma''$. Then equation (6) *possesses a solution $u \in H^{(s)}_{[\gamma',\gamma'']}$ for any $f \in H^{(s)}_{[\gamma',\gamma'']}$, and it satisfies the inequality*

$$\|u\|_{(s),[\gamma',\gamma'']} \leqslant C\|P(y; D)u\|_{(s),[\gamma',\gamma'']}. \tag{10}$$

3.1.4. Now, under the assumption that we know how to prove inequalities (9′) and (10′) for $|s| \leqslant M$ and $\gamma, \gamma', \gamma'' \leqslant \gamma_0(M,\delta)$, $\gamma'' - \gamma' \leqslant \delta$, we complete the proof of the existence of a solution belonging to $H^{(s)}_{[\gamma]+}$.

Fix $s \in \mathbb{R}$ and δ and choose a sufficiently large M such that the numbers s and $s \pm \deg P$ simultaneously belong to the closed interval $[-M, M]$. Furthermore, select $\gamma_0 = \gamma_0(M,\delta)$ such that for $\gamma \leqslant \gamma_0$ the inequalities (9′) and (10′) hold. We show that for $f \in H^{(s)}_{[\gamma]+}$ there exists a solution $u \in H^{(s)}_{[\gamma]+}$.

According to Proposition 1′ in 1.6, the inclusion $f \in H^{(s)}_{[\gamma]+}$ means that $f \in H^{(s)}_{[\rho]}$ for any $\rho \leqslant \gamma$ and that the norm

$$\sup_{f \leqslant \gamma}\|f\|_{(s),[\rho]} = \|f\|_{(s),[\gamma]} \tag{11}$$

is finite. By Proposition 3.1.2, there exists a family of solutions $u_\rho \in H^{(s)}_{[\rho]}$ to equation (6), and we have $\|u_\rho\|_{(s),[\rho]} \leqslant c\|f\|_{(s)[\gamma]} < \infty$. However, in view of (11), in this case

$$\sup_{\rho\leqslant\gamma}\|u_\rho\|_{(s),[\rho]} \leqslant C_s \sup_{\rho\leqslant\gamma}\|Pu_\rho\|_{(s),[\rho]} = C_s \sup_{\rho\leqslant\gamma}\|f\|_{(s),[\rho]} \leqslant C_s\,\|f\|_{(s),[\gamma]}. \tag{12}$$

It follows that if we show that the function u_ρ does not depend on ρ on a ray $(-\infty, \rho_0]$, we prove that $u \in H^{(s)}_{[\rho_0]+}$, and this would imply $u \in H^{(s)}_{[\gamma]+}$. To prove that u_ρ does not depend on ρ it suffices to verify that

$$u_{\rho_1} = u_{\rho_2} \qquad \forall \rho_1 \leqslant \rho_2 \leqslant \rho_0, \quad \rho_1 - \rho_2 \leqslant \delta. \tag{13}$$

To prove (13) we note that if the numbers $\rho_0 < \gamma$, δ, s, and $s - \deg P$ are chosen so that for $\rho_1 = \gamma'$ and $\rho_2 = \gamma''$ the inequality (10′) holds, then, according to Proposition 3.1.3, for $f \in H^{(s)}_{[\gamma]+} \subset H^{(s)}_{[\rho_1,\rho_2]}$ there exists a solution $u_{\rho_1,\rho_2} \in H^{(s)}_{[\rho_1,\rho_2]}$. By virtue of the uniqueness established in 3.3.1, we have $u_{\rho_1} = u_{\rho_1,\rho_2} = u_{\rho_2}$, i. e., relation (13) and, consequently, the fact that $u_\rho = u$ belongs to $H^{(s)}_{[\gamma]+}$ are proved.

3.1.5. It remains to prove that the solution obtained in 3.1.4 satisfies inequality (7), i. e., that the relation (8) holds. To this end, we note first of all that for a fixed s the inequality (7) is extended by continuity to an arbitrary space $H^{(s_1)}_{[\gamma]}$, $s_1 \geqslant s + \deg P$. We now approximate the right-hand side f in $H^{(s)}_{[\gamma]}$ with smoother functions $f_j \in H^{(s_1)}_{[\gamma]+}$. By what has been proved, there are $u_j \in H^{(s_1)}_{[\gamma]+}$ such that $Pu_j = f_j$. By virtue of (7) and because $\{f_j\}$ is a Cauchy sequence, it follows that $\{u_j\}$ is a Cauchy sequence with respect to the norm determined by the left-hand side of (7), i. e., the limit function of the sequence u_j satisfies (8). Since $\lim u_j$ is a solution to (6), the uniqueness implies that $u = \lim u_j$, i. e., (7) holds.

REMARK. The above argument is of a more general character. We have actually proved that if, along with inequality (7′), an additional estimate for the smooth function u in terms of the right-hand side of (7) holds, then this estimate also holds for the solution to equation (6).

3.2. Estimates in the spaces $H^{(s)}_{[\gamma_1,\gamma_2]}$, $\gamma_1 > \gamma_2$. We now show that if (7′) holds, then (10′) also holds. This will complete the proof of Theorem 3.1. To simplify the notation, we will derive from (7) the inequality

$$\|u\|_{(s),[\gamma_1,\gamma_2]} \leqslant C\|Pu\|_{(s),[\gamma_1,\gamma_2]}, \qquad u \in H^{(\infty)}_{[\gamma_1,\gamma_2]}, \qquad \gamma_1 > \gamma_2. \tag{14}$$

By the definition of the norm in $H^{(s)}_{[\gamma_1,\gamma_2]}$ (see (1.29)), we have

$$\|Pu\|_{(s),[\gamma_1,\gamma_2]} = \|\chi Pu\|_{(s),[\gamma_2]} + \|(1-\chi)Pu\|_{(s),[\gamma_1]}.$$

According to the Leibniz formula,

$$\chi Pu = P(\chi u) - \sum_{l=1}^{m} \chi_l P^{(l)} u, \quad \chi_l = D_t^l \chi/\alpha!\,.$$

Writing a similar relation for $(1-\chi)Pu$ and applying the estimates in $H^{(s)}_{[\gamma_2]}$ and $H^{(s)}_{[\gamma_1]}$ to χu and $(1-\chi)u$, respectively, we obtain the inequality

$$
\begin{aligned}
\|Pu\|_{(s),[\gamma_1,\gamma_2]} &\geqslant \|P(\chi u)\|_{(s),[\gamma_2]} + \|P((1-\chi)u)\|_{(s),[\gamma_1]} \\
&\quad - \sum_{l=1}^{m} \|\chi_l P^{(l)} u\|_{(s),[\gamma_2]} - \sum_{l=1}^{m} \|\chi_l P^{(l)} u\|_{(s),[\gamma_1]} \\
&\geqslant \sum_{l=1}^{m} \Big[(\gamma_0-\gamma_2)^l \|P^{(l)}\chi u\|_{(s),[\gamma_2]} + (\gamma_0-\gamma_1)^l \|P^{(l)}\chi u\|_{(s),[\gamma_1]} \\
&\qquad - \|\chi_l P^{(l)} u\|_{(s),[\gamma_2]} - \|\chi_l P^{(l)} u\|_{(s),[\gamma_1]}\Big].
\end{aligned} \tag{15}
$$

To estimate the third and fourth terms on the right-hand side via the first and second terms we note that

$$\|\chi_l P^{(l)} u\|_{(s),[\gamma_2]} \leqslant \|\chi_l P^{(l)}(\chi u)\|_{(s),[\gamma_2]} + \|\chi_l P^{(l)}(u-\chi u)\|_{(s),[\gamma_2]}.$$

Furthermore, the following inequalities hold:

$$\|\chi_l P^{(l)}(\chi u)\|_{(s),[\gamma_2]} \leqslant K_s \|P^{(l)}(\chi u)\|_{(s),[\gamma_2]}, \tag{16}$$

$$\|\chi_l P^{(l)}(u-\chi u)\|_{(s),[\gamma_2]} \leqslant K_s(\gamma_2-\gamma_1) \|P^{(l)}(u-\chi u)\|_{(s),[\gamma_1]}, \tag{17}$$

and similar inequalities hold for the norm $\|\cdot\|_{(s),[\gamma_2]}$:

$$\|\chi_l P^{(l)}(\chi u)\|_{(s),[\gamma_1]} \leqslant K'_s(\gamma_2-\gamma_1) \|P^{(l)}(\chi u)\|_{(s),[\gamma_2]}, \tag{16$'$}$$

$$\|\chi_l P^{(l)}(u-\chi u)\|_{(s),[\gamma_2]} \leqslant K_s \|P^{(l)}(u-\chi u)\|_{(s),[\gamma_1]}. \tag{17$'$}$$

We assume that these estimates are proved and continue the derivation of the desired inequality. Substituting these estimates into the right-hand side of (15) and choosing a number γ_2 such that the difference $\gamma_0-\gamma_2$ is sufficiently large we find a lower bound for this right-hand side:

$$
\begin{aligned}
&\frac{1}{2}\sum_{l=1}^{m} \Big[(\gamma_0-\gamma_2)^l \|P^l(\chi u)\|_{(s),[\gamma_2]} + (\gamma_0-\gamma_2)^l \|P^l(u-\chi u)\|_{(s),[\gamma_1]}\Big] \\
&\qquad \geqslant \frac{m!}{2}(\gamma_0-\gamma_2)^m \left(\|\chi u\|_{(s),[\gamma_2]} + \|u-\chi u\|_{(s),[\gamma_1]}\right) = \text{const}\, \|u\|_{(s),[\gamma_1,\gamma_2]}.
\end{aligned}
$$

Hence, it remains to prove inequalities (16)–(17$'$), which follow from the fact that $H^{(s)}_{[\gamma]}$ is a module over the functions $a(t) \in \mathcal{D}$ and the norm of the corresponding multiplication operator does not depend on γ. Assertions of this type will be proved in the next section (see 4.5). Note that the desired assertion is obvious for the integers $s \geqslant 0$, and, by conjugacy, it is derived for the negative integers s. We now prove (17) (inequality (17$'$) is proved in a similar way).

We define a norm in $H^{(s)}_{[\gamma_2]}$ by means of (1.24) and find

$$
\begin{aligned}
\|\chi_l P^l(u-\chi u)\|_{(s),[\gamma_2]} &= \|\exp(\gamma_2 t)_{[\gamma_2]}\delta_s^+(D)\chi_l P^l(1-\chi)u\| \\
&= \|\delta_s^+(D_x, D_t+i\gamma_2)\exp(\gamma_2 t)\chi_l P^{(l)}(1-\chi)u\| \\
&= \left\|\frac{\delta_s^+(D_x, D_t+i\gamma_2)}{\delta_s^+(D_x, D_t+i\gamma_1)}\delta_s^+(D_x, D_t+i\gamma_1)\{\chi_l e^{(\gamma_2-\gamma_1)t} e^{\gamma_1 t} P^{(l)}((1-\chi)u)\}\right\|.
\end{aligned} \tag{18}
$$

According to the elementary inequality

$$\left|\delta_s^+(\xi,\sigma+\gamma_2)\delta_{-s}^+(\xi,\sigma+\gamma_1)\right| \leqslant (1+|\gamma_2-\gamma_1|)^{|s|},$$

the constant $(1+|\gamma_2-\gamma_1|)^{(s)}$ is an upper bound for the $L_2 \to L_2$ norm of the pseudodifferential operator $\delta_s^+(D_x, D_t+i\gamma_2)\ \delta_{-s}^+(D_x, D_t+i\gamma_2)$. Thus, expression (18) does not exceed

$$(1+|\gamma_2-\gamma_1|)^{(s)}\|\chi_l \exp((\gamma_2-\gamma_1)t)P^{(l)}((1-\chi)u)\|_{(s),[\gamma_1]}.$$

Since the function $\chi_l(t)\exp((\gamma_2-\gamma_1)t)$ belongs to $\mathcal{D}(\mathbb{R})$ and all its derivatives have upper bounds depending on $\gamma_2-\gamma_1$ only, we arrive at inequality (17).

3.3. Energy estimates for differential operators with constant coefficients correct in Petrovskii's sense. Consider the Hermitian form

(19) $$[u,v]_{(s),[\gamma]} = \left[\exp(\gamma t)\delta_s^+(D)u, \exp(\gamma t)\delta_s^+(D)v\right].$$

According to (1.6′), we have

(19′) $$[u,v]_{(s),[\gamma]} = \iint\limits_{\operatorname{Im}\tau=\gamma} |\delta_s^+(\xi,\tau)|^2\widehat{u}(\xi,\tau)\overline{\widehat{u}(\xi,\tau)}\,d\xi\,d\sigma.$$

Replacing u with $P(D)u$ and v with $P^{(1)}(D)v$ in (19′) we obtain (in the notation of 1.1.4)

(20) $$-\operatorname{Im}[P(D)u, P^{(1)}(D)u]_{(s),[\gamma]} = \int\limits_{\operatorname{Im}\tau=\gamma} H_P(\xi,\sigma,\gamma)|\delta_s^+(\xi,\tau)|\,|\widehat{u}(\xi,\tau)|^2\,d\xi\,d\sigma.$$

By (20) and Proposition 1.1.4, we have the following

PROPOSITION. *For a differential operator $P(D_x, D_t)$ with constant coefficients which is resolved with respect to the highest derivative D_t^m the conditions below are equivalent*:

(i) *the correctness condition* (2.3) *holds*;

(ii) *there is γ_0 such that for all $s\in\mathbb{R}$ and $\gamma<\gamma_0$ the inequality*

(21) $$c(m)\sum_{j=1}^{m}(\gamma_0-\gamma)^{(2l-1)}\|P^{(l)}(D)u\|^2_{(s),[\gamma]} \leqslant -\operatorname{Im}[P(D)u, P^{(1)}(D)u]_{(s),[\gamma]} \qquad \forall u\in H^{(\infty)}_{[\gamma]}$$

holds.

By the Schwarz inequality, the right-hand side of (21) is estimated as follows:

$$\|P(D)u\|_{(s),[\gamma]}\|P^{(1)}(D)u\|_{(s),[\gamma]} \leqslant (2\varepsilon(\gamma_0-\gamma))^{-1}\|P(D)u\|^2_{(s),[\gamma]} + \frac{\varepsilon(\gamma_0-\gamma)}{2}\|P^{(1)}(D)u\|^2_{(s),[\gamma]}.$$

Taking $\varepsilon\leqslant c(m)$ and substituting it into (21) we derive

(22) $$c_1(m)\sum_{l=1}^{m}(\gamma_0-\gamma)^{l}\|P^{(l)}(D)u\|_{(s),[\gamma]} \leqslant \|P(D)u\|_{(s),[\gamma]} \quad \forall u\in H^{(\infty)}_{[\gamma]}.$$

If the polynomial $P(\xi,\tau)$ is correct in Petrovskii' sense, then its roots lie in the half-plane $\operatorname{Im}\tau \leqslant \gamma_0$. However, the same half-plane contains the roots of the polynomial $\overline{P}(\xi,-\tau)$ so that inequality (22) holds for the operator $\overline{P}(D_x,-D_t)$ as well. Making change of variable $t \to -t$ in this inequality and replacing s by $-s$ we arrive at an estimate for the adjoint operator:

$$(22') \qquad c_1(m)\sum_{j=1}^{m}(\gamma_0-\gamma)^l\|\overline{P}^{(l)}(D)v\|_{(-s),[-\gamma]} \leqslant \|\overline{P}(D)v\|_{(-s),[-\gamma]}, \quad v \in H^{(\infty)}_{[-\gamma]}.$$

Thus we obtained a new proof of the existence and uniqueness of the solution to the homogeneous Cauchy problem for differential operators with constant coefficients satisfying Petrovskii's correctness condition.

3.4. Remarks on energy estimates for differential operators with constant coefficients. In the subsequent sections, when deriving estimates for hyperbolic and parabolic operators with variable coefficients, we will resort to the fact that, by virtue of the two-sided inequalities for the polynomial H_P in 1.2, in the case of variable coefficients the symbol $H_P(y,\xi,\sigma,\gamma)$ satisfies the following

Condition of constant strength. There are γ_0 and $A > 0$ such that for all y', $y'' \in \mathbb{R}^{n+1}$ the inequalities

$$(23) \qquad A^{-1} \leqslant H_P(y',\eta,\gamma)H_P^{-1}(y'',\eta,\gamma) \leqslant A \qquad \forall\gamma \leqslant \gamma_0, \quad \forall\eta \in \mathbb{R}^{n+1}$$

hold.

The condition of constant strength for the symbol H_P implies a two-sided inequality for the corresponding quadratic differential form:

$$(24) \qquad \begin{aligned} -K^{-1}[P(D)u, P^{(1)}(D)u]_{(s),[\gamma]} &\leqslant -\operatorname{Im}[P(y,D)u, P^{(1)}(y,D)u]_{(s),[\gamma]} \\ &\leqslant -K[P(D)u, P^{(1)}(D)u]_{(s),[\gamma]} \\ \forall u \in H^{(\infty)}_{[\gamma]}, &\quad \gamma \leqslant \gamma_0, \end{aligned}$$

where $P(D) = P(y^0,D)$. In view of the condition of constant strength, we do not have to fix a point $y^0 \in \mathbb{R}^{n+1}$. In the case of operators with variable coefficients correct in Petrovskii's sense, inequalities (23) and (24) imply (7). We will prove the following

PROPOSITION. *Let a symbol* (1) *with coefficients of the type* (2) *possess the properties below*:

(a) *for every* $y^0 \in \mathbb{R}^{n+1}$ *the polynomial* $P(y^0,\xi,\tau)$ *is correct in Petrovskii's sense*;
(b) *the condition of constant strength holds*;
(c) *for some* s *and* γ *inequality* (24) *holds.*
Then inequality (7) *holds for these* s *and* γ.

PROOF. Since the space of the polynomials $P(y,\eta)$ is finite-dimensional, only a finite number among them are linearly independent. Therefore there are $y^1,\dots,y^J$ such that the representation

$$(25) \qquad P(y;\xi,\tau) = \sum b_j(y)P(y^j,\xi,\tau)$$

takes place. The functions $b_j(y)$ result from the solution of an algebraic linear system of equations whose right-hand sides involve the coefficients of the symbol $P(y,\eta)$. If these coefficients have the form (2), then $b_j(y)$ are functions of the same type, and, as was noted above, the operators of multiplication by these functions transform $H^{(s)}_{[\gamma]}$ into itself, and the corresponding norms do not depend on γ.

The l-fold differentiation of (25) with respect to τ yields

$$P^{(l)}(y;\xi,\tau)=\sum D^l_t b_j(y)P^{(l)}(y^j,\xi,\tau). \tag{25'}$$

By (25′) and Proposition 3.3, we have

$$\begin{aligned}\sum_{l=1}^{m}(\gamma_0-\gamma)^{2l-1}\|P^{(l)}(y,D)u\|^2_{(s),\gamma} &\leqslant \text{const}\sum_{l=1}^{m}\sum_{j=1}^{J}(\gamma_0-\gamma)^{2l-1}\|P^{(l)}(y^j,D)u\|^2_{(s),[\gamma]}\\ &\leqslant -\text{const}\sum_{j=1}^{J}\text{Im}[P(y^j,D)u,P^{(1)}(y^jD)u]_{(s),[\gamma]}.\end{aligned}$$

In view of conditions (b) and (c), it follows that

$$\text{const}\sum_{l=1}^{m}(\gamma_0-\gamma)^{2l-1}\|P^{(l)}(y,D)u\|^2_{(s),[\gamma]}\leqslant -\text{Im}[P(y,D)u,P^{(1)}(y,D)u]_{(s),[\gamma]}.$$

Taking

$$(2\varepsilon(\gamma_0-\gamma))^{-1}\|P(y,D)u\|_{(s),[\gamma]}+\frac{\varepsilon}{2}(\gamma-\gamma_0)\|P^{(1)}(y,D)u\|_{(s),\gamma}$$

as an upper bound for the right-hand side of the last inequality and choosing a sufficiently small ε we prove (7).

REMARK 1. If the symbol $P(y,\eta)$ satisfies the conditions of the proposition, then the symbol $\overline{P}(y,\xi,-\tau)$ also satisfies these conditions, which implies that the inequality

$$\text{const}\sum_{l=1}^{m}(\gamma_0-\gamma)^{l}\|\overline{P}^{(l)}(y,D)u\|^2_{(-s),[-\gamma]}\leqslant\|\overline{P}(y,D)u\|^2_{(-s),[-\gamma]}$$

holds. To pass from this inequality to (7′) one should resort to the stability of the estimates in question with respect to the addition of lower terms to the operator. This stability takes place in the case of q-stable correct operators.

REMARK 2. For q-stable correct operators there exists a norm $[\cdot]_{(s),[\gamma]}$ possessing the property that

$$\begin{gathered}[u]^2_{(s),[\gamma]}\leqslant -\text{const Im}\,\|P(D)u,P^{(1)}(D)u\|_{(s),[\gamma]},\\ \|P^{(1)}(D)u\|_{(s),[\gamma]}\leqslant\text{const}[u]_{(s),[\gamma]},\end{gathered}$$

where the constant does not depend on γ. In this case we have

$$[u]^2_{(s),[\gamma]}\leqslant -\text{const Im}\,\|P(D)u,P^{(1)}(D)u\|_{(s),[\gamma]}\leqslant\text{const}\,\|Pu\|_{(s),[\gamma]}[u]_{(s),[\gamma]},$$

i. e., the inequality

$$[u]_{(s),[\gamma]}\leqslant\text{const}\,\|P(y,D)u\|_{(s),[\gamma]} \tag{26}$$

holds. Therefore (see Remark 3.1.5), there exists a solution with a finite norm $[\cdot]_{(s),[\gamma]}$, and it satisfies (26).

For a set of sufficient conditions on a symbol $P(y,\eta)$ that guarantee the validity of an estimate of the type (24) under conditions (a) and (b) for both the operator $P(y,D)$ and its formally adjoint operator see **[GV2**, Chapter VII].

§4. The Cauchy problem for strictly hyperbolic operators with variable coefficients

The Cauchy problem for hyperbolic systems in the case of variable coefficients (and even in the nonlinear case) was studied by Petrovskii in **[P2]**. In particular, by using implicitly pseudodifferential operators Petrovskii constructed energy integrals and obtained a priori estimates for hyperbolic systems. In the case of one equation Leray **[Le]** suggested a simper method for deriving Petrovskii's estimate based on the application of quadratic forms considered in §3. Gårding **[G]** supplemented Leray's estimate with a similar estimate for the formally adjoint operator and derived from them the existence and uniqueness theorems. We also use the Leray-Gårding approach in our presentation. The theory of the Cauchy problem for hyperbolic operators is given following a scheme that will be used later for parabolic and q-hyperbolic operators.

4.1. Statement of the main result. As in §3, we will consider differential operators (3.1) resolved with respect to the highest time derivative D_t^m, and it will be assumed without special stipulation that the coefficients have the form (3.2). By $P_0(y,\eta)$ we will denote the principal homogeneous part of the symbol $P(y,\eta)$:

$$P_0(y,\eta) = \sum_{\alpha_0+\cdots+\alpha_n=m} a_{\alpha_0\alpha_1\ldots\alpha_n}\sigma^{\alpha_0}\xi_1^{\alpha_1}\cdots\xi_n^{\alpha_n}.$$

DEFINITION. A symbol

$$P(y,\eta) = \sigma^m + \sum_{j<m}^{m} a_{\alpha_j}(y)\sigma^j\xi^\alpha \tag{1}$$

is said to be uniformly strictly hyperbolic if

(a) for each $y^0 \in \mathbb{R}^{n+1}$ the polynomial $P(y^0,\eta)$ is strictly hyperbolic, i. e., the roots $\lambda_{0j}(y^0,\xi)$ of the principal homogeneous part $P_0(y^0,\eta)$ are real and distinct;

(b) there is a unified constant δ such that

$$|\lambda_{0j}(y^0;\xi) - \lambda_{0k}(y^0,\xi)| \geqslant \delta|\xi|, \qquad k\neq j \quad \forall y^0\in\mathbb{R}^{n+1}, \quad \forall\xi\in\mathbb{R}^{n+1}. \tag{2}$$

PROPOSITION. *For a symbol* (1) *with coefficients of the type* (3.2) *the following conditions are equivalent*:

(i) *the symbol $P(y,\eta)$ is uniformly strictly hyperbolic*;

(ii) *there is γ_0 and positive constants K_1 and K_2 such that*

$$K_1 \leqslant H_P(y,\xi,\sigma,\gamma)\left[|\gamma|(1+|\gamma|+|\sigma|+|\xi|)^{2m-2})\right]^{-1} \leqslant K_2, \quad \gamma\leqslant\gamma_0;$$

(iii) *there is γ_0 and a constant $C>0$ such that*

$$|\gamma|(1+|\gamma|+|\sigma|+|\xi|)^{m-1} \leqslant c|P(y,\xi,\sigma+i\gamma)|, \quad \gamma\leqslant\gamma_0.$$

PROOF. (i)$\Rightarrow$(ii). To prove this assertion one should carefully analyze the derivation of (1.2.11), after which it becomes clear that the constant c in this inequality depends on the maximum of the moduli of the coefficients in the lower terms of P and on

$$\max_{1\leqslant j\leqslant m}\sup_{|\xi|=1}|\lambda_{0j}(y,\xi)|,$$

where $\lambda_{0j}(y,\xi)$ are the roots of the principal homogeneous part. This quantity can also be estimated in terms of the maximum of the moduli of the coefficients of the principal homogeneous part. By conditions (3.2), all the coefficients of the symbol (1) are uniformly bounded for $y\in\mathbb{R}^{n+1}$, which proves the existence of the constant K_2. As to the lower bound, the corresponding constant depends on the maximum of the moduli of the coefficients in the lower terms and on

$$\min_{j\neq k}\inf_{|\xi|=1}|\lambda_{0j}(y,\xi)-\lambda_{0k}(y,\xi)|.$$

In view of condition (b) in the definition of uniform strict hyperbolicity, we can choose the desired constant K_1.

(ii)$\Rightarrow$(iii)$\Rightarrow$(ii). The first implication is obvious. The above inequality and Theorem 1.2.3 imply that for each $y\in\mathbb{R}^{n+1}$ the roots of the polynomial $P_0(y,\eta)$ are real and distinct. It remains to show that in the inequality (2) a uniform constant can be chosen. Assume the contrary. Then there is a sequence $y^{(\tau)}\in\mathbb{R}^{n+1}$ and an element $\xi^{(\tau)}\in\mathbb{R}^n$, $|\xi^{(\tau)}|=1$, such that for some j and k we have

$$\left|\lambda_{0j}(y^{(\tau)},\xi^{(\tau)})-\lambda_{0k}(y^{(\tau)},\xi^{(\tau)})\right|\longrightarrow 0,\qquad \tau\to\infty.$$

Because the symbol (1) is resolved with respect to the highest power τ^m, the roots $\lambda_{0j}(y,\xi)$ are continious functions of the variables y and ξ. By virtue of the stabilization condition (3.2), $\lambda_{0j}(y,\xi)$ are continuous functions on a compact set $\{|y|\leqslant R\}\times\{|\xi|=1\}$, where R is sufficiently large. We have thus found the limiting point $(y^{(0)},\xi^{(0)})$ where the roots $\lambda_{0j}(y^{(0)},\xi^{(0)})$ and $\lambda_{0k}(y^{(0)},\xi^{(0)})$ coincide, which contradicts the condition of strong hyperbolicity of the polynomial $P(y^0,\eta)$.

An operator $P(y;D)$ is said to be uniformly strictly hyperbolic if its symbol possesses this property.

We note that the property of strict hyperbolicity of a polynomial is determined by its principal homogeneous part only. Therefore, if an operator $P(y;D)$ is uniformly strictly hyperbolic, then all operators $P(y,D)+Q(y,D)$, where $\deg Q<\deg P$, possess this property.

The present section is devoted to the proof of the following

THEOREM. *For a differential operator* (3.1), (3.2) *the conditions below are equivalent*:

(I_+) *For every $M>0$ there is $\gamma_0=\gamma_0(M)$ such that for $|s|\leqslant M$ and for each $\gamma\leqslant\gamma_0$ the equation*

$$P(y;D)u=f \tag{3}$$

has a unique solution $u\in H^{(s+m-1)}_{[\gamma]+}$ for any $f\in H^{(s)}_{[\gamma]+}$, and the inequality

$$|\gamma|\,\|u\|_{(s+m-1),[\gamma]}\leqslant C\|f\|_{(s),[\gamma]} \tag{4}$$

holds, where C does not depend on γ.

(II) *For all* $M > 0$ *there is* $\gamma_0 = \gamma_0(M)$ *such that for* $|s| \leqslant M$ *and for each* $\gamma \leqslant \gamma_0$ *the a priori estimate*

$$(5) \qquad |\gamma| \, \|u\|_{(s+m-1),[\gamma]} \leqslant C \|P(y, D)u\|_{(s),[\gamma]} \quad \forall u \in H^{(\infty)}_{[\gamma]}$$

holds.

(II_+) *is obtained from* (II) *by replacing* $H^{(\infty)}_{[\gamma]}$ *with* $H^{(\infty)}_{[\gamma]+}$.

(III) *The operator* $P(y, D)$ *is uniformly strictly hyperbolic.*

REMARK. 1) Condition (I), which is obtained from (I_+) by replacing the spaces $H^{(s)}_{[\gamma]+}$ with $H^{(s)}_{[\gamma]}$, can be included in the set of equivalent conditions of the theorem.

2) Condition (I_+) can be formulated in operator form, namely the operator

$$\mathcal{P} \colon H^{(s+m-1)}_{[\gamma]+} \longrightarrow H^{(s)}_{[\gamma]+},$$

which is originally defined on $H^{(s+m)}_{[\gamma]+}$, is closable, and its closure has a bounded inverse operator whose norm does not exceed $c|\gamma|^{-1}$, where c does not depend on $|s| \leqslant M$, $\gamma \leqslant \gamma_0(M)$.

4.2. The proof of Theorem 4.1. It is based on the following two assertions.

THEOREM. *Let* $P(y, D)$ *be a uniformly strictly hyperbolic operator. Then for each* $M > 0$ *there is* $\gamma_0 = \gamma_0(M)$ *and a constant* $K > 0$ *such that for an arbitrary* $y^0 \in \mathbb{R}^{n+1}$ *and for* $P(D) = P(y^0; D)$ *the two-sided inequality*

$$(6) \qquad \begin{aligned} -K^{-1} \operatorname{Im}[P(D)u, P^{(1)}(D)u]_{(s),[\gamma]} &\leqslant - \operatorname{Im}[P(y, D)u, P^{(1)}(y, D)u]_{(s),[\gamma]} \\ &\leqslant K \operatorname{Im}[P(D)u, P^{(1)}(D)u]_{(s),[\gamma]} \\ \forall u \in H^{(\infty)}_{[\gamma]}, \quad |s| \leqslant M, \quad &\gamma \leqslant \gamma_0(M), \end{aligned}$$

holds.

PROPOSITION. *Let for some* s *and* $\gamma \leqslant \gamma_0 = \gamma_0(s)$ *the differential operator* $P(y; D)$ *satisfy* (5) *with a constant* c *not depending on* γ. *Then for the homogeneous operator* $P_0(D) = P(y^0, D)$ *corresponding to an arbitrary point* $y^0 \in \mathbb{R}^{n+1}$ *the inequality*

$$(7) \qquad \begin{aligned} |\gamma|^2 \int_{\operatorname{Im}\tau=\gamma} &(|\tau|^2 + |\xi|^2)^{s+m-1} |\widehat{u}(\xi, \tau)|^2 \, d\xi \, d\tau \\ &\leqslant c^2 \int_{\operatorname{Im}\tau=\gamma} (|\tau|^2 + |\xi|^2)^{s} |P_0(\xi, \tau)\widehat{u}(\xi, \tau)|^2 \, d\xi \, d\tau \quad \forall u \in H^{(\infty)}_{[\gamma]} \end{aligned}$$

holds with the same constant c *for any* $\gamma < 0$. *If* (5) *holds only on the subspace* $H^{(\infty)}_{[\gamma]+}$, *then* (7) *remains valid for any points* y^0 *belonging to the subspace* $t \geqslant 0$.

Assuming that the above results are proved we complete the derivation of Theorem 4.1.

(I_+)$\Rightarrow$(II_+). Fix $M > 0$. Let us take an arbitrary function $u \in H^{(\infty)}_{[\gamma]+}$, $\gamma \leqslant \gamma_0(M)$ and set $f = P(y; D)u$. Then, by virtue of the uniqueness, the function u is the solution of (3) corresponding to the right-hand side $f \in H^{(\infty)}_{[\gamma]+}$ regarded as an element of $H^{(s)}_{[\gamma]+}$. Hence, (4) implies (5).

(II_+), $(\mathrm{II})\Rightarrow(\mathrm{III})$. By the proposition, inequality (5) on $H^{(\infty)}_{[\gamma]}$ or on $H^{(\infty)}_{[\gamma]+}$ implies inequality (7) on the corresponding space. By virtue of Theorem 2.2, the polynomials $P(y, D)$ satisfy the conditions of Theorem 1.2.3 with constants not depending on y^0. Therefore, according to Proposition 4.1, the symbol (1) is uniformly strictly hyperbolic.

$(\mathrm{III})\Rightarrow(\mathrm{I}_+)$, (II). First of all we note that Theorem 1.2.3 implies that for every fixed $y \in \mathbb{R}^{n+1}$ the symbol $H_P(y; \xi, \sigma, \gamma)$ has upper and lower bounds expressed in terms of $|\gamma|(\sigma^2 + \gamma^2 + \xi^2)^{2m-2}$, and, generally, the constant c in (1.2.11) depends on γ. A careful analysis of the derivation of inequality (1.2.11) shows that the constant c in this inequality depends on the constant δ in (2) and on the maximum of the moduli of the coefficients of the symbol. Thus, the conditions of uniform strict hyperbolicity and condition (3.2) imply that there exists a unified constant $c > 0$ such that

$$(8)\quad c^{-1}|\gamma| \leqslant H_P(y, \xi, \sigma, \gamma)(|\tau|^2 + |\xi|^2)^{-2m+2} \leqslant c|\gamma|, \qquad \gamma \leqslant \gamma_0, \quad (\xi, \sigma) \in \mathbb{R}^{n+1}.$$

Inequality (8) guarantees the fulfilment of the condition of constant strength for the symbol $H_P(y, \sigma, \gamma)$. Combining this condition with the theorem we prove, according to Proposition 3.4, the validity of inequality (3.7) for a uniformly strictly hyperbolic operator.

By virtue of (8) and the Parseval relation, any operator $P(D) = P(y^0, D)$ satisfies the inequality

$$c_1|\gamma|\,\|u\|_{(s+m-1),[\gamma]} \leqslant -\operatorname{Im}[P(D)u, P^{(1)}(D)u]_{(s),[\gamma]},$$

which implies (3) (cf. Remark 2 in 3.4). Thus, the theorem together with the proposition implies all assertions of Theorem 4.1.

4.3. Proof of Proposition 4.2. Take an arbitrary $\gamma < 0$ and an arbitrary function $u(y) \in \mathcal{D}$ and set

$$u_\varepsilon(y) = u\left(\frac{y - y^0}{\varepsilon}\right).$$

Suppose that for some $s \in \mathbb{R}$ and for $\gamma \leqslant \gamma_0(s)$ inequality (5) holds with a constant C independent of γ. If ε is sufficiently small, then this inequality can be applied to elements of the space $H^{(s)}$ with weight $\exp(\gamma t/\varepsilon)$. Replacing γ by γ/ε and u by u_ε in (5) we obtain

$$(5')\qquad \gamma\varepsilon^{-1}\|u_\varepsilon\|_{(s+m-1),[\gamma/\varepsilon]} \leqslant C\|P(y, D)u_\varepsilon\|_{(s),[\gamma/\varepsilon]}.$$

We now show that by multiplying both sides of (5′) by a suitable normalization factor:

$$(9)\qquad K(\varepsilon) = \varepsilon^{(m+s)-(n+1)/2}\exp(-t^0\gamma/\varepsilon)$$

and making ε tend to zero we derive inequality (5). So, we pass to the calculation of the left- and right-hand sides of (5′). By the definition of the Fourier transform,

$$\widehat{u}_\varepsilon(\xi, \tau) = (2\pi)^{-(n+1)/2}\int \exp(-i\langle x, \xi\rangle - it\tau)u((y - y^0)/\varepsilon)\,dx\,dt.$$

Making change of variable $y \to y^0 + \varepsilon y$ we derive

$$(10)\qquad \widehat{u}_\varepsilon(\xi, \tau) = \exp(i\langle x^0, \xi\rangle + it^0\tau)\varepsilon^{n+1}\widehat{u}(\varepsilon\xi, \varepsilon\tau).$$

Therefore for an arbitrary real r we have

$$\begin{aligned}
&\|u_\varepsilon\|_{(r),[\gamma/\varepsilon]}\\
&= \varepsilon^{n+1}\exp(t^0\gamma/\varepsilon)\left[\iint |\widehat{u}(\varepsilon\xi,\varepsilon\sigma+i\gamma)|^2(1+|\xi|^2+|\sigma+i\gamma/\varepsilon|^2)^r\,d\xi\,d\sigma\right]^{1/2}\\
&= \varepsilon^{(n+1)/2-r}\exp(t^0\gamma/\varepsilon)\left[\iint |\widehat{u}(\xi,\tau)|^2(\varepsilon^2+|\xi|^2+|\tau|^2)^r\,d\xi\,d\sigma\right]^{1/2}\\
&= \varepsilon^{(n+1)/2-r}\exp(t^0\gamma/\varepsilon)\left(\iint |\widehat{u}(\xi,\tau)|^2(|\tau|^2+|\xi|^2)^r\,d\xi\,d\sigma+O(\varepsilon^2)\right)^{1/2}.
\end{aligned}$$

Replacing r in this formula by $s+m-1$ we obtain
(11)

$$\begin{aligned}
\|u_\varepsilon\|_{(s+m-1),[\gamma/\varepsilon]} &= \varepsilon^{(n+1)/2-(m+s-1)}\exp(t^0\gamma/\varepsilon)\\
&\quad\times\left(\int_{\operatorname{Im}\tau=\gamma} |\widehat{u}(\xi,\tau)|^2(|\tau|^2+|\xi|^2)^{s+m-1}\,d\xi\,d\sigma+O(\varepsilon^2)\right)^{1/2}.
\end{aligned}$$

Now we consider the right-hand side of $(5')$. Take the auxiliary operator

$$P^{(\varepsilon)(y,D)} = P(y^0+\varepsilon y,\varepsilon^{-1}D)$$

and set

$$g_\varepsilon(y) = \varepsilon^m P^{(\varepsilon)}(y,D)u.$$

As can easily be seen, we have

$$f_\varepsilon(y) \overset{\text{def}}{=} P(y,D)u_\varepsilon = \varepsilon^{-m}g_\varepsilon\left(\frac{y-y^0}{\varepsilon}\right).$$

Repeating the above estimation of the norm of the function u_ε we find

$$\begin{aligned}
\|f_\varepsilon\|_{(s),[\gamma/\varepsilon]} &= \varepsilon^{(n+1)/2-s-m}\exp(t^0\gamma/\varepsilon)\\
&\quad\times\left(\int_{\operatorname{Im}\tau=\gamma} |\widehat{g}_\varepsilon(\xi,\tau)|^2(\varepsilon^2+|\tau|^2+|\xi|^2)^s\,d\xi d\sigma\right)^{1/2}.
\end{aligned} \tag{12}$$

Furthermore,

$$\begin{aligned}
g_\varepsilon(y) = P_0(D)u &+ \sum_{|\alpha|=m}(a_\alpha(y^0+\varepsilon y)-a_\alpha)D^\alpha u\\
&+ \sum_{|\alpha|<m}\varepsilon^{m-|\alpha|}a_\alpha(y^0+\varepsilon y)D^\alpha u.
\end{aligned}$$

Since $u\in\mathcal{D}$, we have $g_\varepsilon(y) = g_0(y)+O(\varepsilon)$, where $g_0(y) = P_0(D)u$, and therefore similar expansions can also be written for the Fourier transform: $\widehat{g}_\varepsilon(y) = \widehat{g}_0(y)+O(\varepsilon)$. The substitution into (12) results in

$$\begin{aligned}
\|f_\varepsilon\|_{(s),[\gamma/\varepsilon]} &= \varepsilon^{(n+1)/2-(s+m)}\exp(t^0\gamma/\varepsilon)\\
&\quad\times\left(\int_{\operatorname{Im}\tau=\gamma} |P_0(\xi,\tau)\widehat{u}(\xi,\tau)|^2\,(|\tau|^2+|\xi|^2)^s\,d\xi\,d\sigma+O(\varepsilon)\right)^{1/2}.
\end{aligned} \tag{12'}$$

Substituting (11) and (12′) into (5′), multiplying both sides by the function (9), and passing to the limit for $\varepsilon \to 0$ we obtain (7).

If inequality (5) holds originally for functions with supports in the region $t > 0$, then (7) also holds for these functions. Since inequality (7) is invariant with respect to the shift of the argument, it holds on the functions belonging to $H^{(\infty)}_{[\gamma]}$ whose supports are bounded from the right. However, these functions are dense in $H^{(\infty)}_{[\gamma]}$, which completes the proof of the desired assertion.

4.4. The proof of Theorem 4.2. Since it is rather lengthy, we split it into several stages.

4.4.1. Auxiliary assertions. Fix a point $y^0 \in \mathbb{R}^{n+1}$ and set

$$\{u\}^2_{(s),[\gamma]} = -\operatorname{Im}[P(D)u, P^{(1)}(D)u]_{(s),[\gamma]}. \tag{13}$$

In the case $s = 0$ the corresponding index will be omitted. As was noted in 4.2, expression (13) defines a norm, and we have

$$\{u\}_{(s),[\gamma]} \approx |\gamma|^{1/2}\|u\|_{(s+m-1),[\gamma]}, \quad \gamma \leqslant \gamma_0, \tag{14}$$

where the equivalence constants do not depend on γ. Relation (14) implies the following inequalities, which will be used later:

$$\|u\|_{(s+m-1-j),[\gamma]} \leqslant \operatorname{const}|\gamma|^{-j-1/2}\{u\}_{(s),[\gamma]}, \quad \gamma \leqslant \gamma_0, \tag{15}$$

where the constant does not depend on γ.

PROPOSITION. *Let $A(y, D)$ and $B(y, D)$ be differential operators with coefficients of the type* (3.2). *Assume that*

$$\deg A \leqslant m, \quad \deg B \leqslant m-1, \quad \deg A + \deg B = 2m - 2 - \lambda, \quad \lambda \geqslant 0. \tag{16}$$

Then the quadratic form

$$J = [A(y, D)u, B(y, D)u]_{(s),[\gamma]} \tag{17}$$

satisfies the inequality

$$|J| \leqslant \operatorname{const}|\gamma|^{-\lambda}\|u\|^2_{(m-1-s),[\gamma]}. \tag{18}$$

If $-\gamma$ is sufficiently large, then, in view of (15), *we get*

$$|J| \leqslant \varepsilon(\gamma)\{u\}^2_{(s),[\gamma]} \qquad \varepsilon(\gamma) = \operatorname{const}|\gamma|^{-1-\lambda} \to 0, \quad \gamma \to -\infty.$$

The proof of the proposition is based on the integration by parts in (17). Since the pseudodifferential operators $\delta^+_s(D)$ commute with the operators of differentiation, the formulas for $s \neq 0$ do not differ from the standard formulas for the case $s = 0$. For further references, we present these formulas:

$$[a(y)D_k v, w]_{(s),[\gamma]} = [av, D_k w]_{(s),[\gamma]} - [(D_k a)v, w]_{(s),[\gamma]}, \quad k = 1, \ldots, n, \tag{19}$$

(19′)

$$[a(y)D_t v, D_t w]_{(s),[\gamma]} = [av, D_t w]_{(s),[\gamma]} - [(D_t a)v, w]_{(s),[\gamma]} - 2i\gamma[av, w]_{(s),[\gamma]}.$$

While proving (14), we can assume, without loss of generality, that each of the operators consists of a single monomial, i. e.,

(17′)

$$J = [aD^\alpha u, bD^\beta u], |\alpha| = m - \lambda_1, \quad |\beta| = m - 1 - \lambda_2, \quad \lambda_1, \lambda_2 \geqslant 0, \quad \lambda_1 + \lambda_2 \geqslant 1.$$

If $\lambda_1 \geqslant 1$, then, by the Schwarz inequality, we have

$$|J| \leqslant \|aD^\alpha u\|_{(s),[\gamma]}\|bD^\beta u\|_{(s),[\gamma]} \leqslant \text{const}\,\|u\|_{(s+m-\lambda_1),[\gamma]}\|u\|_{(s+m-1-\lambda_2),[\gamma]}$$
$$\leqslant \text{const}\,|\gamma|^{-(\lambda_1+\lambda_2-1)}\|u\|^2_{(s+m-1)}.$$

(Here we have used the fact that the operator of multiplication by a function of the type (3.2) is bounded in $H^{(s)}_{[\gamma]}$, and its norm does not depend on γ.)

If $\lambda_1 = 0$, then $\lambda_2 \geqslant 1$. By integrating by parts we transfer one derivative to the second term in the form and reduce the case $\lambda_1 = 0$ to the already considered case $\lambda_1 \geqslant 1$.

4.4.2. Localization of inequality (6). We now show that for $s = 0$ it suffices to prove (6) for functions $u(y)$ with compact supports belonging to a ball of small radius. Let us prove the following

PROPOSITION. *Assume that for some $\delta > 0$ there is $\gamma_0(\delta)$ such that (6) holds when the support of $u(y)$ belongs to a ball of a small radius no greater than δ. Then there is γ_1 such that for $\gamma \leqslant \gamma_1$ inequality (6) holds with $s = 0$ for all $u \in H^{(\infty)}_{[\gamma]}$.*

PROOF. 1) Since the coefficients of the operator P are constant outside some ball, there exists a finite covering

$$\mathbb{R}^{n+1} \in U_0 \cup U_1 \cup \cdots \cup U_J,$$

and hence for the functions whose supports belong to U_j, $j = 1, \dots, J$, inequality (6) has already been proved. As to U_0, in this region the operator has constant coefficients, and, consequently, for functions with supports lying in U_0 inequality (6) is trivial.

2) Consider a partition of unity $\{\varphi_j(y)\}$ subordinate to the covering $\{U_j\}$, i. e., assume that

(i) $\varphi_j(y) \in \mathcal{D}(U_j)$;
(ii) $\varphi_j(y) \geqslant 0$ for all $y \in \mathbb{R}^{n+1}$;
(iii) $\sum_{j=0}^J \varphi_j^2(y) \equiv 1$.

3) According to the condition of the proposition, the inequalities

$$K^{-1}\{\varphi_j u\}_{(s),[\gamma]} \leqslant -\operatorname{Im}\Big[P(y,D)(\varphi_j u), P^{(1)}(y,D)(\varphi_j u)\Big]_{(s),[\gamma]}$$
$$\leqslant K\{\varphi_j u\}_{(s),[\gamma]}, \quad \gamma \leqslant \gamma_0, \quad \forall u \in H^{(\infty)}_{[\gamma]} \tag{20}$$

hold. Using Proposition 4.4.1, we will prove that

$$-\operatorname{Im}\big[P(y,D)u, P^{(1)}(y,D)u\big]\sum_{j=0}^J \operatorname{Im}\Big[P(y,D)(\varphi_j u), P^{(1)}(y,D)(\varphi_j u)\Big]_{(s),[\gamma]}$$
$$= O(\varepsilon(\gamma)\{u\}^2_{[\gamma]}), \qquad \varepsilon(\gamma) \to 0, \quad \gamma \to -\infty. \tag{21}$$

A similar result is true in the case of forms involving operators with constant coefficients:

$$\{u\}^2_{[\gamma]} - \sum_{j=1}^J \{\varphi_j u\}^2_{[\gamma]} = O(\varepsilon(\gamma)\{u\}^2_{[\gamma]}). \tag{22}$$

Adding together estimates (20) for $j = 0, \dots, J$ and using (21) and (22) we find γ_1 such that inequality (6) holds for $\gamma \leqslant \gamma_1$.

4) Now we prove (21) and (22). Since these cases are considered in an identical manner, we will simply write P for operators with constant or variable coefficients.

Conditions (iii) for a function $\varphi_j(y)$ imply that

$$-\operatorname{Im}[P(u), P^{(1)}(u)]_{[\gamma]} = -\sum_{j=0}^{J} \operatorname{Im}[\varphi_j Pu, \varphi_j P]_{[\gamma]}.$$

By the Leibniz-Hörmander formula, we have

$$P(\varphi_j u) = \varphi_j Pu + \sum \varphi_{j\alpha} P^{(\alpha)} u, \quad \varphi_{j\alpha} = D^\alpha \varphi_j / \alpha!.$$

In view of the above two relations, the left-hand sides of (21) and (22) are equal to

$$-\operatorname{Im} \sum_{j=0}^{J} \sum_{\alpha+\beta>0} \left[\varphi_{j\alpha} P^{(\alpha)} u, \varphi_{j\beta} P^{(1)(\beta)} u\right]_{[\gamma]}.$$

Since $|\alpha| + |\beta| > 0$, all quadratic forms under the double summation sign satisfy the conditions of Proposition 4.4.1. The application of this proposition results in (21) and (22).

4.4.3. Proof of inequality (6) (the case $s = 0$). For a differential operator $P(y, D) = \sum a_\alpha(y) D^\alpha$ we define the kth norm of the coefficients by setting

$$[P]_k = \max_{|\beta| \leqslant k} \max_{\alpha} \sup_{y \in \mathbb{R}^{n+1}} |D^\beta a_\alpha(y)|.$$

The index will be omitted in the case $k = 0$. The proof of inequality (6) is based on the following auxiliary

PROPOSITION. *Let $A(y, D)$ and $B(y, D)$ be differential operators with real coefficients of orders m and $m-1$, respectively. Let A_0 and B_0 be their principal homogeneous parts. Then the inequality*

$$|\operatorname{Im}[A(y, D)u, B(y, D)u]_{[\gamma]}| \leqslant \left(\kappa [A_0][B_0] + c|\gamma|^{-1}\right) \{u\}^2_{[\gamma]} \tag{23}$$

holds.

Assuming that (23) is proved we complete the derivation of (6) for the case $s = 0$. According to the results in the previous subsection, it can be assumed that the supports of the functions u belong to a ball

$$S_\delta(y^0) = \{y \in \mathbb{R}^{n+1}, |y - y^0| \leqslant \delta\},$$

where y^0 is a point in $\mathbb{R}^{n+1}$. Here we can assume, without loss of generality, that the operator P has the form

$$P(y; D) = P(D) + P_1(y, D), \quad P(D) = P(y^0, D), \tag{24}$$

where the coefficients of the operator P_1 have compact supports belonging to a neighborhood of y^0, and that

$$[P_1] \leqslant K\delta. \tag{25}$$

Indeed, choose a function $\chi_\delta(y) \in \mathcal{D}(S_{2\delta}(y^0))$ that is equal to 1 in the ball $S_\delta(y^0)$. If the support of u belongs to $S_\delta(y^0)$, we have

$$P(y, D)u = P(D)u + \chi_\delta(P(y, D) - P(y^0, D))u. \tag{26}$$

Denoting the operator on the right by P_1, we obtain (24) and (25). We now define the norm $\{\cdot\}_{[\gamma]}$ with the aid of $P(D) = P(y^0, D)$ and find

$$\begin{aligned} &-\{u\}^2_{[\gamma]} - \operatorname{Im}[P(y,D)u, P^{(1)}(y,D)u]_{[\gamma]} \\ &\quad = \operatorname{Im}[P(D)u, P^{(1)}(y,D)u]_{[\gamma]} - \operatorname{Im}[P_1(y,D)u, P^{(1)}(y,D)u]_{[\gamma]}. \end{aligned}$$

By the assumption, the modulus of the right-hand side does not exceed

$$(K_1\delta + K_2|\gamma|^{-1})\{u\}^2_{[\gamma]}.$$

Choosing δ satisfying the condition $K_1\delta < 1/4$ and γ_0 satisfying $K_2|\gamma_0|^{-1} \leqslant 1/4$ we derive for $\gamma \leqslant \gamma_0$ the inequality

$$\frac{1}{2}\{u\}^2_{[\gamma]} \leqslant -\operatorname{Im}[P(y,D)u, P^{(1)}(y,D)u]_{[\gamma]} \leqslant \frac{3}{2}\{u\}^2_{[\gamma]}.$$

4.4.4. Proof of Proposition 4.4.3. 1) We set $A = A_0 + A_1$ and $B = B_0 + B_1$, where A_0 and B_0 are the principal homogeneous parts, and rewrite the form on the left-hand side of (23) as a sum of three forms:

$$\operatorname{Im}[A_0u, B_0u]_{[\gamma]} + \operatorname{Im}[A_1u, Bu]_{[\gamma]} + \operatorname{Im}[A_0u, B_1u]_{[\gamma]}.$$

According to Proposition 4.4.1, the second and third forms can be estimated in terms of const $|\gamma|^{-1}\{u\}^2_{[\gamma]}$.

2) We now assume that the symbols $A(y;\eta)$ and $B(y;\eta)$ are homogeneous polynomials of degrees m and $m-1$, respectively. Setting $u = \exp(\gamma t)v$, $A_\gamma = A(y, D_x, D_t + i\gamma)$, and $B_\gamma = B(y, D_x, D_t + i\gamma)$ we rewrite the form under consideration as follows:

$$\operatorname{Im}[Au, Bu]_{[\gamma]} = \operatorname{Im}[A_\gamma v, B_\gamma v] = \left(\frac{1}{2}(B^*_\gamma A_\gamma - A^*_\gamma B_\gamma)v, \overline{v}\right). \tag{27}$$

3) Denote by $H(y, D, \gamma)$ the differential operator on the right-hand side of (27), i. e.,

$$\begin{aligned} H(y,D,\gamma) = \frac{1}{2i}\Big(&B^*(y, D_x, D_t + i\gamma)\cdot A(y, D_x, D_t + i\gamma) \\ &- A^*(y, D_x, D_t + i\gamma)\cdot B(y, D_x, D_t + i\gamma)\Big). \end{aligned} \tag{28}$$

LEMMA. *Operator* (28) *is representable in the form*

$$H(y,D,\gamma) = \gamma H^{(0)}(y,D,\gamma) + H^{(1)}(y,D,\gamma), \tag{29}$$

where the symbol $H^{(0)}(y,\eta,\gamma)$ is a homogeneous real function of degree $2m-2$, and the moduli of the coefficients of $H^{(0)}$ do not exceed $\kappa[A_0][B_0]$. The symbol $H^{(1)}(y,\eta,\gamma)$ is a polynomial in η and γ of degree no higher than $2m-2$.

4) Assuming that the lemma is proved we complete the proof of the proposition. Integration by parts of the form

$$(H^{(0)}(y,D,\gamma)v, \overline{v}) \overset{\text{def}}{=} \sum_{\substack{j+|\alpha+\beta|=2m-2 \\ |\alpha|,|\beta|\leqslant m-1}} \gamma^j (h_{\alpha\beta j}(y)D^{\alpha+\beta}v, \overline{v})$$

results in

$$\sum_{\substack{j+|\alpha+\beta|=2m-2\\ |\alpha|,|\beta|\leqslant m-1}} \gamma^j (h_{\alpha\beta j}(y) D^\alpha v, \overline{D^\beta v}) + \sum_{\substack{k+|\alpha'+\beta'|<2m-2\\ |\alpha'|,|\beta'|\leqslant m-1}} \gamma^k (c_{\alpha\beta k} D^{a'} v, \overline{D^{\beta'} v}).$$

The modulus of the first form has an upper bound

$$\sum \max_y |h_{\alpha\beta j}(y)|\, |\gamma|^j \, \|D^\alpha v\| \, \|D^\beta v\| \leqslant \kappa_1 [A_0][B_0] \|u\|^2_{(m-1),[\gamma]},$$

and the modulus of the other has an upper bound

$$K_1 |\gamma|^{-1} \|u\|^2_{(m-1),[\gamma]}.$$

Thus,

$$|\gamma (H^{(0)} v, v)| \leqslant (\kappa_1 [A_0][B_0] + K_1 |\gamma|^{-1}) \{u\}^2_{[\gamma]}.$$

A similar inequality holds for the second form:

$$|(H^{(1)} v, v)| \leqslant K_2 \|u\|^2_{(m-1),[\gamma]} \leqslant K_2 |\gamma|^{-1} \{u\}^2_{[\gamma]}.$$

Combining these estimates we prove the desired proposition.

4.4.5. Proof of Lemma 4.4.4. To prove expansion (29) we calculate the symbol of the operator (28) in explicit form. Recall once again that if A has a symbol $A(y,\eta)$, then the symbol of A^* is equal to $\sum \overline{A}^{(\alpha)}_{(\alpha)}/\alpha!$. Furthermore, if A and B are two arbitrary operators, then the symbol of their composition $A \cdot B$ is $\sum A^{(\alpha)} B_{(\alpha)}/\alpha!$. Thus, the symbol of the operator (28) is equal to

$$\begin{aligned}
&\frac{1}{2i} \sum \frac{1}{\alpha!\beta!} \Big[\overline{B^{(\alpha+\beta)}_{(\beta)}(y,\xi,\sigma+i\gamma)} A_{(\alpha)}(y,\xi,\sigma+i\gamma) \\
&\qquad - \overline{A^{(\alpha+\beta)}_{\beta}(y,\xi,\sigma+i\gamma)} B_{(\alpha)}(y,\xi,\sigma+i\gamma) \Big] \\
&\quad = \frac{1}{2i} \sum \frac{1}{\alpha!} \Big[\overline{B^{(\alpha)}(y,\xi,\sigma+i\gamma)} A(y,\xi,\sigma+i\gamma) \\
&\qquad - \overline{A^{(\alpha)}(y,\xi,\sigma+i\gamma)} B(y,\xi,\sigma+i\gamma) \Big]_{(\alpha)} \\
&\quad = \frac{1}{2i} \Big[\overline{B(y,\xi,\sigma+i\gamma)} A(y,\xi,\sigma+i\gamma) \\
&\qquad - \overline{A(y,\xi,\sigma+i\gamma)} B(y,\xi,\sigma+i\gamma) \Big] + H^{(1)}(y,\xi,\sigma+i\gamma).
\end{aligned}$$

It is clear that $H^{(1)}$ is a polynomial in ξ, σ, and γ of degree no higher than $2m-2$. As to the first term on the right-hand side, it is a homogeneous polynomial in ξ, σ, and γ of degree $2m-1$. Since the coefficients of the symbols A and B are real, this polynomial is identically equal to zero for $\gamma = 0$. Hence, the polynomial is divisible by γ and can be represented as $\gamma H^{(0)}(y,\xi,\sigma,\gamma)$. Homogeneity considerations imply that $H^{(0)}(y,\xi,\sigma,\gamma)$ is a homogeneous polynomial of degree $2m-2$. The explicit representation of $\gamma H^{(0)}$ shows that the coefficients of this symbol are bilinear forms in the coefficients of $\overline{A}$ and B or $\overline{B}$ and A. It follows that these coefficients can be estimated in terms of $[A_0][B_0]$.

4.4.6. Proof of inequality (6) (the case $s \neq 0$). Set

$$w = {}_{[\gamma]}\delta_s^+(D) u \overset{\text{def}}{=} e^{-\gamma t} \delta_s^+(D_x, D_t + i\gamma) e^{\gamma t} u.$$

Then we have

$$u = \delta_{-s}^{+}(D)w,$$

and the form in question can be rewritten as

$$-\operatorname{Im}[Pu, P^{(1)}u]_{(s),[\gamma]} = -\operatorname{Im}[\widetilde{P}w, \widetilde{P}^{(1)}w]_{[\gamma]}, \tag{30}$$

where we set

$$\widetilde{P} = \delta_s^+ P \delta_{-s}^+, \quad \widetilde{P}^{(1)} = \delta_s^+ P^{(1)} \delta_{-s}^+.$$

The right-hand side of (30) is representable in the form

$$\begin{aligned}-\operatorname{Im}[Pw, P^{(1)}w]_{[\gamma]} &- \operatorname{Im}[Pw, (\widetilde{P}^{(1)} - P^{(1)}w]_{[\gamma]} - \operatorname{Im}[(\widetilde{P} - P)w, P^{(1)}w]_{[\gamma]} \\ &- \operatorname{Im}[(\widetilde{P} - P)w, (\widetilde{P}^{(1)} - P^{(1)}w]_{[\gamma]} \\ &= J_0 + J_1 + J_2 + J_3.\end{aligned}$$

According to 4.4.3, we have

$$J_0 \geqslant K_0\{w\}_{[\gamma]}^2 = K_0\{u\}_{(s),[\gamma]}^2,$$

and the desired assertion will follow from the inequalities

$$|J_j| \leqslant \operatorname{const}|\gamma|^{-1}\{w\}_{[\gamma]}^2, \quad j = 1, 2, 3. \tag{31}$$

We first consider the form J_1. We emphasize that the proof of inequalities (31) will be based on the fact that the operator P is resolved with respect to the highest derivative D_t^m, and the derivative itself enters the operator with a constant coefficient. We write

$$P^{(1)}(y, D) = mD_t^{m-1} + \sum_{\alpha, j>0} b_{\alpha j}(y) D_t^{m-j} D_x^\alpha.$$

Then

$$\begin{aligned}P^{(1)}(y, D) - \widetilde{P}^{(1)}(y, D) &= \sum (b_{\alpha j} - \delta_s^+ b_{\alpha j} \delta_{-s}^+) D_t^{m-j-1} D_x^\alpha \\ &= \sum_{j=1}^{m-1} \sum_{|\alpha|=j} + \sum_{j=1}^{m-1} \sum_{|\alpha|<j} = Q_1 + Q_2.\end{aligned}$$

Let the multiindex α correspond to the first sum. Then it can be represented as a sum of two multiindices:

$$\alpha = \alpha' + \alpha'', \qquad |\alpha''| = 1, \quad |\alpha'| = j - 1 \geqslant 0.$$

Hence, the operator Q_1 is a linear combination of operators having the form

$$T_{\alpha,\alpha'',j} D_t^{m-j-1} D_x^{\alpha'} = T_{\alpha,\alpha'',j} D^\beta, \quad |\beta| \leqslant m - 2, \tag{32}$$

where

$$T_{\alpha,\alpha'',j} = (b_{\alpha j} - \delta_s^+ b_{\alpha j} \delta_{-s}^+) D_x^{\alpha''}, \quad |\alpha''| = 1. \tag{33}$$

The operator Q_2 is a linear combination of operators of the type

$$(b_\beta - \delta_s^+ b_\beta \delta_{-s}^+) D^\beta, \quad |\beta| \leqslant m - 2, \tag{32$'$}$$

i. e., operators (33) corresponding to $\alpha'' = 0$. Substitute these expressions into the form J_1. This results in a form differing from the one in Proposition 4.4.1 in that some of the coefficients are replaced by operators of the type (33). Integrating by parts in the same way as in Proposition 4.4.1 we finally represent the forms Q_1 and Q_2 as linear combinations of the forms

$$[T_1 D^\alpha w, T_2 D^\beta w]_{[\gamma]}, \qquad |\alpha| \leqslant m-1, \quad |\beta| \leqslant n-1,$$

where T_1 and T_2 are either operators of the type (33) or some functions. If we manage to prove that the $H_{[\gamma]} \to H_{[\gamma]}$ norms of these operators are bounded and do not depend on γ, this will imply inequality (31) and thus complete the proof of (6). The estimation of the norms of operators of the type (33) is considered in the next subsection.

4.5. The spaces $H^\mu_{[\gamma]}$ and estimates for the commutators of pseudodifferential operators and operators of multiplication by a function.

4.5.1. Let $\mu(\eta)$ be a positive function satisfying the condition that there are constants N_μ and $K_\mu > 0$ such that

$$\mu(\eta')\mu^{-1}(\eta'') \leqslant K_\mu(1+|\eta'-\eta''|)^{N_\mu} \quad \forall \eta', \eta'' \in \mathbb{R}^{n+1}. \tag{34}$$

With every such function μ we associate the space of distributions

$$H^\mu = \{u \in H^{-\infty}, \mu(D)u \in L_2\} \tag{35}$$

and endow it with the norm

$$\|u\|_\mu = \|\mu(D)u\|. \tag{36}$$

Definition (36) can also be rewritten in the form

$$\{u \in H^\mu\} \leftrightarrow \{u = \mu^{-1}(D)v, \ v \in L_2, \ \|v\| = \|u\|_\mu\}. \tag{35'}$$

Proposition. *The space H^μ is a module over $\mathcal{D}$, and the inequality*

$$\|au\|_\mu \leqslant c_\mu(a)\|u\|_\mu \quad \forall a \in D, \quad \forall u \in H^\mu \tag{37}$$

holds, where

$$c_\mu = K_\mu \int |\widehat{a}(\eta)|(1+|\eta|)^{N_\mu}\, d\eta. \tag{37'}$$

Proof. According to (35′), on the left-hand side of (37) the function $w = \mu(D)(a\mu^{-1}(D)v)$ stands under the sign of the L_2 norm. By the Parseval relation, the function w can be replaced by its Fourier transform

$$\widehat{w}(\eta) = \int \widehat{a}(\eta-\eta')\mu(\eta)\mu^{-1}(\eta')\widehat{v}(\eta')\, d\eta' \stackrel{\text{def}}{=} \int K(\eta,\eta')\widehat{v}(\eta')\, d\eta'.$$

In view of condition (34), we have

$$|K(\eta,\eta')| \leqslant K_\mu(1+|\eta-\eta'|)^{N_\mu}|\widehat{a}(\eta-\eta')|.$$

Since the function $\widehat{a}(\eta)$ decreases faster than any power of η, the kernel $K(\eta,\eta')$ is absolutely integrable with respect to the two variables, and we have

$$\int |K(\eta,\eta')|\, d\eta, \quad \int |K(\eta,\eta')|\, d\eta' \leqslant C_\mu(a).$$

Applying the Schur lemma, we derive inequality (37).

COROLLARY 1. *H^μ is a module over the space of functions $a(y) = a + a'(y)$, $a \in \mathbb{C}$, $a'(y) \in \mathcal{D}$, i. e.,*

$$\|au\|_\mu \leqslant (|a| + C_\mu(a'))\|u\|_\mu. \tag{38}$$

COROLLARY 2. *If $a \in \mathcal{D}$, then the norm of the operator of multiplication by $a(y)$ regarded as an operator from H^μ into H^μ can be estimated from above in terms of the maximum of the moduli of the derivatives $D^\beta a$, where $|\beta| \leqslant N_\mu + n + 2$.*

PROOF. If $a \in \mathcal{D}$, then, in view of the inversion formula for the Fourier transform, we have

$$\begin{aligned} |\widehat{a}(\eta)| &= \left|(2\pi)^{-(n+1)/2}\int \exp(i\langle y, \eta\rangle)a(y)\,dy\right| \\ &\leqslant (2\pi)^{-(n+1)/2}\operatorname{Vol}(\operatorname{supp} a)\max|a(y)| \leqslant L(a)|\max a(y)|, \end{aligned}$$

whence

$$\sum_{|\beta|\leqslant N_\mu+n+1} |\eta^\beta|\,|\widehat{a}(\eta)| \leqslant L'(a) \max_{|\beta|\leqslant N_\mu+n+1} |D^\beta a|$$

(for simplicity, here N_μ is assumed to be an integer). Thus,

$$(1 + |\eta|)^{N_\mu}|\widehat{a}(\eta)| \leqslant L''(a) \max_{|\beta|\leqslant N_\mu+n+2} |D^\beta a|(1 + |\eta|)^{-n-2}.$$

By integrating this inequality we estimate the constant $C_\mu(a)$ in terms of the maximum of the moduli of the derivatives of a up to order $N_\mu + n + 2$.

4.5.2. LEMMA. *Let a function $\mu(\eta)$ satisfy condition (34) and belong to C^2, and let*

$$|(\partial_k\mu)(\eta)| \leqslant m(1 + |\eta|)^{-1}\mu(\eta), \quad k = 0, \dots, n. \tag{39}$$

Then for any function $a \in \mathcal{D}$ the inequality

$$\|(a - \mu a\mu^{-1})D_k v\| \leqslant d_\mu(a)\|v\|, \quad k = 0, \dots, n, \tag{40}$$

holds, where

$$d_\mu a = K_\mu m\sqrt{n+1}\int (1 + |\eta|)^{N_\mu+2}|\widehat{a}(\eta)|\,d\eta. \tag{41}$$

This lemma implies the following

PROPOSITION. *Let a function $\mu(\eta)$ satisfy the conditions of the lemma and let $a(y) = a + a'(y)$, where $a \in \mathbb{C}$ and $a'(y) \in \mathcal{D}$. Then the inequality*

$$\|au\|_\mu \leqslant (\max|a| + d_\mu(a'))\|u\|_\mu \tag{42}$$

holds.

PROOF. Using (35′) we rewrite the left-hand side in the form

$$\begin{aligned} \|au\|_\mu &= \|\mu a\mu^{-1}v\| \leqslant \|av\| + \|(\mu a'\mu^{-1} - a')v\| \\ &\leqslant \max|a|\,\|v\| + d_\mu(a')\|v\| = (\max|a| + d_\mu(a'))\|u\|_\mu. \end{aligned}$$

PROOF OF THE LEMMA. As in Proposition 4.5.1, passing to the Fourier transforms under the sign of the norm on the left-hand side we obtain an integral operator

$$\widehat{u}(\eta) \to \int H(\eta, \eta')\widehat{u}(\eta')\, d\eta'$$

with kernel

$$H(\eta, \eta') = \widehat{a}(\eta - \eta')(1 - \mu(\eta)\mu^{-1}(\eta'))\eta'_k. \tag{43}$$

Let us estimate this kernel. By the Lagrange formula,

$$\mu(\eta) = \mu(\eta') + \sum_{j=0}^{n}(\eta_j - \eta'_j)\partial_j\mu(\eta' + \Theta(\eta - \eta')), \quad 0 \leqslant \Theta \leqslant 1.$$

Therefore

$$\begin{aligned} |1 - \mu(\eta)\mu^{-1}(\eta')| &= \left|\sum_{j=0}^{n}(\eta_j - \eta'_j)\partial_j\mu(\eta' + \Theta(\eta - \eta'))\mu^{-1}(\eta')\right| \\ &\leqslant \left(\sum(\eta_j - \eta'_j)^2\right)^{1/2}\left(\sum|\partial_j\mu(\eta' + \Theta(\eta - \eta'))|^2\right)^{1/2}\mu^{-1}(\eta') \\ &\leqslant |\eta - \eta'|m\sqrt{n+1}(1 + |\eta' + \Theta(\eta - \eta')|)^{-1}\mu(\eta' + \Theta(\eta - \eta'))\mu^{-1}(\eta') \\ &\leqslant \sqrt{n+1}K_\mu(1 + |\eta - \eta'|)^{N_\mu + 2}(1 + |\eta'|)^{-1}. \end{aligned}$$

Using this inequality, we estimate the expressions

$$\int |H(\eta, \eta')|\, d\eta, \quad \int |H(\eta, \eta')|\, d\eta' \leqslant d_\mu(a).$$

The application of the Schur lemma now proves the lemma.

4.5.3. Denote by $H^\mu_{[\gamma]}$ the space of distributions u such that $\exp(\gamma t) \in H^\mu$, i. e., the integral

$$\int |\mu(D)(\exp(\gamma t)u)|^2\, dy = \int \mu^2(\eta)|\widehat{u}(\xi, \sigma + i\gamma)|^2\, d\eta$$

converges. In what follows we will deal with spaces $H^\mu_{[\gamma]}$ corresponding to weight functions $\mu(\xi, \tau)$ defined for $(\xi, \operatorname{Re}\tau) \in \mathbb{R}^{n+1}$, $\operatorname{Im}\tau \leqslant \gamma_0 = \gamma_0(\mu)$, and satisfying the inequalities

$$\mu(\xi', \tau')\mu^{-1}(\xi'', \tau'') \leqslant K_\mu(1 + |\xi' - \xi''| + |\tau' - \tau''|)^{N_\mu}. \tag{44}$$

If condition (44) holds, then for the norm in $H^\mu_{[\gamma]}$ we take

$$\begin{aligned} \|u\|_{\mu,[\gamma]} &\overset{\text{def}}{=} \|\exp(\gamma t)_{[\gamma]}\mu(D)u\| = \|\mu(D_x, D_t + i\gamma)\exp(\gamma t)u\| \\ &= \left(\int_{\operatorname{Im}\tau = \gamma} |\mu(\xi, \tau)\widehat{u}(\xi, \tau)|^2\, d\xi\, d\tau\right)^{1/2}, \end{aligned} \tag{45}$$

and the space $H^\mu_{[\gamma]}$ consists of functions of the type

$$u = {}_{[\gamma]}\mu^{-1}(D)v, \quad v \in H_{[\gamma]}, \quad \|u\|_{\mu,[\gamma]} = \|\exp(\gamma t)v\|. \tag{46}$$

REMARK. If the symbol $\mu(\xi,\tau)$ is holomorphic, then the index $[\gamma]$ in (46) can be omitted. However, when deriving energy estimates we may encounter norms for which the symbol $\mu(\xi,\tau)$ is not a holomorphic function.

Note that if $\operatorname{Im}\tau' = \operatorname{Im}\tau'' = \gamma$, then the right-hand side of inequality (44) does not depend on γ, hence a trivial modification of the proof of Proposition 4.5.1 shows that the following proposition holds.

PROPOSITION 1. *If condition* (44) *holds, then the operator*

$$H^{\mu}_{[\gamma]} \to H^{\mu}_{[\gamma]} \quad (u \longmapsto au,\, a \in \mathbb{C} + \mathcal{D})$$

is bounded, and its norm does not depend on γ.

By modifying the argument in Lemma 4.5.2 we prove the following

LEMMA. *Let a symbol* $\mu(\xi,\tau) \in C^2$ *satisfy* (44) *and the conditions*

$$(47) \qquad |\partial^{\alpha}\mu(\xi,\tau)| \leqslant m(1+|\xi|+|\operatorname{Re}\tau|+|\operatorname{Im}\tau|)^{-1}\mu(\xi,\tau), \qquad |\alpha| = 1.$$

Let $a(y) \in \mathcal{D}$. *Then the inequalities below hold*:

$$(48) \qquad \|(a - {}_{[\gamma]}\mu(D)a_{[\gamma]}\mu^{-1}(D))D_k v\|_{[\gamma]} \leqslant d_{\mu}(a)\|v\|_{[\gamma]},$$

$$(49) \qquad \|(a - {}_{[\gamma]}\mu(D)a_{[\gamma]}\mu^{-1}(D))v\|_{[\gamma]} \leqslant |\gamma|^{-1} d_{\mu}(a)\|v\|_{[\gamma]},$$

where the constant $d_{\mu}(a)$ *is defined in* (41).

A consequence of (49) is

PROPOSITION 2. *Let* $a(y) = a + a'(y)$, $a' \in \mathcal{D}$, *and let the conditions of the lemma hold. Then*

$$(50) \qquad \|au\|_{\mu,[\gamma]} \leqslant (\max|a| + |\gamma|^{-1} d_{\mu}(a'))\|u\|_{\mu,[\gamma]}.$$

4.5.4. Now we consider the symbol (1.2.3):

$$\delta_s^{+}(\xi,\tau) = \left(i\tau + \gamma_0 + \sqrt{1+|\xi|^2}\right)^{s}.$$

As can easily be shown, this symbol and also the symbol

$$(51) \qquad \delta_s(\xi,\tau) = (1+|\xi|^2+\sigma^2+\gamma^2)^{s/2}$$

satisfy conditions (44) and (47). Therefore, if the norm in $H^{(s)}_{[\gamma]}$ is defined using pseudodifferential operators with such symbols, then the norm of the multiplication operator

$$H^{(s)}_{[\gamma]} \to H^{(s)}_{[\gamma]} \quad (u \to a(y)u,\ a(y) = a + a'(y),\ a' \in \mathcal{D})$$

is uniformly bounded with respect to γ, and, moreover, we have

$$\|au\|_{(s),[\gamma]} \leqslant (\max|a| + |\gamma|^{-1} d_s(a))\|u\|_{(s),[\gamma]}.$$

§5. The Cauchy problem for q-parabolic operators with variable coefficients

In this section we prove the existence and uniqueness theorem for q-parabolic differential operators with variable coefficients. Combining the basic theorem of this section with that in §4 we obtain a general result on the Cauchy problem for q-stable correct operators. Most of the proofs in the present section follow the scheme presented in §4. Therefore we will dwell only on those few characteristic features where the specificity of the parabolic case manifests itself.

5.1. Uniformly q-parabolic operators. As in §§3 and 4, we consider differential operators (3.1) resolved with respect to the highest time derivative D_t^m, which enters the operator with a coefficient equal to 1. It is assumed without special stipulation that the coefficients of the operator in question have the form (3.2). For a fixed $y \in \mathbb{R}^{n+1}$ the symbol $P(y,\eta)$ is a polynomial in η, and for a given $q \geqslant 1$ the q-degree of the polynomial can be defined, which will be denoted $\deg_q P(y)$. By the q-degree of the symbol we mean

$$\deg_q P = \sup_{y \in \mathbb{R}^{n+1}} \deg_q P(y).$$

By $P_0(y,\eta)$ we denote the principal q-homogeneous part of the symbol $P(y;\eta)$:

$$P_0(y,\eta) = \sum_{q|\alpha|+j=\deg_q P} a_{\alpha j} \xi^\alpha \sigma^j.$$

DEFINITION. Let $q \geqslant 2$ be an even integer. A symbol $P(y,\eta)$ of the type (4.2) is said to be uniformly q-parabolic if

(a) for each $y^0 \in \mathbb{R}^{n+1}$ the polynomial $P(y^0,\eta)$ is q-parabolic, i. e., the roots $\lambda_{0j}(y^0,\eta)$ of the principal q-homogeneous part $P_0(y,\eta)$ have positive imaginary parts for $\xi \neq 0$;

(b) there is a constant $\kappa > 0$ such that

$$\operatorname{Im} \lambda_{0j}(y^0,\xi) \geqslant \kappa|\xi|^q \quad \forall \xi \in \mathbb{R}^{n+1}, \quad \forall y^0 \in \mathbb{R}^{n+1}.$$

An operator $P(y,D)$ is said to be uniformly q-parabolic if its symbol $P(y,\eta)$ possesses this property.

By analogy with Proposition 4.1, Theorem 1.2.4 can be applied to prove the following

PROPOSITION. *For a polynomial* (4.1) *with coefficients satisfying the stabilization condition* (3.2) *and for an even* $q \geqslant 2$ *the conditions below are equivalent*:

(i) $P(y,\xi,\tau)$ *is a uniformly q-parabolic symbol*;

(ii) *there is* γ_0 *and some numbers* K_1 *and* $K_2 > 0$ *such that*

$$K_1 \leqslant H_P(y,\xi,\sigma,\gamma) \left[(|\gamma| + |\xi|^q)(1 + |\sigma| + |\gamma| + |\xi|^q)^{2m-2}\right]^{-1} \leqslant K_2,$$
$$\gamma \leqslant \gamma_0, \quad y \in \mathbb{R}^{n+1};$$

(iii) *there are* γ_0 *and* $C > 0$ *such that*

$$(|\gamma| + |\xi|^q)(1 + |\sigma| + |\gamma| + |\xi|^q)^{m-1} \leqslant C|P(y,\xi,\sigma + i\gamma)|, \quad \gamma \leqslant \gamma_0, \quad y \in \mathbb{R}^{n+1}.$$

We will need the spaces $H^{(s|q),\lambda}$ of functions characterized by an "additional" degree λ of smoothness with respect to the spatial variables. Namely, we define the space $H^{(s|q),\lambda}$ as a set of functions $u \in H^{(-\infty)}$ for which the norm

$$\|u\|_{(s|q),\lambda} = \left(\int (1 + |\sigma|^2 + |\xi|^{2q})^s (1 + |\xi|^2)^\lambda |\widehat{u}(\xi, \sigma)|^2 \, d\eta \right)^{1/2}$$

is finite; for $\lambda = 0$ we obtain the norm (2.6). The space $H^{(s|q),\lambda}_{[\gamma]}$ is defined in a natural manner; as a norm in this space we will take

$$\|u\|_{(s|q),\lambda,[\gamma]} = \left(\int (1 + |\tau|^2 + |\xi|^{2q})^s (1 + |\xi|^2)^\lambda |\widehat{u}(\xi, \tau)|^2 \, d\xi \, d\tau \right)^{1/2}. \tag{1}$$

In this section we prove the following

THEOREM. *For a differential operator $P(y, D)$ and an even integer $q \geqslant 2$ the conditions below are equivalent.*

(I_+) *For every $M > 0$ there is $\gamma_0 = \gamma_0(M)$ such that for $|s| \leqslant M$ and $\gamma \leqslant \gamma_0$ equation* (4.3) *has a unique solution $u \in H^{(s+m-1|q),q}_{[\gamma]+}$ for any $f \in H^{(s|q)}_{[\gamma]+}$, and the inequality*

$$|\gamma| \, \|u\|_{(s+m-1|q),[\gamma]} + \|u\|_{(s+m-1|q)q,[\gamma]} \leqslant C\|f\|_{(s|q),[\gamma]} \tag{2}$$

holds, where the constant C does not depend on γ.

(II) *For each $M > 0$ there is $\gamma_0 = \gamma_0(M)$ such that for $|s| < M$ and $\gamma \leqslant \gamma_0(M)$ the a priori estimate*

$$|\gamma| \, \|u\|_{(s+m-1|q),[\gamma]} + \|u\|_{(s+m-1|q),q,[\gamma]} \leqslant C\|P(y, D)u\|_{(s|q),[\gamma]} \quad \forall u \in H^{(\infty)}_{[\gamma]}. \tag{3}$$

holds.

(II$'$) *For every $M > 0$ there is $\gamma_0 = \gamma_0(M)$ such that for $|s| \leqslant M$ and $\gamma \leqslant \gamma_0(M)$ the inequality*

$$|[u]|_{(s+m-1|q),[\gamma]} \leqslant C\|P(y, D)u\|_{(s|q),[\gamma]} \quad \forall u \in H^{(\infty)}_{[\gamma]} \tag{4}$$

holds, where (cf. (2.9))

$$|[u]|_{(s+m-1|q),[\gamma]} = |\gamma| \, \|u\|_{(s+m-1|q),[\gamma]} + |\gamma|^{1/q} |[u]|_{(s+m-1|q),q-1,[\gamma]}. \tag{5}$$

Conditions (II_+) and (II'_+) *are obtained from* (II) *and* (II$'$) *by replacing $H^{(\infty)}_{[\gamma]}$ by $H^{(\infty)}_{[\gamma]+}$.*

(III) *The symbol $P(y; \eta)$ is uniformly q-parabolic.*

5.2. Scheme of the proof of Theorem 5.1. As in the case of hyperbolic equations, the proof is based on the following two assertions.

THEOREM. *Let $P(y, \eta)$ be a uniformly q-parabolic symbol, where q is an even integer. Then for all $M > 0$ and for all $L > 0$ there is $\gamma_0(M, L)$ and a number $K > 0$ such that for an arbitrary $y^0 \in \mathbb{R}^{n+1}$ and for $P(D) = P(y^0, D)$ the two-sided inequality*

$$\begin{aligned} -K^{-1} \operatorname{Im} \left[P(D)u, P^{(1)}(D)u \right]_{(s|q),\lambda,[\gamma]} &\leqslant - \operatorname{Im} \left[P(y, D)u, P^{(1)}(y, D)u \right]_{(s|q),\lambda,[\gamma]} \\ &\leqslant - K \operatorname{Im} \left[P(D)u, P^{(1)}(D)u \right]_{(s|q),\lambda,[\gamma]}, \end{aligned} \tag{6}$$

$$|s| \leqslant M, \quad |\lambda| \leqslant L, \quad \gamma \leqslant \gamma_0(M, L), \quad u \in H^{(\infty)}_{[\gamma]},$$

holds.

PROPOSITION. *Suppose that for some $s \in \mathbb{R}$ and $\gamma \leqslant \gamma_0(s)$ a differential operator $P(y, D)$ satisfies inequality (5) with a constant C not depending on γ. Then for the q-homogenous operator $P_0(D) = P_0(y^0, D)$ corresponding to an arbitrary point $y^0 \in \mathbb{R}^{n+1}$ the inequality*

$$\int\limits_{\operatorname{Im}\tau=\gamma} (\gamma^2 + |\gamma|^{2/q}|\xi|^{2q-2})(|\tau|^2 + |\xi|^{2q})^{s+m-1}|\widehat{u}(\xi, \tau)|^2\, d\xi\, d\tau \tag{7}$$
$$\leqslant C^2 \int\limits_{\operatorname{Im}\tau=\gamma} (|\tau|^2 + |\xi|^{2q})^s |P_0(\xi, \tau)\widehat{u}(\xi, \tau)|^2\, d\xi\, d\tau \quad \forall u \in H^{(\infty)}_{[\gamma]}$$

holds with the same constant C. If (4) holds only in the subspace $H^{(\infty)}_{[\gamma]+}$, then (7) remains valid for any points y^0 belonging to the half-space $t \geqslant 0$.

Assuming that the above assertions are proved we will complete the proof of the theorem.

$(\mathrm{I}_+)\Rightarrow(\mathrm{II}_+)\Rightarrow(\mathrm{II}'_+)$. The first implication is proved in the same way as $(\mathrm{I}_+)\Rightarrow(\mathrm{II}_+)$ in Theorem 4.1. The other implication follows from the elementary inequality

$$|\gamma|^{2/q}|\xi|^{2q-2} \leqslant \frac{1}{q}|\gamma|^2 + \frac{q-1}{q}|\xi|^{2q}$$

and the inequality for the norms

$$|[u]|_{(s+m-1|q),\gamma} \leqslant |\gamma|\, \|u\|_{(s+m-1|q),[\gamma]} + |\gamma|^{1/q}\|u\|_{(s+m-1|q),q-1,[\gamma]} \tag{8}$$

implied by the former.

(II'), $(\mathrm{II}'_+)\Rightarrow(\mathrm{III})$. According to the proposition, inequality (4) on $H^{(\infty)}_{[\gamma]}$ or $H^{(\infty)}_{[\gamma]+}$ implies inequality (7) in the same space. By virtue of Theorem 2.2, the equivalent conditions of Theorem 1.2.4 hold with constants not depending on $y^0 \in \mathbb{R}^{n+1}$. Applying Proposition 5.1 we conclude that the symbol $P(y, \eta)$ is uniformly q-parabolic.

$(\mathrm{III})\Rightarrow(\mathrm{II})$, (I_+). The condition of uniform q-parabolicity and Theorem 1.2.4 imply that there exists a unified constant $K_0 > 0$ such that for $\gamma \leqslant \gamma_0$ we have

$$K_0^{-1} \leqslant H_P(y, \xi, \sigma, \gamma)\left[(|\gamma| + |\xi|^q)(|\xi|^q + |\tau| + |\gamma|)^{2m-2}\right]^{-1} \leqslant K_0. \tag{9}$$

Inequality (9) guarantees the fulfilment of the condition of constant strength for the symbol $H_P(y, \xi, \sigma, \gamma)$. By virtue of the theorem and Proposition 3.4, this implies inequality (3.7). Since the operator with symbol $P^*(x, -t, \xi, -\tau)$ is also q-parabolic, it satisfies (3.7). Making the change of variable $t \to -t$ we arrive at (3.7′). To prove assertion (I_+) to the full extent it remains to derive (3) from (III).

Thus, in view of (9) and the theorem, we have

$$\begin{aligned}
|\gamma|\, \|u\|_{(s+m-1|q),[\gamma]} &\leqslant -K_0 \operatorname{Im}\left[P(D)u, P^{(1)}(D)u\right]_{(s|q),[\gamma]} \\
&\leqslant -K_0 K \operatorname{Im}\left[P(y, D)u, P^{(1)}(y, D)u\right]_{(s|q),[\gamma]} \\
&\leqslant K_0 K \|P(y, D)u\|_{(s|q),[\gamma]} \|P^{(1)}(y, D)u\|_{(s|q),[\gamma]} \\
&\leqslant K_1 \|P(y, D)u\|_{(s|q),[\gamma]} \|u\|_{(s+m-1|q),[\gamma]},
\end{aligned}$$

whence

$$|\gamma|\, \|u\|_{(s+m-1|q),[\gamma]} \leqslant K_1 \|P(y, D)u\|_{(s|q),[\gamma]}. \tag{3′}$$

To estimate the second term on the right-hand side of (3) we need inequality (6) involving the additional parameter λ. According to (9), we have

$$\begin{aligned}\|u\|^2_{(s+m-1|q),\lambda+q/2,[\gamma]} &\leqslant -K_0 \operatorname{Im}\left[P(D)u, P^{(1)}(D)u\right]_{(s|q),\lambda,[\gamma]} \\ &\leqslant -K_0K\left[P(y,D)u, P^{(1)}(D)u\right]_{(s|q),\lambda,[\gamma]} \\ &\leqslant K_0K\|P(y,D)u\|_{(s|q),\lambda-q/2,[\gamma]}\|P^{(1)}(y,D)u\|_{(s|q),\lambda+q/2,[\gamma]} \\ &\leqslant K_2\|P(y,D)u\|_{(s|q),\lambda-q/2,[\gamma]}\|u\|_{(s+m-1|q),\lambda+q/2,[\gamma]},\end{aligned}$$

which implies the inequality

$$\|u\|_{(s+m-1|q),\lambda+q/2,[\gamma]} \leqslant K_2\|P(y,D)u\|_{(s|q),\lambda-q/2,[\gamma]}.$$

Setting $\lambda = q/2$ in this inequality we obtain

$$\|u\|_{(s+m-1|q),q,[\gamma]} \leqslant K_2\|P(y,D)u\|_{(s|q),[\gamma]}. \tag{3''}$$

Comparing (3′) and (3″) we derive (3).

5.3. The proof of Proposition 5.2. It is a simple modification of the proof of Proposition 4.2. Consider the family of functions

$$u_\varepsilon(y) = u\left(\frac{t-t^0}{\varepsilon^q}, \frac{x-x^0}{\varepsilon}\right), \quad (x^0,t^0)\in\mathbb{R}^{n+1},\ u\in\mathcal{D},$$

for which the inequality

$$|[u_\varepsilon]|_{(s+m-1|q),[\gamma\varepsilon^{-q}]} \leqslant C\|P(y,D)u_\varepsilon\|_{(s|q),[\gamma\varepsilon^{-q}]}$$

can be written. The passage to the limit for $\varepsilon\to 0$ is carried out following the scheme of Proposition 4.2, which results in inequality (7) for the principal q-homogeneous part $P_0(y^0,D)$.

5.4. The proof of Theorem 5.2. It follows the scheme of the proof of Theorem 4.2. In a sense, it is simpler because, in view of (9), we need not the considerations relating to the fact that the coefficients are real (an exception is a simple estimate for the highest time derivative). To preserve the parallelism with Theorem 4.2, the proof of Theorem 5.2 will also be split into similar stages.

5.4.1. Auxiliary assertions. As in (4.13), we set

$$\{u\}^2_{(s|q),\lambda,[\gamma]} = -\operatorname{Im}\left[P(y^0,D)u, P^{(1)}(y^1,D)u\right]_{(s|q),\lambda,[\gamma]}. \tag{10}$$

In view of (9), the replacement of the point y^0 by another point belonging to $\mathbb{R}^{n+1}$ leads to an equivalent norm, and we have

$$\{u\}_{(s|q),\lambda,[\gamma]} \approx |\gamma|^{1/2}\|u\|_{(s+m-1|q),\lambda,[\gamma]} + \|u\|_{(s+m-1|q),\lambda+q/2,[\gamma]} \tag{11}$$

with an equivalence constant not depending on γ. We will need an analog of Proposition 4.4.1.

PROPOSITION. *Let $q \geqslant 2$ be an even integer and let $A(y, D)$ and $B(y, D)$ be differential operators with coefficients of the type* (3.2). *Assume that*

$$\deg_q A \leqslant mq, \qquad \deg_q B \leqslant mq - q, \quad \deg_q A + \deg_q B \leqslant 2mq - q - 1. \tag{12}$$

Then the quadratic form

$$J = [A(y, D)u, B(y, D)u]_{(s|q),\lambda,[\gamma]} \tag{13}$$

satisfies the inequality

$$J \leqslant \text{const}\, |\gamma|^{-\varepsilon} \{u\}^2_{(s+m-1|q),\lambda,[\gamma]}. \tag{14}$$

PROOF. It can be assumed, without loss of generality, that J is written as

$$\begin{gathered} J = [aD_t^i D_x^\alpha u, D_t^j D_x^\beta u]_{(s|q),\lambda,[\gamma]}, \\ i \leqslant m, \quad j \leqslant m-1, \quad iq + |\alpha| \leqslant mq, \quad jq + |\beta| \leqslant mq - q, \\ (i+j)q + |\alpha| + |\beta| \leqslant 2mq - q - 1. \end{gathered} \tag{13'}$$

We first consider the case $i \leqslant m - 1$. Using integration by parts with respect to x, we represent J as a linear combination of forms of the type (13′), where

$$iq + |\alpha| \leqslant mq - q + q/2, \quad jq + |\beta| \leqslant mq - q + (q-2)/2.$$

Under these conditions on the multiindices the form (13′) has an upper bound

$$\text{const}\, \|u\|_{(s+m-1|q),q/2,[\gamma]} \|u\|_{(s+m-1|q),(q-2)/2,[\gamma]}.$$

Applying the interpolation inequality

$$\begin{aligned} \|u\|_{(s+m-1|q),(q-2)/2,[\gamma]} &\leqslant |\gamma|^{-1/q} \left(|\gamma|^{1/2} \|u\|_{(s+m-1|q),[\gamma]} + \|u\|_{(s+m-1|q),q/2,[\gamma]} \right) \\ &\leqslant \text{const}\, |\gamma|^{-1/q} \{u\}_{(s+m-1|q),\lambda,[\gamma]}, \end{aligned}$$

we prove (14).

If $i = m$ (in this case $\alpha = 0$), then we transfer the derivative with respect to t to the right-hand side of the form and the q derivatives with respect to x to the left-hand side and thus arrive at the already considered forms. The proposition is proved.

5.4.2. Localization of the estimate. Based on Proposition 5.4.1 we can repeat literally the argument in 4.4.2 to prove the following

PROPOSITION. *Assume that for some $\delta > 0$ there is $\gamma_0(\delta)$ such that inequality* (6) *with $s = \lambda = 0$ holds when the support of $u(y)$ belongs to a ball of a radius no greater than δ. Then there is γ_1 such that for $\gamma \leqslant \gamma_1$ the inequality* (6) *with $s = \lambda = 0$ holds for all $u \in H^{(\infty)}_{[\gamma]}$.*

5.4.3. Proof of inequality (6) (the case $s = \lambda = 0$). An analog of Proposition 4.4.3 for the case of even values of q is stated as the following

PROPOSITION. (i) *Let $q \geqslant 2$ be an even integer. Let $A(y, D)$ and $B(y, D)$ be differential operators of q-orders mq, $mq - q$, respectively, where either the operator A does not contain D_t^m or the operator B does not contain D_t^{m-1}. Then the following inequality holds (cf. (4.23)):*

$$|[A(y, D)u, B(y, D)u]_{[\gamma]}| \leqslant \left(\kappa[A_0][B_0] + C|\gamma|^{-1/q}\right) \{u\}^2_{[\gamma]}. \tag{15}$$

(ii) *Let $q \geqslant 2$ be an even integer and let the q-orders of the operators $A(y, D)$ and $B(y, D)$ not exceed mq and $mq - q$, respectively. Assume that the highest derivatives with respect to t, i. e., D_t^m and D_t^{m-1}, enter the corresponding operators A and B with real coefficients. Then*

$$|\mathrm{Im}[A(y, D)u, B(y, D)u]_{[\gamma]}| \leqslant \left(\kappa[A_0][B_0] + C|\gamma|^{-1/q}\right). \tag{16}$$

PROOF. (i) is proved using integration by parts. The form on the left-hand side is represented as a linear combination of the forms

$$\left[a\gamma^{i_1} D_t^{i_2} D_x^{\alpha} u, b\gamma^{j_1} D_t^{j_2} D_x^{\beta} u\right]_{[\gamma]},$$
$$(i_1 + i_2)q + |\alpha| + (j_1 + j_2)q + |\beta| = (m - 1)q + q/2,$$

and some forms satisfying the conditions of Proposition 5.4.1. Here $a(y)$ and $b(y)$ are some of the coefficients of the principal q-homogeneous parts of A and B. This implies (15).

(ii) We write down $A = a(y)D_t^m + A'$ and $B = b(y)D_t^{m-1} + B'$ and represent the form on the left-hand side of (16) as

$$\mathrm{Im}[aD_t^m u, bD_t^{m-1} u]_{[\gamma]} + \mathrm{Im}[aD_t^m u, B'u]_{[\gamma]} + \mathrm{Im}[A'u, Bu]_{[\gamma]}.$$

Inequality (16) for the second and third forms follows from the already proved assertion (i). As for the first form, it satisfies the conditions of Proposition 4.4.3 and, consequently, is estimated in terms of

$$(\max|a| \max|b| + C|\gamma|^{-1})\|u\|^2_{(m-1|1),[\gamma]}.$$

The proposition is proved.

Using the above propositions and repeating literally the argument in 4.4.3 we prove (6) for $s = \lambda = 0$.

5.4.4. Proof of inequality (6) in the general case. As in 4.4.4, we apply pseudodifferential operators to reduce inequality (6) for the general case to the case $s = \lambda = 0$. Assuming that q is a positive integer we set

$$\delta_{(s|q)}(\xi, \tau) = (1 + |\tau|^2 + |\xi|^{2q})^{s/2}, \quad \delta_\lambda(\xi) = (1 + |\xi|^2)^{\lambda/2}.$$

We have $u \in H^{(s|q),\lambda}_{[\gamma]}$ if and only if

$$u = {}_{[\gamma]}\delta_{(-s|q)}(D)\delta_{-\lambda}(D_x)w, \quad \|w\|_{[\gamma]} = \|u\|_{(s|q),\lambda,[\gamma]}.$$

Using this representation we rewrite the form we are interested in as

$$-\mathrm{Im}[Pw, P^{(1)}w]_{[\gamma]} - \mathrm{Im}[Pw, (\widetilde{P}^{(1)} - P^{(1)})w]_{[\gamma]}$$
$$- \mathrm{Im}[(\widetilde{P} - P)w, P^{(1)}w]_{[\gamma]} = J_0 + J_1 + J_2,$$

where

$$\widetilde{P} = \delta^{-}_{(s|q)}\delta_\lambda P \delta^{-}_{(-s|q)}\delta_{-\lambda}, \quad \widetilde{P}^{(1)} = \delta^{-}_{(s|q)}\delta_\lambda P^{(1)}\delta^{-}_{(-s|q)}\delta_{-\lambda}.$$

As in §4, we will assume that the derivative D_t^m enters the operator P with a constant coefficient.

According to the previous subsection, we have

$$J_0 \geqslant K_0\{w\}^2_{[\gamma]} = K_0\{u\}_{(s|q),\lambda,[\gamma]},$$

and to prove the desired assertion it suffices to show that

$$(17) \qquad |J_j| \leqslant \text{const}\,|\gamma|^{-1/q}\{w\}^2_{[\gamma]}.$$

Repeating the argument in 4.4.6 and employing integration by parts we rewrite the forms J_1 and J_2 as linear combinations of forms of the type

$$[T_1 D_t^i D_x^\alpha w, T_2 D_t^j D_x^\beta w]_{[\gamma]},$$

where

$$iq + |\alpha|, iq + |\beta| \leqslant mq - q/2, \quad (i+j)/q + |\alpha| + |\beta| \leqslant 2mq - q - 1,$$

and T_1 and T_2 are either some functions or operators of the type

$$\left(b(y) - \delta^{+}_{(s|q)}(D)\delta_\lambda(D_x)b(y)\delta^{+}_{(-s|q)}(D)\delta_{-\lambda}(D_x)\right)\frac{\partial}{\partial x_k}, \quad k = 1, \ldots, n.$$

According to 4.5, the operators T_1, T_2: $H_{[\gamma]} \to H_{[\gamma]}$ are bounded, and their norms are uniformly bounded for $\gamma \leqslant \gamma_0$. This implies the desired estimates.

5.5. The Cauchy problem for q-stable correct differential operators with variable coefficients. Here we state the general result by combining the results of the present and foregoing sections as a unified theorem.

THEOREM. *Let $P(y, D)$ be a differential operator of the type* (3.1) *with coefficients satisfying conditions* (3.2) *and let $q \geqslant 1$ be a real number. Then the following conditions are equivalent.*

(I) *For any differential operator $Q(y, D)$, with coefficients of the type* (3.2), *whose q-order is less than that of P and for any $M > 0$ there is $\gamma_0 = \gamma_0(M, Q)$ such that for $|s| \leqslant M$, $\gamma \leqslant \gamma_0$ the equation*

$$(18) \qquad (P(y; D) + Q(y; D))u = f$$

possesses a unique solution $u \in H^{(s+m-1|q),q-1}_{[\gamma]+}$ for any $f \in H^{(s|q)}_{[\gamma]+}$, and this solution satisfies the inequality

$$(19) \qquad \begin{aligned} |[u]|_{(s+m-1|q),[\gamma]} &\overset{\text{def}}{=} |\gamma|\,\|u\|_{(s+m-1|q),\gamma} + |\gamma|^{1/q}\|u\|_{(s+m-1|q),q-1,[\gamma]} \\ &\leqslant C\|f\|_{(s|q),[\gamma]}. \end{aligned}$$

(II) *The inequality*

$$(20) \qquad |[u]|_{(s+m-1|q),[\gamma]} \leqslant C\|P(y, D)u\|_{(s|q),[\gamma]} \quad \forall u \in H^{(\infty)}_{[\gamma]+}$$

holds.

(II_+) *Inequality* (20) *holds on the subspace $H^{(\infty)}_{[\gamma]+}$.*

(III) *The number q is either equal to 1 or is an even integer. For $q = 1$ the operator $P(y, D)$ is uniformly strictly hyperbolic, and for even values of q it is uniformly q-parabolic.*

The implication (I)$\Rightarrow$(II$_+$) is obvious. The implications (II) and (II$_+$)$\Rightarrow$(III) are proved by means of Proposition 5.2 (in whose proof the fact that q was even was not used) and Theorem 2.2. The main implications (III)$\Rightarrow$(I) and (III)$\Rightarrow$(II)$\Rightarrow$(II$_+$) follow from Theorems 4.1 and 5.1.

REMARKS. 1) For $q = 1$ the above theorem coincides with Theorem 4.1.

2) In the case of even values of q we obtain a less precise result as compared to Theorem 5.1 in which a solution belonging to $H_{[\gamma]+}^{(s+m-1|q),q}$ was obtained (and not to $H_{[\gamma]+}^{(s+m-1|q),(q-1)}$ as in the theorem of the present subsection), and inequality (3) is stronger than estimate (20), which coincides with (4).

3) We note that the argument in Theorems 5.1 and 5.2 can be applied to derive a stronger version of inequality (3). To this end we introduce the norms

$$[u]^2_{(s|q),\lambda,[\gamma]} = \int_{\operatorname{Im}\tau=\gamma} (|\gamma| + |\xi|^q)(1 + |\xi|^2)^{\lambda/2}(|\tau|^2 + |\xi|^{2q})^s |\widehat{u}(\xi,\tau)|^2 \, d\xi \, d\tau, \tag{21}$$

$$'[u]^2_{(s|q),\lambda,[\gamma]} = \int_{\operatorname{Im}\tau=\gamma} (|\gamma| + |\xi|^q)^{-1}(1 + |\xi|^2)^{\lambda/2}(|\tau|^2 + |\xi|^{2q})^s |\widehat{u}(\xi,\tau)|^2 \, d\xi \, d\tau. \tag{21'}$$

It is clear that the following version of the Schwarz inequality holds:

$$[f, g]_{(s|q),\lambda,[\gamma]} \leqslant {}'[f]_{(s|q),\lambda,[\gamma]}[g]_{(s|q),\lambda,[\gamma]}. \tag{22}$$

We also note that, by virtue of (9) and Theorem 5.2, we have

$$\begin{aligned}
[u]^2_{(s+m-1|q),\lambda,[\gamma]} &\leqslant -\operatorname{const}\left[P(y^0, D), P^{(1)}(y^0, D)u\right]_{(s|q),\lambda,[\gamma]} \\
&\leqslant -\operatorname{const}\left[P(y, D), P^{(1)}(y, D)u\right]_{(s|q),\lambda,[\gamma]} \\
&\leqslant \operatorname{const}'\left[P(y, D)u\right]_{(s|q),\lambda,[\gamma]}\left[P^{(1)}(y^0, D)u\right]_{(s|q),\lambda,[\gamma]} \\
&\leqslant \operatorname{const}'\left[P(y, D)u\right]_{(s|q),\lambda,[\gamma]}[u]_{(s+m-1|q),\lambda,[\gamma]}
\end{aligned}$$

which implies the inequality

$$[u]_{(s+m-1|q),\lambda,[\gamma]} \leqslant \operatorname{const}'\left[P(y, D)u\right]_{(s|q),\lambda,[\gamma]}. \tag{23}$$

In the case $\lambda = q/2$ this inequality is equivalent to (3); however, in (23) the dependence of the right-hand side on the large parameter $|\gamma|$ is traced more thoroughly.

4) With the aid of the parametrix method of §2 we can strengthen inequality (3) by estimating $\|u\|_{(s+m|q),\lambda,[\gamma]}$ in terms of $\|Pu\|_{(s|q),\lambda,[\gamma]}$. In this case the solution corresponding to an arbitrary $f \in H_{[\gamma]+}^{(s|q),\lambda}$ belongs to $H_{[\gamma]+}^{(s+m|q),\lambda}$.

§6. The Cauchy problem for q-hyperbolic operators with constant and variable coefficients

The aim of the present section is to include q-hyperbolic operators in the theory presented in §§2–5. The main technique in the consideration of these operators is the replacement of spaces corresponding to the weight $\exp(\gamma t)$ by spaces related to the weight $\exp(\gamma t + \nu\langle N, x\rangle)$, where N is a direction in $\mathbb{R}^n$. We first introduce the necessary spaces and formulate the main result of 1.4 in terms of the Cauchy problem for differential operators with constant coefficients.

The main part of this section is devoted to the Cauchy problem for q-hyperbolic operators with variable coefficients, which is studied according to the scheme in §§4 and 5. The central point is a two-sided inequality for Leray's form

$$-\operatorname{Im}\left[P(y, D)u, P^{(1)}(y, D)u\right]_{(s|q),\lambda,[\gamma,\nu N]}$$

in terms of similar forms corresponding to operators with constant (frozen) coefficients.

At the end of the section we state the basic result of the present chapter that includes Theorems 4.1, 5.1, and 6.1 as special cases.

6.1. For $N \in \mathbb{R}^n$ we set

$$H^{(s|q),\lambda}_{[\gamma,\nu N]} = \left\{u \in \mathcal{D}', \exp(\gamma t + \nu\langle N, x\rangle)u \in H^{(s|q),\lambda}\right\}$$

and endow this space with the natural norm

$$(1) \qquad \|u\|_{(s|q),\lambda,[\gamma,\nu N]} = \left(\int\limits_{\substack{\operatorname{Im}\tau=\gamma\\ \operatorname{Im}\zeta=\nu N}} (1+|\tau|^2+|\zeta|^{2q})^s(1+|\zeta|^2)^\lambda|\widehat{u}(\zeta,\tau)|^2 d\zeta\, d\tau\right)^{1/2}.$$

Introducing the pseudodifferential operators

$${}_{[\gamma,\nu N]}a(D)u = \exp(-\gamma t - \nu\langle N, x\rangle)a(D_x + i\nu N, D_t + i\gamma)(\exp(\gamma t + \nu\langle N, x\rangle)u)$$

in the space $H^{(-\infty)}_{[\gamma,\nu N]}$ we can rewrite the norm (1) in the form

$$(1') \qquad \|u\|_{(s|q),\lambda,[\gamma,\nu N]} = \left\|{}_{[\gamma,\nu N]}(\delta_{(s|q)}(D)\delta_\lambda(D_x))u\right\|_{[\gamma,\nu N]},$$

where

$$\delta_{(s|q)} = (1+|\tau|^2+|\xi|^{2q})^{s/2}, \quad \delta_\lambda(\xi) = (1+|\xi|^2)^{\lambda/2}.$$

In this case we have

$$(2) \qquad \left\{u \in H^{(s|q),\lambda}_{[\gamma,\nu N]}\right\} \Longleftrightarrow \left\{u = {}_{[\gamma,\nu N]}\delta_{(s|q)}(D)\delta_\lambda(D)w,\right.$$
$$\left. w \in H_{[\gamma,\nu N]}, \quad \|w\|_{[\gamma,\nu N]} = \|u\|_{(s|q),\lambda,[\gamma,\nu N]}\right\}.$$

THEOREM. *For $q \geqslant 1$, $N \in \mathbb{R}^n$, and a polynomial $P(\xi,\tau)$ the following conditions are equivalent*:

(I_+) *For all $Q(\xi,\tau)$, $\deg_q Q < \deg_q P$, there is $\nu_0 = \nu_0(Q)$ and $\sigma = \sigma(Q)$ such that the equation*

$$(3) \qquad (P+Q)(D_x, D_t)u = f$$

possesses a unique solution $u \in H^{(s-\sigma|q)}_{[\gamma,\nu N]+}$ for any right-hand side $f \in H^{(s|q)}_{[\gamma,\nu N]}$.

(II) *There is* $\nu_0 > 0$ *such that for all* $s \in \mathbb{R}$, $\nu \geqslant \nu_0$, *and* $\gamma \leqslant \gamma_0$,

$$\begin{aligned} |[u]|_{(s+m-1|q),[\gamma,\nu N]} &\overset{\text{def}}{=} \gamma|\,\|u\|_{(s+m-1|q),[\gamma,\nu N]} \\ &\leqslant \nu\|u\|_{(s+m-1|q),q-1,[\gamma,\nu N]} \\ &\leqslant C\|P(D)u\|_{(s+m-1|q),[\gamma,\nu N]}, \quad \forall u \in H^{(\infty)}_{[\gamma,\nu N]}. \end{aligned} \tag{4}$$

(II_+) *Inequality* (4) *holds in the space* $H^{(\infty)}_{[\gamma,\nu N]+}$.
(III) *Condition* (III) *of Theorem* 1.4.3 *holds.*

PROOF. The transformation $u = \exp\langle -\nu N, x\rangle v$, $f = \exp\langle -\nu N, x\rangle g$ reduces (3) to the family of equations

$$(P_{\nu N} + Q_{\nu N})(D_x, D_t)v = g,$$

where

$$P_{\nu N}(\xi,\tau) = P(\xi + i\nu N, \tau), \quad Q_{\nu N}(\xi,\tau) = Q(\xi + i\nu N, \tau),$$

and the subspaces $H^{(s|q)}_{[\gamma,\nu N]}$ are replaced by $H^{(s|q)}_{[\gamma]}$. Application of Proposition 2.1 allows us to conclude that if (I) holds, then the polynomials $P_{\nu N}$ are correct in Petrovskii's sense, i. e., condition (I) of Theorem 1.4.3 holds. Using this theorem we conclude that condition (III) holds. Similarly, by the Parseval relation and Theorem I.4.3, it follows from (II) and (II_+) that (III) holds (cf. Proposition 2.1). Furthermore, if (III) holds, then equation (3) is solved with the aid of the Fourier transform, and, in view of the Parseval relation, the lower estimate for the polynomial $P(\xi,\tau)$ implies inequality (4).

6.2. We will deal with symbols

$$P(y,\xi,\tau) = \tau^m + \sum a_{j\alpha}(y)\tau^j\xi^\alpha \tag{5}$$

whose coefficients $a_{j\alpha}(y)$ satisfy conditions (3.2).

DEFINITION. Let $q \geqslant 1$ be an odd integer. Symbol (5) is said to be uniformly q-hyperbolic if

(a) for every $y \in \mathbb{R}^{n+1}$ the polynomial $P(y^0,\xi,\tau)$ is q-hyperbolic, i. e., the roots $\lambda_{0j}(y^0,\xi)$ satisfy the conditions of Definition 1 in 1.3.6;

(b) there is a constant δ such that

$$|\lambda_{0j}(y^0,\xi) - \lambda_{0k}(y^0,\xi)| \geqslant \delta|\xi|^q, \qquad k \neq j, \quad \forall \xi \in \mathbb{R}^n, \quad \forall y^0 \in \mathbb{R}^{n+1}; \tag{6}$$

(c) let $q > 1$ and let $V(P)$ be the intersection of the cones of q-hyperbolicity of the polynomials $P(y^0,\xi,\tau)$. Then for each $N \in V(P)$, $|N| = 1$, there is a unified constant $\kappa(N) > 0$ for all $y^0 \in \mathbb{R}^{n+1}$ such that

$$\langle N, \operatorname{grad}\lambda_{0j}(\xi^0,\xi)\rangle \geqslant \kappa(N)|\xi|^{q-1}, \quad \forall y^0 \in \mathbb{R}^{n+1}, \quad \forall \xi \in \mathbb{R}^n. \tag{7}$$

The cone $V(P)$ is called the cone of q-hyperbolicity of the symbol (4.1).

By combining an analysis of the equivalent definitions of q-hyperbolic polynomials and the compactness considerations according to the plan of Proposition 4.1 we can prove the following

PROPOSITION. *Let the coefficients in* (4.1) *have the form* (3.2), *let* $q > 1$ *be an odd integer, and let* $N \in \mathbb{R}^n$. *Then the conditions below are equivalent.*

(i) *The symbol* $P(y; \xi, \tau)$ *is uniformly* q*-hyperbolic and* N *belongs to the cone* $V(P)$.

(ii) *There is* ν_0 *and some numbers* $K_1, K_2 > 0$ *such that for* $\nu \geqslant \nu_0$ *and* $\gamma \leqslant -R\nu^q - R_1$ *the inequality*

$$K_1(|\gamma| + \nu|\xi|^{q-1})(|\gamma| + |\sigma| + (|\xi| + \nu)^q)^{m-1} \leqslant H_P(y, \xi, \nu, \sigma, \gamma)$$
$$\leqslant K_2(|\gamma| + \gamma|\xi|^{q-1})(|\gamma| + |\sigma| + (|\xi| + \nu)^q)^{m-1}$$

holds, where

$$H_P(y, \xi, \nu, \sigma, \gamma) = -\operatorname{Im}\left[P(y, \xi + i\nu N, \sigma + i\gamma)\overline{P^{(1)}(y, \xi + i\nu N, \sigma + i\gamma)}\right].$$

(iii) *There is* ν_0 *such that for* $\nu \geqslant \nu_0$ *and* $\gamma < -R\nu^q - R_1$ *the inequality*

$$(|\gamma| + \nu|\zeta|^{q-1})(1 + |\tau| + |\zeta|^q)^{m-1} \leqslant C|P(y, \zeta, \tau)|,$$
$$\zeta = \xi + i\nu N, \quad \tau = \sigma + i\gamma, \quad \forall y \in \mathbb{R}^{n+1}$$

holds.

THEOREM. *For a given odd integer* $q > 1$, *a vector* $N \in \mathbb{R}^{n+1}$, *and a differential operator* $P(y, D)$ *with symbol* (4.1) *the following conditions are equivalent.*

(I_+) *For all* $M > 0$ *there is* $\nu_0 = \nu_0(M)$ *and some numbers* $R > 0$ *and* R_1 *such that for* $|s| \leqslant M$, $\nu \geqslant \nu_0$, *and* $\gamma \leqslant -R\nu^q - R_1$ *the equation*

$$P(y, D)u = f \tag{8}$$

possesses a unique solution $u \in H^{(s+m-1|q),q-1}_{[\gamma,\nu N]+}$ *for any* $f \in H^{(s|q)}_{[\gamma,\nu N]+}$, *and this solution satisfies the inequality*

$$|[u]|_{(s+m-1|q),[\gamma,\nu N]} \leqslant C\|f\|_{(s|q),[\gamma,\nu N]}$$

with a constant C *not depending on* ν *and* γ.

(II) *For every* $M > 0$ *there is* $\nu_0 = \nu_0(M)$ *such that the inequality*

$$\begin{gathered} |[u]|_{(s+m-1|q),[\gamma,\nu N]} \leqslant C\|P(y, D)u\|_{(s|q),[\gamma,\nu N]}, \\ |s| \leqslant M, \quad \nu \geqslant \nu_0, \quad \gamma \leqslant -R\nu^q - R_1, \quad \forall u \in H^{(\infty)}_{[\gamma,\nu N]} \end{gathered} \tag{9}$$

holds.

(II_+) *is obtained from* (II) *by replacing* $H^{(\infty)}_{[\gamma,\nu N]}$ *with* $H^{(\infty)}_{[\gamma,\nu N]+}$.

(III) *The symbol* $P(y; \eta)$ *is uniformly* q*-hyperbolic.*

6.3. The proof of the implications (II), (II_+)$\Rightarrow$(III) is based on an analog of Propositions 4.2 and 5.2. We will prove the following

PROPOSITION. *For a given* $q \geqslant 1$, $N \in \mathbb{R}^n$, *and a differential operators* $P(y, D)$ *with coefficients of the type* (3.2) *let the inequality* (9) *hold for some* $s \in \mathbb{R}$, $\nu \geqslant \nu_0(s)$, *and*

$\gamma < -R\nu^q - R_1$. *Then for this s, each* $\nu > 0$, *each* $\gamma < -R\nu^q$, *and* $y^0 \in \mathbb{R}^{n+1}$ *the differential operators* $P(D) = P(y^0, D)$ *with constant coefficients satisfy the inequality*

$$\tag{10} \begin{aligned} &\int\limits_{\substack{\operatorname{Im}\tau=\gamma\\ \operatorname{Im}\tau=\nu N}} (\gamma^2+\nu^2|\zeta|^{2q-2})(|\tau|^2+|\zeta|^{2q})^{m-1+s}|\widehat{u}(\zeta,\tau)|^2\,d\zeta\,d\tau \\ &\qquad\leqslant C^2 \int\limits_{\substack{\operatorname{Im}\tau=\gamma\\ \operatorname{Im}\zeta=\nu N}} (|\tau|^2+|\zeta|^{2q})^s\,\left|P_0(\xi,\tau)\widehat{u}(\xi,\tau)\right|^2\,d\zeta\,d\tau. \end{aligned}$$

PROOF. As in Proposition 5.2, we take $u(y) \in \mathcal{D}$ and set

$$u_\varepsilon(y) = u\left(\frac{x-x^0}{\varepsilon}, \frac{t-t^0}{\varepsilon^q}\right).$$

Let $\nu > 0$ and $\gamma < -R\nu^q$. If ε is sufficiently small, then $\varepsilon^{-1}\nu \geqslant \nu_0(s)$ and $\gamma\varepsilon^{-q} \leqslant -R(\varepsilon^{-1}\nu)^q - R_1$. Therefore, according to the condition of the proposition, we can write the inequality

$$\begin{aligned} &|\gamma|\varepsilon^{-q}\|u_\varepsilon\|_{(s+m-1|q),[\gamma\varepsilon^{-q},\nu\varepsilon^{-1}N]} + \nu\varepsilon\|u_\varepsilon\|_{(s+m-1|q),q-1,[\gamma\varepsilon^{-q},\nu\varepsilon^{-1}N]} \\ &\quad\leqslant C\|P(y,D)u_\varepsilon\|_{(s|q),[\gamma\varepsilon^{-q},\nu\varepsilon^{-1}N]}. \end{aligned}$$

A modification of the argument in Proposition 4.2 shows that by multiplying the last inequality by

$$K(\varepsilon) = \varepsilon^{-(n+q)/2+(m+s)q}\exp(-t^0\gamma\varepsilon^{-q} - \langle N, x^0\rangle\nu\varepsilon^{-1})$$

and passing to the limit as $\varepsilon \to 0$ we derive (10).

6.4. As in §§4 and 5, the central point in the proof of the solvability of the Cauchy problem is Leray's two-sided estimate. We have the following

THEOREM. *Let* $q \geqslant 1$ *be an integer and let the coefficients of an operator* $P(y,D)$ *with symbol* (4.1) *satisfy conditions* (3.2). *Let* $N \in V(P)$. *Then for all* $M > 0$ *and* $L > 0$ *there is* $\nu_0 = \nu_0(M,L)$ *such that for* $|s| \leqslant M$, $|\lambda| \leqslant L$, $\nu \geqslant \nu_0$, *and* $\gamma < -R\nu^q - R_1$ *there is a constant* $K > 0$ *such that for every* $y^0 \in \mathbb{R}^{n+1}$ *the two-sided inequality*

$$\tag{11} \begin{aligned} &-K^{-1}\operatorname{Im}\left[P(D)u, P^{(1)}(D)u\right]_{(s|q),\lambda,[\gamma,\nu N]} \\ &\qquad\leqslant -\operatorname{Im}\left[P(y,D)u, P^{(1)}(y,D)u\right]_{(s|q),\lambda,[\gamma,\nu N]} \\ &\qquad\leqslant -K\operatorname{Im}\left[P(D)u, P^{(1)}(D)u\right]_{(s|q),\lambda,[\gamma,\nu N]} \qquad \forall u \in H^{(\infty)}_{[\gamma,\nu N]} \end{aligned}$$

holds, where $P(D) = P(y^0, D)$.

Assuming that the theorem is proved we will complete the proof of Theorem 6.2.

6.4.1. First we prove (9). To simplify the notation, we define the norm

$$\tag{12} \{u\}^2_{(s|q),\lambda,[\gamma,\nu N]} = -\operatorname{Im}\left[P(y^0,D)u, P^{(1)}(y^0,D)u\right]_{(s|q),\lambda,[\gamma,\nu N]}.$$

By Proposition 6.2, replacing the point y^0 with another point belonging to $\mathbb{R}^{n+1}$ results in an equivalent norm, and the equivalence constant does not depend on ν and γ. Furthermore, by the same proposition,

$$\{u\}_{(s|q),\lambda,[\gamma,\nu N]} \approx |\gamma|^{1/2}\|u\|_{(s+m-1|q),\lambda,[\gamma,\nu N]} + \nu^{1/2}\|u\|_{(s+m-1|q),\lambda+(q-1)/2,[\gamma,\nu N]}. \tag{13}$$

According to (13) and the theorem, we have

$$\begin{aligned}|\gamma|\,\|u\|^2_{(s+m-1|q),[\gamma,\nu N]} &\leqslant C\{u\}^2_{(s+m-1|q),[\gamma,\nu N]}\\ &\leqslant -CK\operatorname{Im}\left[P(y,D)u, P^{(1)}(y,D)\right]_{(s|q),[\gamma,\nu N]}\\ &\leqslant CK\|Pu\|_{(s|q),[\gamma,\nu N]}\,\|P^{(1)}u\|_{(s|q),[\gamma,\nu N]}\\ &\leqslant K_1\|Pu\|_{(s|q),[\gamma,\nu N]}\,\|u\|_{(s+m-1|q),[\gamma,\nu N]},\end{aligned}$$

which implies the inequality

$$|\gamma|\,\|u\|_{(s+m-1|q),[\gamma,\nu N]} \leqslant K_1\|P(y,D)u\|_{(s|q),[\gamma,\nu N]}. \tag{9$'$}$$

Furthermore, as in the case of q-parabolic equations, we have

$$\begin{aligned}\nu\|u\|^2_{(s+m-1|q),q-1,[\gamma,\nu N]} &\leqslant C^2\{u\}^2_{(s|q),(q-1)/2,[\gamma,\nu N]}\\ &\leqslant -C^2K\operatorname{Im}[Pu, P^{(1)}u]_{(s|q),(q-1)/2,[\gamma,\nu N]}\\ &\leqslant C^2K\|Pu\|_{(s|q),0,[\gamma,\nu N]}\,\|P^{(1)}u\|_{(s|q),q-1,[\gamma,\nu N]}\\ &\leqslant K_2\|Pu\|_{(s|q),[\gamma,\nu N]}\,\|u\|_{(s|q),q-1,[\gamma,\nu N]},\end{aligned}$$

whence

$$\nu\|u\|_{(s+m-1|q),q-1,[\gamma,\nu N]} \leqslant K_2\|P(y,D)u\|_{(s|q),[\gamma,\nu N]}. \tag{9$''$}$$

Adding together the inequalities (9$'$) and (9$''$) we obtain (9).

6.4.2. As is seen from the proof of Theorem 1.1.3, the constant C in (1.1.12) depends on m only, i. e., on the degree of the polynomial P with respect to the variable τ, and does not depend on the coefficients of the polynomial. Therefore, if for $\nu \geqslant \nu_0$ and $\gamma < -R\nu^q - R_1$ the polynomial $P(\xi + i\nu N, \sigma + i\gamma)$ is nonzero, then it satisfies the inequality

$$\begin{aligned}&\sum_{l=1}^{m}(\gamma_0-\gamma)^{2l-1}|P^{(l)}(\xi+i\nu N,\sigma+i\gamma)|^2\\ &\qquad\leqslant -c(m)\operatorname{Im}\left[P(\xi+i\nu N,\sigma+i\gamma), \overline{P^{(1)}(\xi+i\nu N,\sigma+i\gamma)}\right] = c(m)H_P(\xi,\nu,\sigma,\gamma).\end{aligned}$$

By the Parseval relation, it follows that

$$\sum_{l=1}^{m}(\gamma_0-\gamma)^{2l-1}\|P^{(l)}(D)u\|^2_{(s|q),[\gamma,\nu N]} \leqslant c(m)\{u\}^2_{(s|q),[\gamma,\nu N]}.$$

Now, using representation (3.25) and repeating the argument in 3.4 we show that

$$c^{-1}m\sum_{l=1}^{m}(\gamma_0-\gamma)^{2l-1}\|P^{(l)}(y;D)u\|^2_{(s|q),[\gamma,\nu N]}$$
$$\leqslant -\operatorname{Im}[P(y,D)u,P^{(1)}(y,D)u]_{(s|q),[\gamma,\nu N]}$$
$$\leqslant (2\varepsilon(\gamma_0-\gamma))^{-1}\|P(y,D)u\|_{(s|q),[\gamma,\nu N]}$$
$$+\frac{\varepsilon}{2}(\gamma-\gamma_0)\|P^{(1)}(y,D)u\|_{(s|q),[\gamma,\nu N]}.$$

Taking a sufficiently small ε we arrive at the inequality

$$\sum_{l=1}^{m}(\gamma_0-\gamma)^l\|P^{(l)}(y;D)u\|_{(s|q),[\gamma,\nu N]}\leqslant \operatorname{const}\|P(y,D)u\|_{(s|q),[\gamma,\nu N]}. \tag{14}$$

6.4.3. If $q>1$ and a symbol $P(x,t,\xi,\tau)$ satisfies condition (III) of Theorem 6.2, then the symbol $P^*(x,-t,\xi,-\tau)$ also satisfies it provided that the cone of q-hyperbolicity $V(P)$ is replaced by the cone $-V(P)$. Indeed, the principal q-homogeneous part of the symbol $P^*(x,-t,\xi,-\tau)$ is written

$$P_0(x,-t,\xi,-\tau)=(-1)^{mq}P_0(x,-t,-\xi,\tau).$$

Thus, we can write inequality (10) for the operator $P^*(x,-t,D_x,-D_t)$ with replacement of s and N by $-s$ and $-N$, respectively. After the transformation $t\to -t$ we obtain the inequality

$$\sum_{l=1}^{m}(\gamma_0-\gamma)^l\|P^{*(l)}(y,D)u\|_{(-s|q),[-\gamma,-\nu N]}\leqslant \operatorname{const}\|P^*(y,D)u\|_{(-s|q),[-\gamma,-\nu N]}. \tag{14'}$$

A simple modification of the argument in §3 shows that (9), (14), and (14′) imply the solvability in the spaces $H^{(s|q)}_{[\gamma,\nu N]}$, i. e., assertion (I$_+$) of the theorem.

6.5. Now we prove Theorem 6.4. Since it does not practically differ from the proof of similar theorems in §§4 and 5, our presentation will be very sketchy.

6.5.1. An analog of Proposition 4.1 is the following

PROPOSITION. *Let $A(y,D)$ and $B(y,D)$ be two differential operators such that $\deg_q A\leqslant mq$ and $\deg_q B\leqslant(m-1)q$, where one of the inequalities is strict, and q is an odd integer. Let $\nu>0$ be sufficiently large and let $\gamma<-R\nu^q$. Then the quadratic form*

$$J=[A(y,D)u,B(y,D)u]_{(s|q),\lambda,[\gamma,\nu N]}$$

satisfies the inequality

$$|J|\leqslant \operatorname{const}\nu^{-1}\{u\}^2_{(s|q),\lambda,[\gamma,\nu N]}. \tag{15}$$

The proof is based on formulas for integration by parts of the type (4.19), (4.19′), where the weight $\exp(\gamma t)$ should be replaced by $\exp(\gamma t+\nu\langle N,x\rangle)$, and in (4.19) the term $2\nu\nu_k[av,w]_{(s|q),\lambda,[\gamma,\nu N]}$ should be added, where $N=(\nu_1,\dots,\nu_n)$. Without loss of generality, it can be assumed that the form J is written as

$$J=[aD_t^iD_x^\alpha u,bD_t^jD_x^\beta u]_{(s|q),\lambda,[\gamma,\nu N]},$$
$$iq+|\alpha|\leqslant mq,\qquad jq+|\beta|\leqslant mq-q.$$

We first consider the case $i \leqslant m-1$. Here, with the aid of integration by parts, we transfer the $(q+1)/2$ derivatives with respect to x to the second term of the form. As a result, we obtain a linear combination of forms of the type

$$[a\nu^k D_t^i D_x^\alpha u, b\nu^l D_t^i D^\beta u]_{(s|q),\lambda,[\gamma,\nu N]},$$

where $qi+k+|\alpha|$ and $jq+l+|\beta| \leqslant (m-1)q+(q-1)/2$. We thus arrive at (15). In the case $i=m$ and $|\alpha|=0$ one of the derivatives with respect to t should be transferred to the second term of the form, after which we arrive at the already considered case.

6.5.2. Using Proposition 6.5.1 and repeating literally the argument in 4.4.2, we will show that for $s=\lambda=0$ it suffices to prove inequality (11) for functions with support of a sufficiently small diameter.

6.5.3. When proving (11) in the case $s=\lambda=0$ for functions whose supports lie in a ball of a small diameter, we can assume, without loss of generality, that the operator P differs slightly from an operator with constant coefficients. In this case the proof is based on the following analog of Proposition 4.4.3.

PROPOSITION. *Let $q \geqslant 1$ be an odd integer, let $A(y,D)$ and $B(y,D)$ be differential operators with real coefficients in the q-homogeneous principal part, and let the q-orders of these operators not exceed mq and $(m-1)q$, respectively. Then for $\gamma < -R\nu^q - R_1$ the inequality*

$$\left|\operatorname{Im}[A(y,D)u, B(y,D)u]_{[\gamma,\nu N]}\right| \leqslant (\kappa[A_0][B_0]+c\nu^{-1})\{u\}^2_{[\gamma,\nu N]} \tag{16}$$

holds. Here A_0 and B_0 are the principal q-homogeneous parts of the symbols of A and B, respectively, κ does not depend on γ, ν, and the coefficients of the operators A and B, and c do not depend on ν.

The proof follows the scheme in Proposition 4.4.3. Setting

$$\begin{aligned} u &= \exp(\gamma t + \nu\langle N,x\rangle)v, \\ A_{\gamma,\nu N} &= A(y, D_x+i\gamma N, D_x+i\gamma), \\ B_{\gamma,\nu N} &= B(y, D_x+i\nu N, D_t+i\gamma) \end{aligned}$$

we rewrite the form in question as

$$\operatorname{Im}[Au,Bu]_{[\gamma,\nu N]} = (H(y,D_x,\nu,D_t,\gamma)v,\overline{v}),$$

where

$$H(y,D_x,\nu,D_t,\gamma) = A^*_{\gamma,\nu N}B_{\gamma,\nu N} - B^*_{\gamma,\nu N}A_{\gamma,\nu N},$$

and the symbol of this operator is written

$$\begin{aligned} &\frac{1}{2i}\Big[\overline{A_0(y,\xi+i\nu N,\sigma+i\gamma)}B_0(y,\xi+i\nu N,\sigma+i\gamma) \\ &\quad - \overline{B_0(y,\xi+i\nu N,\sigma+i\gamma)}A_0(y,\xi+i\nu N,\sigma+i\gamma)\Big] + H^{(1)}(y,\xi,\nu,\sigma,\gamma). \end{aligned} \tag{17}$$

The symbol $H^{(1)}$ is a polynomial in the variables ξ, ν, σ and γ of degree no higher than $(2m-2)q$ (here ξ and ν are taken with weight 1, and σ and γ are taken with weight q). With the aid of integration by parts the form

$$(H^{(1)}(y,D_x,\nu,D_t,\gamma)v,\overline{v})$$

is represented as a linear combination of the forms considered in Proposition 6.5.1, whence follows

$$|(H^{(1)}v, \overline{v})| \leqslant \text{const}\, \nu^{-1}\{u\}^2_{[\gamma,\nu N]}.$$

Since the coefficients of A_0 and B_0 are real, the first term in (17) is identically equal to zero for $\gamma = \nu = 0$, and therefore it can be represented in the form

$$\gamma L(y, \xi, \nu, \sigma, \gamma) + \nu M(y, \xi, \nu, \sigma, \gamma),$$

where L and M are q-homogeneous polynomials in ξ, ν, σ, and γ of degrees $2(m-1)q$ and $2(m-1)q + q - 1$, respectively, and their coefficients are bilinear forms in the coefficients of A and B. Integrating by parts we prove that

$$|(Lv, \overline{v})| \leqslant (\kappa_1[A_0][B_0] + c_1\nu^{-1})\|u\|^2_{m-1|q,0,[\gamma,\nu N]},$$
$$|(Mv, v)| \leqslant (\kappa_2[A_0][B_0] + c_2\nu^{-1})\|u\|^2_{m-1|q,(q-1)/2,[\gamma,\nu N]},$$

which implies the desired assertion.

6.5.4. In the case of arbitrary s and λ the inequality (11) is reduced to the case $s = \lambda = 0$ following the scheme of 4.4.4 and 5.5.4. Setting

$$\begin{aligned}\mu(\zeta, \tau) &= \delta_{s|q}(\zeta, \tau)\delta_\lambda(\zeta),\\ u &= {}_{[\gamma,\nu N]}\mu^{-1}(D)w\\ &= \exp(-\gamma t - \nu\langle N, x\rangle)\mu^{-1}(D_x + i\nu N, D_t + i\gamma)\exp(\gamma t + \nu\langle N, x\rangle)w\end{aligned}$$

we rewrite the middle term in (11) in the form

$$-\operatorname{Im}\left[\widetilde{P}w, \widetilde{P^{(1)}}w\right]_{[\gamma,\nu N]},$$

where

$$\widetilde{P} = \mu^{-1}(D)P\mu(D), \qquad \widetilde{P^{(1)}} = \mu^{-1}(D)P\mu(D).$$

The subsequent exact derivation of the inequality requires no new ideas as compared to 4.4.4 and 5.4.4, and we leave this to the reader as a simple exercise. It should only be taken into account that because we have

$$\mu(\xi, \nu, \sigma, \gamma)\mu^{-1}(\xi', \nu', \sigma', \gamma) \leqslant 2^{|s|+|\lambda|}(1 + |\xi - \xi'| + |\sigma - \sigma'|)^{|s|}(1 + |\xi - \xi'|)^{|\lambda|},$$
$$\left|\frac{\partial}{\partial \xi_k}\mu(\xi, \nu, \sigma, \nu)\right| \leqslant 2(|\lambda| + |s|)\mu(\xi, \nu, \sigma, \nu)(1 + |\xi + i\nu N|)^{-1},$$

the results in 4.5 imply estimates for the commutators of the operators $\mu(D)$ and $\mu^{-1}(D)$ with operators of multiplication by a function, and the constants in these estimates do not depend on ν and γ.

REMARK (cf. Remark 3 in 5.5). Inequality (9) can be refined by introducing the norms

$$[u]^2_{(s|q),\lambda,[\gamma,\nu N]} = \iint\limits_{\substack{\operatorname{Im}\tau=\gamma\\ \operatorname{Im}\zeta=\nu N}} (|\gamma| + \nu|\zeta|^{q-1})(1 + |\zeta|^2)^{\lambda/2}(|\tau|^2 + |\zeta|^{2q})^s|\widehat{u}(\tau, \zeta)|^2\, d\tau\, d\zeta$$

and

$$'[u]^2_{(s|q),\lambda,[\gamma,\nu N]} = \iint\limits_{\substack{\operatorname{Im}\tau=\gamma\\ \operatorname{Im}\zeta=\nu N}} (|\gamma| + \nu|\zeta|^{q-1})^{-1}(1 + |\zeta|^2)^{\lambda/2}(|\tau|^2 + |\zeta|^{2q})^s|\widehat{u}(\tau, \zeta)|^2\, d\tau\, d\zeta.$$

Repeating the argument presented for the case of q-parabolic equations in Remark 3 of 5.5, we derive the inequality

$$[u]_{(s+m-1|q),\lambda,[\gamma,\nu N]} \leqslant \text{const}'[Pu]_{(s|q),\lambda,[\gamma,\nu N]}.$$

For $\lambda = (q-1)/2$ this inequality is a refinement of (9).

6.6. Theorems 4.1, 5.1, and 6.2 lead to the following assertion, which can be regarded as the main result of the present chapter.

THEOREM. *For a given differential operator $P(y, D)$ of order m with respect to $\partial/\partial t$ with coefficients satisfying conditions* (3.2), *a vector $N \in \mathbb{R}^n$, and a number $q \geqslant 1$ the following conditions are equivalent.*

(I) *For any differential operator $Q(y, D)$ with coefficients of the type* (3.2), $\deg_q Q < \deg_q P$, *and any $M > 0$ there is R and a number $\nu_0 = \nu_0(M, Q)$ such that for $|s| \leqslant M$, $\nu \geqslant \nu_0$, and $\gamma \leqslant -R\nu^q$ the equation*

$$(P(y,D) + Q(y,D))u = f$$

has a unique solution $u \in H^{(s+m-1|q),q-1}_{[\gamma,\nu N]+}$ for any $f \in H^{(s|q)}_{[\gamma,\nu N]+}$, and this solution satisfies the inequality

$$\begin{aligned}|[u]|_{(s+m-1|q),[\gamma,\nu N]} &\overset{\text{def}}{=} |\gamma|\, \|u\|_{(s+m-1|q),0,[\gamma,\nu N]} \\ &\quad + \nu\|u\|_{(s+m-1|q),q-1,[\gamma,\nu N]} \leqslant C_Q\|f\|_{(s+m-1|q),[\gamma,\nu N]}.\end{aligned}$$

(II) *The inequality*

$$|[u]|_{(s+m-1|q),[\gamma,\nu N]} \leqslant \|P(y,D)u\|_{(s|q),[\gamma,\nu N]} \quad \forall u \in H^{(\infty)}_{[\gamma,\nu N]+} \tag{18}$$

holds.

(II_+) *Inequality* (18) *holds on the subspace $H^{(\infty)}_{[\gamma,\nu N]+}$.*

(III) *q is an integer and*

(a) *for $q = 1$ the operator P is uniformly strictly hyperbolic, and N is arbitrary;*

(b) *for even $q \geqslant 2$ the operator P is uniformly q-parabolic, and N is arbitrary;*

(c) *for odd $q > 1$ the operator P is q-hyperbolic, and N belongs to the cone of q-hyperbolicity of the symbol $P(y, \zeta, \tau)$.*

§7. The Cauchy problem for systems of differential equations

The Cauchy problem for hyperbolic systems was investigated by Petrovskii in his classical memoir **[P2]**. In the first part of the memoir Petrovskii considered first-order systems, and in the second part the investigation of a high-order system was reduced to the case of a first-order system by means of a transformation of the unknown functions. The conditions imposed by Petrovskii on the first-order systems guarantee the diagonalizability of their symbols. By constructing the calculus of pseudodifferential operators in a rather implicit form Petrovskii diagonalized in fact the system in question and derived an energy estimate, and then combined it with the Cauchy-Kovalevskaya theorem to prove the existence and uniqueness theorem.

In his Princeton lectures, Leray **[Le]** derived similar estimates for strictly hyperbolic operators. Furthermore, he noted that they can be used for establishing estimates (and, consequently, existence and uniqueness theorems) for a broad class of hyperbolic systems. Although in application to hyperbolic systems Leray's approach is somewhat less general than Petrovskii's method, it can readily be applied to q-hyperbolic systems as well **[V4]**.

The presentation of the material in this section is organized in the following way. In 7.1 we present a general method for defining the principal part of a polynomial matrix. In 7.2 hyperbolic systems are defined for which the Petrovskii-Leray solvability theorem for the Cauchy problem is proved. Finally, in 7.3 we define q-hyperbolic systems and prove an analog of the Petrovskii-Leray theorem for them.

As for the general q-parabolic systems, the Cauchy problem and some other initial value problems were investigated for them by Eidel'man **[E]**, Solonnikov **[So]**, and some other authors with the aid of different versions of the parametrix method. For completeness, we present the Cauchy problem for parabolic systems following the same approach as in the case of hyperbolic and q-hyperbolic systems.

In Appendix 1 we give a different definition of hyperbolic systems, which was also suggested by Petrovskii (see **[P1]**). In Appendix 2, proceeding from this definition, we define a class of q-hyperbolic systems distinct from those in §7; these results are due to the second author **[V4]**.

7.1. q-nondegenerate systems of partial differential operators. In the present section we will consider systems of the general type

$$\mathcal{P}(y, D)u = f. \tag{1}$$

Here $u = (u_1, \dots, u_M)$ and $f = (f_1, \dots, f_M)$ are vector-valued functions and $\mathcal{P}(y, D)$ is a square matrix of differential operators:

$$\mathcal{P}(y, D) = \|P_{ij}(y, D)\|_{i,j=1,\dots,M}, \tag{2}$$

where

$$P_{ij}(y, D) = \sum_{\alpha} a_{\alpha ij} D^{\alpha}. \tag{2'}$$

By $\mathcal{P}(y, \eta) = \|P_{ij}(y, \eta)\|$ we denote the symbol of operator (2).

We set a weight $q \geqslant 1$ that will be assigned to the variable $\sigma = \eta_{n+1}$. Since we are not interested in converse theorems, in what follows the number q is assumed to be an integer.

Denote by $m_{ij}(q)$, $i, j = 1, \dots, M$, the q-orders of the elements of the matrix operator (2). If $P_{ij} \equiv 0$, then we formally set $m_{ij}(q) = -\infty$.

With every permutation $\pi = \binom{1,\dots,M}{i_1,\dots,i_M}$ of the natural numbers $1, \dots, M$ we associate the natural number

$$R(\pi, q) = m_{1 i_1}(q) + \dots + m_{M i_M}(q). \tag{3}$$

The maximum of the numbers (3) over all permutations π of the numbers $1, \dots, M$ will be called the q-order of system (1) or, which is the same, of the matrix operator (2):

$$R(q) = \max_{\pi} R(\pi, q). \tag{4}$$

We now define the q-principal part of the symbol (1). The definition is based on the following

PROPOSITION ([**V2,3**]). *There are nonnegative numbers $t_j(q)$ and $s_j(q)$, $j = 1, \dots, M$, such that the following conditions hold*:

$$\text{(i)} \qquad m_{ij}(q) \leqslant t_j(q) - s_j(q),$$

$$\text{(ii)} \qquad \sum_{j=1}^{M} (t_j(q) - s_j(q)) = R(q).$$

This proposition makes it possible to define the principal part (in the Leray [**Le**] and Douglis-Nirenberg [**DN**] sense) of the system. We set

$$\mathcal{P}_q(y,\eta) = \|P_{qij}(y,\eta)\|, \quad i, j = 1, \dots, M,$$

$$\text{(5)} \qquad P_{qij}(y,\eta) = \begin{cases} \sum_{\alpha_1+\dots+\alpha_n+q\alpha_{n+1}=t_j(q)-m_j(q)} a_\alpha(y)\eta^\alpha, \\ 0 \qquad \text{for} \quad m_{ij}(q) < t_j(q) - s_j(q). \end{cases}$$

Note that the matrix (5) is q-homogeneous:

$$\text{(6)} \qquad \mathcal{P}_q(y, c\xi, c^q\sigma) = S_q^{-1}(c)\mathcal{P}_q(y,\eta)T_q(c),$$

where $S_q(c)$ and $T_q(\sigma)$ are diagonal matrices:

$$\text{(6')} \qquad S_q(c) = \|\delta_{ij}c^{s_i(q)}\|, \quad T_q(c) = \|\delta_{ij}c^{t_j(q)}\|.$$

DEFINITION ([**V2**]). System (1) is said to be q-nondegenerate if its q-order $R(q)$ coincides with the q-order of the determinant of its symbol (3):

$$\text{(7)} \qquad R(q) = \deg_q \det \mathcal{P}(y;\eta).$$

By this definition, the calculation of the determinant shows that in the case of a nondegenerate system (1) we have

$$\text{(8)} \qquad \det \mathcal{P}_q(y;\eta) = (\det \mathcal{P}(y;\eta))_q,$$

where the right-hand side is the principal q-homogeneous part of the symbol $\det \mathcal{P}(y;\eta)$. For $q = 1$ the number $R(q)$ will be called the order of the system, the principal q-homogeneous part will simply be called the principal part, and (7) will be called the nondegeneracy condition.

7.2. The Cauchy problem for hyperbolic systems.

7.2.1. Definition of hyperbolic systems. Thus, we consider system (1). We assume the following conditions to hold.

(a) System (1) is nondegenerate in the sense of 7.1, i. e., there are nonnegative integers t_j, s_j, $j = 1, \dots, M$, such that $\deg P_{ij} \leqslant t_j - s_j$ and the determinant of the principal part

$$\text{(9)} \qquad \mathcal{P}_0(y,\eta) = \|P_{0ij}(y,\eta)\|, \quad P_{0ij}(y,\eta) = \sum_{|\alpha|=t_j-s_i} a_\alpha(y)\eta^\alpha$$

defined using these numbers is not identically zero.

(b) The direction σ dual to time t is not characteristic for system (1), i. e.,

$$\det \mathcal{P}_0(y, 0, \dots, 1) = P_0(y, 0, \dots, 1) \neq 0,$$

where $P_0(y,\eta)$ is the principal homogeneous part of the polynomial $\det P(y,\eta)$.

Hence, the polynomial $P_0(y, \xi, \tau)$ is resolved with respect to the highest power τ^R, $R = \sum t_j - s_j$, and therefore we can separate R branches of the roots $\lambda_{0j}(y, \xi)$ that depend continuously on the parameters y and ξ and are homogeneous functions of degree 1 in ξ.

(c) The roots $\lambda_{0j}(y, \xi)$ are real.

Conditions (a), (b), and (c) are not sufficient for the correctness of the Cauchy problem even in the case of one scalar equation. Earlier we supplemented them with the condition that the roots $\lambda_{0j}(y, \xi)$ are simple for $\xi \neq 0$. In the case of a system of equations the condition that the roots of the determinant $P_0(y, \xi, \sigma)$ are pairwise distinct appears unnatural. Indeed, let system (1) possess this property. We can double the number of the unknown functions and consider the system

$$\begin{aligned} &\mathcal{P}_0(y, D)u + \cdots = f, \\ &\mathcal{P}_0(y, D)v + \cdots = g, \end{aligned} \tag{1'}$$

where the dots symbolize the lower terms. It is natural to require that, along with (1), system (1′) also belong to the class of hyperbolic systems. Proceeding from Petrovskii's definition **[P2]**, we impose the following condition on the system.

(d) $\mathcal{P}_0(y, D)$ is a block diagonal matrix (in particular, it may consist of a single block):

$$\mathcal{P}_0(y, \eta) = \begin{pmatrix} \mathcal{P}_0^{(1)} & \dots & 0 \\ \vdots & \ddots & \vdots \\ 0 & \dots & \mathcal{P}_0^{(\kappa)} \end{pmatrix}, \tag{10}$$

and the determinants

$$P_0^{(\lambda)}(y, \eta) = \det \mathcal{P}_0^{(\lambda)}(y, \eta) \tag{11}$$

are strictly hyperbolic symbols.

Note that condition (d) implies (b) and (c).

DEFINITION. System (1) satisfying conditions (a) and (d) is called a hyperbolic system.

REMARK. 1) In the special case $s_1 = \cdots = s_M$ this definition is an unimportant generalization of the original Petrovskii definition **[P1]**.

2) In the special case $t_j = s_j + 1$ this definition becomes the Leray definition **[Le]**.

3) Note that in the simplest case of a first-order system

$$\frac{\partial u}{\partial t} + A \frac{\partial u}{\partial x} + Bu = f$$

with one spatial variable, the above definition means that A is a block diagonal matrix:

$$A = \begin{pmatrix} A^{(1)} & \dots & 0 \\ \vdots & \ddots & \vdots \\ 0 & \dots & A^{(\lambda)} \end{pmatrix},$$

where each of the matrices $A^{(l)}$, $l = 1, \dots, \lambda$, has simple real roots.

The presented definitions imply that the matrix is representable as

$$A = C^{-1} \Lambda C,$$

where the determinant of C is nonzero and Λ is a real diagonal matrix. In this special case the assumption that A is a block matrix is an unnatural constraint because by means of the transformation $u = Cv$ of the unknown functions the system is reduced to a system with a diagonal principal part:

$$\frac{\partial v}{\partial t} + \Lambda v + C^{-1}Bv = C^{-1}f.$$

In the case of several spatial variables,

$$\frac{\partial u}{\partial t} + \sum A_j \frac{\partial u}{\partial x_j} + Bu = f,$$

such a simple reduction is no longer possible.

4) In Appendix 1 to this section we will discuss a different version of the definition of hyperbolic systems, which is suggested by Petrovskii's definition **[P1]**. It is based on the diagonalizability of the principal part.

7.2.2. Basic estimate for hyperbolic systems.

THEOREM. *Let system* (1) *satisfy conditions* (a) *and* (d) *in* 7.2.1. *Then for each L there is* $\gamma_0(L)$ *such that for* $|s| + \max\{s_1, \dots, t_M\} \leqslant L$ *and* $\gamma \leqslant \gamma_0(L)$ *the inequality*

$$C|\gamma| \sum_{j=1}^{M} \|u_j\|_{(s+t_j-1),[\gamma]} \leqslant \sum_{i=1}^{M} \|(\mathcal{P}(y,D)u)_i\|_{(s+s_i),[\gamma]}, \tag{12}$$

$$\forall u = (u_1, \dots, u_M), \quad u_j \in H^{(\infty)}_{[\gamma]}, \quad j = 1, \dots, M,$$

holds, where $(\mathcal{P}u)_i$ *denotes the* i*th component of the vector* $\mathcal{P}u$.

The proof will be carried out in three stages.

1) First we consider the case when all blocks in (9) are one-dimensional, i. e., the principal part $\mathcal{P}_0$ is a diagonal matrix. In other words, all $P_{ii}(y,\eta)$'s are strictly hyperbolic symbols of degrees $t_i - s_i$, and we have $t_i - s_i \geqslant 0$, $\deg P_{ij} < t_j - s_i$, for $i \neq j$.

Denote by $f = (f_1, \dots, f_M)$ the right-hand side of (1). Then the jth equation can be rewritten in the form

$$P_{jj0}(y,D)u_j = f_j + (P_{jj}(y,D) - P_{jj0}(y,D))u_j + \sum_{k \neq j} P_{jk}(y,D)u_k.$$

Because the left-hand side is a uniformly strictly hyperbolic operator of order $t_j - s_j$, the results presented in §4 imply[1]

$$c|\gamma|\,\|u_j\|_{(s+t_j-s_j-1),[\gamma]} \leqslant \|f_j\|_{(s),[\gamma]} + \text{const}\sum_{k=1}^{M} \|u_k\|_{(s+t_k-s_j-1),[\gamma]}.$$

Replacing s with $s + s_j$ in this inequality we obtain

$$c|\gamma|\,\|u_j\|_{(s+t_j-1),[\gamma]} \leqslant \|f_j\|_{(s+s_j),[\gamma]} + \text{const}\sum_{k=1}^{M} \|u_k\|_{(s+t_k-1),[\gamma]}.$$

[1] Note that we do not exclude the possibility that $t_j - s_j = 0$. In this case we have $P_{jj} \equiv$ const, and the inequality in question is trivial.

The summation with respect to j in the last inequality results in

$$c|\gamma|\sum\|u_j\|_{(s+t_j-1),[\gamma]}\leqslant\sum\|f_i\|_{(s+s_i),[\gamma]}+c_1\sum\|u_j\|_{(s+t_j-1),[\gamma]}.$$

Selecting γ_0 from the condition $c_1 < c|\gamma_0|/2$, we arrive at (12).

2) We now consider the case when $\det\mathcal{P}(y,\eta)$ has no multiple roots, i. e., it can be assumed that the principal part $\mathcal{P}_0$ consists of a single block. This case will be reduced to the above.

Denote by $B_{ij}(y,\eta)$ the complement to the element $P_{ji0}(y,\eta)$ in the matrix $\mathcal{P}_0$. By the conditions in Proposition 7.1, we have

$$\deg B_{ij}\leqslant\sum(t_k-s_k)-t_j+s_i=R-t_j+s_i. \tag{13}$$

The application of the matrix differential operator

$$\mathcal{B}(y,D)=\|B_{ij}(y,D)\|,\quad i,j=1,\dots,M,$$

on the left of the system results in a new system

$$\sum Q_{ij}(y,D)u_j=g_i, \tag{14}$$

where

$$g_i=\sum_{k=1}^{M}B_{ik}(y,D)f_k, \tag{15}$$

$$Q_{ij}(y,D)=\sum B_{ik}(y,D)\cdot P_{kj}(y,D). \tag{16}$$

LEMMA. *The symbol $Q_{ij}(y,\eta)$ of the differential operator* (16) *has the form*

$$Q_{ij}(y,\eta)=\delta_{ij}\det\mathcal{P}_0(y,\eta)+C_{ij}(y,\eta), \tag{17}$$

$$\deg C_{ij}(y,\eta)<R+t_j-t_i\quad(R=\deg\det\mathcal{P}_0). \tag{17'}$$

PROOF. By the Leibniz-Hörmander composition formula, the symbol of the operator (16) is equal to

$$\begin{aligned}&\sum_k\sum_{|\alpha|\geqslant0}B_{ik}^{(\alpha)}D_y^\alpha P_{kj}/\alpha!\\&\quad=\sum_k B_{ik}P_{0kj}+\sum_k B_{ik}(P_{kj}-P_{0kj})+\sum_k\sum_{|\alpha|\geqslant1}B_{jk}^{(\alpha)}D_y^\alpha P_{kj}\Big/\alpha!\\&\quad\overset{\text{def}}{=}\sum B_{ik}P_{0kj}+C_{ij}.\end{aligned}$$

By virtue of (13), inequalities (17′) hold. According to the well-known formula of the theory of determinants, we have

$$\sum B_{ik}P_{0kj}=\delta_{ij}\det\mathcal{P}_0,$$

and the lemma is proved.

The lemma shows that the principal part of system (14) is a diagonal matrix so that it satisfies an inequality of the type (12) if f_j in (12) is replaced by g_j and t_j and s_j are replaced by $R + t_j$ and t_j, respectively. This implies the inequality

$$C|\gamma| \sum_j \|u_j\|_{(s+R+t_j-1),[\gamma]} \leqslant \sum_i \|g_i\|_{(s+t_i),[\gamma]} \leqslant \sum_{i,k} \|B_{ik}(y,D) f_k\|_{(s+t_i),[\gamma]}$$
$$\leqslant \text{const} \sum \|f_k\|_{(s+R+s_i),[\gamma]}.$$

Replacing $s + R$ by s in these inequalities we arrive at (12).

3) We now consider the general case. Let

$$\mathcal{B}(y,\eta) = \begin{pmatrix} \mathcal{B}^{(1)}(y,\eta) & \dots & 0 \\ \vdots & \ddots & \vdots \\ 0 & \dots & \mathcal{B}^{(\kappa)}(y,\eta) \end{pmatrix}$$

be a block diagonal matrix, where $\mathcal{B}^{(l)}(y,\eta)$ is a matrix composed of the algebraic adjuncts of the elements of the matrix $\mathcal{P}^{(l)}(y,\eta)$. Applying the matrix operators $\mathcal{B}(y,D)$ to the left of system (1) we again arrive at the case considered above. We leave the remaining details of the proof to the reader as a simple and instructive example. The theorem is proved.

Let us discuss the necessity of the conditions in 7.2.1. Using a natural modification of the argument in Proposition 4.2 we derive from (12) an inequality for the matrix operator $\mathcal{P}_0(D) = \mathcal{P}_0(y^0, D)$, where y^0 is an arbitrary point belonging to $\mathbb{R}^{n+1}$:

$$\text{(18)}\quad \begin{aligned} &C^2|\gamma|^2 \sum_{j=1}^{M} \int_{\operatorname{Im}\tau=\gamma} (|\tau|^2 + |\xi|^2)^{s+t_j} |\widehat{u}_j(\xi,\tau)|^2 \, d\xi \, d\tau \\ &\leqslant \sum_{j=1}^{M} \int_{\operatorname{Im}\tau=\gamma} (|\tau|^2 + |\xi|^2)^{s+s_j} \left|(\widehat{\mathcal{P}_0(D)u})_j\right|^2 d\xi \, d\tau. \end{aligned}$$

Inequality (18) must hold for all $\gamma < 0$. By homogeneity, it readily follows that it holds for $\gamma = \operatorname{Im}\tau > 0$ as well. In turn, (18) is equivalent to the following condition on the polynomial matrix $\mathcal{P}_0(\xi,\tau)$:

$$\text{(18')}\quad \|\mathcal{P}_0^{-1}(\xi,\tau)\| \leqslant C|\operatorname{Im}\tau|^{-1}, \qquad \operatorname{Im}\tau \neq 0, \quad |\xi|^2 + |\tau|^2 = 1.$$

This condition implies that $\det \mathcal{P}_0 \not\equiv 0$, i. e., the nondegeneracy condition of 7.1 holds. Furthermore, by virtue of (18′), the polynomial $\det \mathcal{P}_0(\xi,\tau)$ can have only real roots. Hence, for systems of equations (in contrast to the scalar case) there is a notable "gap" between the necessary and sufficient conditions for the validity of (12).

REMARK. Ivrii and Petkov **[IP]** showed that in the case of first-order systems, i. e., for matrices of the type

$$\mathcal{P}_0(\xi,\tau) = \tau E + \sum A_j \xi_j,$$

condition (18′) implies a condition for diagonalizability of the matrix $\sum \xi_j A_j$, i. e., the existence of a matrix $C(\xi)$ such that $C^{-1}(\xi) \sum \xi_j A_j C(\xi)$ is a diagonal matrix. However, even in the case of first-order systems, for the derivation of (12) by Petrovskii's method **[P2]** it is required that the matrix $C(y,\xi)$ is a smooth function of the variables (y,ξ) (see **[VI]**).

7.2.3. Dual estimate. We define the natural duality for vector-valued functions $u = (u_1, \dots, u_M)$ and $v = (v_1, \dots, v_M)$:

$$((u, v)) = \sum_{j=1}^{M} (u_j, v_j).$$

It is also possible to consider the Hermitian form

$$[[u, v]] = \sum_{j=1}^{m} [u_j, v_j] \stackrel{\text{def}}{=} \sum_{j=1}^{m} (u_j, \overline{v}_j). \tag{19}$$

For an $M \times M$ matrix differential operator $\mathcal{P}(y, D)$ with elements $P_{ij}(y, D)$, $i, j = 1, \dots, M$, we denote by $\mathcal{P}^*(y, D)$ the dual matrix differential operator of $\mathcal{P}$ with respect to the form (19):

$$[[\mathcal{P}(y, D)u, v]] = [[u, \mathcal{P}^*(y, D)v]].$$

It is easy to show that the symbol of $\mathcal{P}^*$ has the form

$$\mathcal{P}^*(y, \eta) = {}^t\overline{\mathcal{P}}(y, \eta) + \sum \frac{1}{\alpha!} {}^t\mathcal{P}^{(\alpha)}_{(\alpha)}(y, \eta), \tag{20}$$

where ${}^t\mathcal{P}$ is the transpose of the matrix $\mathcal{P}$. We also consider the dual matrix operator ${}^t\mathcal{P}(y, D)$ of $\mathcal{P}(y, D)$ with respect to the form $((\,,\,))$:

$$((\mathcal{P}(y, D)u, v)) = ((u, {}^t\mathcal{P}(y, D)v)).$$

It is clear that $\mathcal{P}^*(y, D) = \overline{{}^t\mathcal{P}(y, D)}$. Formula (20) immediately implies the following

LEMMA. *Let $\mathcal{P}(y, \eta)$ be the symbol of a hyperbolic system (in the sense of* 7.2.1) *for which the principal part of $\mathcal{P}$ is defined with the aid of numbers $t_1, \dots, t_M$ and $s_1, \dots, s_M$. Then $\mathcal{P}^*(y, \eta)$ is also the symbol of a hyperbolic system in the same sense, and the principal part of $\mathcal{P}^*$ is defined via the numbers $-s_1, \dots, -s_M$ and $-t_1, \dots, -t_M$.*

This lemma together with the results of 7.2.2 implies that the operator $\mathcal{P}^*$ satisfies the inequality

$$C|\gamma| \sum_{j=1}^{M} \|v_j\|_{(-s-s_j-1),[\gamma]} \leqslant \sum_{j=1}^{M} \|(\mathcal{P}^*u)_j\|_{(-s-t_j),[\gamma]}, \qquad v_j \in H^{(\infty)}_{[\gamma]}, \quad j = 1, \dots, M \tag{21}$$

(for convenience, $s \in \mathbb{R}$ is replaced by $-s$).

We note that if a symbol $\mathcal{P}^*(t, x, \xi, \tau)$ satisfies the conditions of 7.2.1, then the symbol $\mathcal{P}^*(-t, x, \xi, -\tau)$ also possesses this property. Writing out inequality (21) for the matrix operator corresponding to this symbol and making change of variable $t \to -t$ we arrive at an estimate that is dual to (12):

$$C|\gamma| \sum_{j=1}^{M} \|v_j\|_{(-s-s_j-1),[-\gamma]} \leqslant \sum_{j=1}^{M} \|(\mathcal{P}^*u)_j\|_{(-s-t_j),[-\gamma]}, \qquad v_j \in H^{(\infty)}_{[-\gamma]}, \quad j = 1, \dots, M. \tag{22}$$

7.2.4. The Cauchy problem. Generally, the systems defined in 7.2.1 are not Kovalevskaya's systems for which the Cauchy problem can be stated (see **[P1,2]**). In **[Le]** the following generalization of the Cauchy problem is suggested.

Since the numbers $t_1, \dots, s_M$ in Proposition 7.1 are defined up to an arbitrary additive constant, we normalize them by means of the condition $\min s_j = 0$. Then we have $t_k \geqslant 0$, $k = 1, \dots, M$.

Let $f = (f_1, \dots, f_M)$ and $g = (g_1, \dots, g_M)$ be some vector-valued functions. A vector-valued function $u = (u_1, \dots, u_M)$ is said to be a solution to the generalized Cauchy problem (1) with initial data g if

$$\begin{aligned} &u_j - g_j \quad \text{has a zero of order } t_j \text{ at } t = 0, \text{ i. e.,} \\ &D^k u_j\big|_{t=0} = D^k g_j\big|_{t=0}, \quad k = 0, \dots, t_j - 1. \end{aligned} \tag{23}$$

In this case for a smooth solution to exist it is necessary that the following compatibility conditions hold:

$$f_i - \sum P_{ij}(y, D)v_j \quad \text{has a zero of order } s_i \text{ at } t = 0. \tag{24}$$

If all the numbers s_j are equal, i. e., $s_1 = \dots = s_M = 0$, then the compatibility condition disappears, and we arrive at the traditional Cauchy problem.

If the functions g_j are identically zero, then conditions (23) ((24)) say that the functions u_j have a zero of order t_j (accordingly, s_j) at $t = 0$. Under these conditions, u and f in (1) can be extended as zero to $t < 0$, and (1) can be regarded as an equation on the space $\mathbb{R}^{n+1}$.

Thus, by the (generalized) Cauchy problem for system (1) we mean the problem of determining the vector-valued functions $u = (u_1, \dots, u_M)$, $u_j \in H^{(s+t_j-1)}_{[\gamma]+}$, $j = 1, \dots, M$, from the given right-hand side $f = (f_1, \dots, f_M)$, $f_j \in H^{(s+t_j)}_{[\gamma]+}$, $j = 1, \dots, M$. We will prove the following

THEOREM. *Let* (1) *be a hyperbolic system (in the sense of* 7.2.1). *Then for an arbitrary* $S > 0$ *there is* $\gamma_0 = \gamma_0(S)$ *such that for* $|s| \leqslant S$, $\gamma \leqslant \gamma_0(S)$, *and any*

$$f = (f_1, \dots, f_M), \qquad f_j \in H^{(s+s_j)}_{[\gamma]+}, \quad j = 1, \dots, M,$$

there is a unique solution

$$u = (u_1, \dots, u_M), \qquad u_j \in H^{(s+t_j-1)}_{[\gamma]+}, \quad j = 1, \dots, M,$$

to system (1), *and it satisfies inequality* (12).

PROOF. 1) We do not dwell on the proof of the uniqueness which is derived from (12) in the same way as in Theorem 3.1.

2) As in Theorem 3.1, we first prove the solvability of the problem in question for functions whose support is the entire space. This proof is based on the dual estimate (22) and requires in fact only the reformulation of the arguments of Proposition 3.1 in terms of systems of equations.

Thus, we introduce the spaces of vector-valued functions

$$\mathcal{U}^{(s)}_{[\gamma]} = \prod_{j=1}^{M} H^{(s+t_j)}_{[\gamma]}, \quad \mathcal{F}^{(s)}_{[\gamma]} = \prod_{j=1}^{M} H^{(s+s_j)}_{[\gamma]}$$

and endow them with the natural norms

$$\left|u, \mathcal{U}_{[\gamma]}^{(s)}\right|^2 = \sum \|u_j\|^2_{(s+t_j),[\gamma]},$$
$$\left|u, \mathcal{F}_{[\gamma]}^{(s)}\right|^2 = \sum \|f_j\|^2_{(s+t_j),[\gamma]}.$$

The dual spaces with respect to the form $((\ ,\ ))$ are the spaces

$$\left(\mathcal{U}_{[\gamma]}^{(s)}\right)' = \prod_{j=1}^{M} H_{[-\gamma]}^{(-s-t_j)}, \qquad \left(\mathcal{F}_{[\gamma]}^{(s)}\right)' = \prod_{j=1}^{M} H_{[-\gamma]}^{(-s-s_j)}$$

endowed with the natural norms

$$\left|\psi, (\mathcal{U}_{[\gamma]}^{(s)})'\right|^2 = \sum \|\psi_j\|^2_{(-s-t_j),[-\gamma]}, \tag{25}$$

$$\left|\varphi, (\mathcal{F}_{[\gamma]}^{(s)})'\right|^2 = \sum \|\varphi_j\|^2_{(-s-s_j),[-\gamma]}. \tag{25$'$}$$

As in Proposition 3.2.1, we consider the lineal

$$\left\{\psi, \quad \psi = {}^t\mathcal{P}\varphi, \varphi \in \prod_{j=1}^{M} H_{[-\gamma]}^{(\infty)}\right\}$$

and define the linear functional $L(\psi) = ((f, \varphi))$ on it. Then

$$\begin{aligned} |L(\psi)| \leqslant \sum |(f_j, \varphi_j)| &\leqslant \sum \|f_j\|_{(s+t_j),[\gamma]} \|\varphi_j\|_{(-s-t_j),[-\gamma]} \\ &\leqslant \left|f, \mathcal{F}_{[\gamma]}^{(s)}\right| \left|\varphi, (\mathcal{F}_{[\gamma]}^{(s)})'\right|. \end{aligned} \tag{26}$$

In view of (25) and (25$'$), inequality (22) can be rewritten as

$$\left|\varphi, (\mathcal{F}_{[\gamma]}^{(s)})'\right| \leqslant C|\gamma|^{-1} \left|{}^t\mathcal{P}\varphi, (\mathcal{U}_{[\gamma]}^{(s-1)})'\right|.$$

Substituting this inequality into the right-hand side of (26) and recalling the definition of ψ we conclude that

$$|L(\psi)| \leqslant C|\gamma|^{-1} |f, \mathcal{F}_{[\gamma]}^{(s)}| \left|\psi, (\mathcal{U}_{[\gamma]}^{(s-1)})'\right|. \tag{27}$$

By the Hahn-Banach theorem, there is a vector-valued function $u \in \mathcal{U}_{[\gamma]}^{(s-1)}$ such that

$$L(\psi) = ((u, \psi)), \qquad \text{i. e.} \quad ((u, {}^t\mathcal{P}\varphi)) = (f, \varphi) \quad \forall \varphi \in \prod H_{[-\gamma]}^{(\infty)},$$

and in this case we have

$$\gamma |u, \mathcal{U}_{[\gamma]}^{(s-1)}| \leqslant c |f, \mathcal{F}_{[\gamma]}^{(s)}|,$$

i. e., inequality (12) holds.

3) We now show that if $\operatorname{supp} f \subset \{(x,t) \in \mathbb{R}^{n+1}, t \geqslant 0\}$, i. e., $f \in \mathcal{F}_{[\gamma]+}^{(s)}$, then the solution u also belongs to $\mathcal{U}_{[\gamma]+}^{(s-1)}$. In view of the arguments in 3.1.4, this fact can be proved if not only inequality (21) but also the more general inequality

$$\text{const} \sum \|v_j\|_{(-s-s_j-1),[-\gamma'-\gamma'']} \leqslant \sum \|\mathcal{P}^* v_j\|_{(-s-s_j-1),[-\gamma'-\gamma'']}$$

is applied, where $\gamma' \leqslant \gamma'' < \gamma_0$. In turn, the latter inequality is derived from the estimate

$$(28) \qquad C|\gamma|\,\|u\|_{(s+m-1),[\gamma',\gamma'']} \leqslant \|P(y,D)u\|_{(s),[\gamma',\gamma'']}, \quad \gamma'' < \gamma' \leqslant \gamma \leqslant \gamma_0,$$

for strictly hyperbolic operators by means of literal repetition of the argument in 7.2.2 and 7.2.3.

7.2.5. Proof of inequality (28). This inequality is derived using a modification of the argument in 3.2. According to the definition of the norm in $H^{(s)}_{[\gamma',\gamma'']}$ and the basic estimate for strictly hyperbolic operators $P(y,D)$, we have

$$\begin{aligned}\|u\|_{(s+m-1),[\gamma',\gamma'']} &= \|\chi u\|_{(s+m-1),[\gamma']} + \|(1-\chi)u\|_{(s+m-1),[\gamma'']}\\ &\leqslant C\left(|\gamma'|^{-1}\|P(\chi u)\|_{(s),[\gamma']} + |\gamma''|^{-1}\|P((1-\chi)u)\|_{(s),[\gamma'']}\right)\\ &\leqslant C|\gamma|^{-1}\left(\|\chi Pu\|_{(s),[\gamma']} + \|(1-\chi)Pu\|_{(s),[\gamma'']}\right)\\ &\quad + C|\gamma|^{-1}\sum_{k=1}^{m}\left(\|\chi_k P^{(k)}u\|_{(s),[\gamma']} + \|\chi_k P^{(k)}u\|_{(s),[\gamma'']}\right).\end{aligned}$$

The first term on the right-hand side is nothing other than

$$C|\gamma|^{-1}\|Pu\|_{(s),[\gamma',\gamma'']}, \quad \gamma'' \leqslant \gamma' < \gamma < 0.$$

Repeating the argument in 3.2 we prove that the second term on the right-hand side has an upper bound

$$C_1|\gamma|^{-1}\|u\|_{(s+m-1),[\gamma',\gamma'']}.$$

If $\gamma < \gamma_0$ and $-\gamma_0$ is a sufficiently large number, then we arrive at (28). The proof of the theorem is completed.

7.3. The Cauchy problem for q-hyperbolic systems. In this subsection q is an odd positive integer. We will say that (1) is a q-hyperbolic system if its symbol $\mathcal{P}(y,\eta)$ satisfies the following conditions.

(i) System (1) in nondegenerate, i. e., there exist integers $t_1,\dots,t_M$ and $s_1,\dots,s_M$ related to the q-orders $m_{ij}(q)$ of the operators P_{ij} and the q-order $R(q)$ of the system by inequalities (i) and (ii) in Proposition 7.1. We denote by $\mathcal{P}_q(y,\eta) = \|P_{qij}(y,\eta)\|$ the q-principal part of the matrix $\mathcal{P}(y,\eta)$.

(ii) $\mathcal{P}_q(y,\eta)$ is a block diagonal matrix of the type (9):

$$\mathcal{P}_q = \begin{pmatrix} \mathcal{P}^{(1)} & & \\ & \ddots & \\ & & \mathcal{P}^{(\kappa)} \end{pmatrix},$$

where

$$(29) \qquad P^{(\lambda)}(y,\eta) = \det \mathcal{P}_q(y,\eta)$$

are uniformly q-hyperbolic symbols.

(iii) The (open) cones of q-hyperbolicity $V(P^{(\lambda)})$ of symbols (29) have a nonempty intersection, i. e., there is a nonzero ray $\{vN\}$ such that

$$\{vN\} \subset \bigcap_{\lambda=1}^{\kappa} V(P^{(\lambda)}).$$

To illustrate the above definition we consider a q-hyperbolic system with one spatial variable:

$$\frac{\partial u}{\partial t} + A\frac{\partial^q u}{\partial x^q} + B\left(\frac{\partial}{\partial x}\right)u = f,$$

where B is a matrix of differential operators whose orders do not exceed $q-1$. In this case condition (i) means that the matrix A has real characteristic numbers and splits into blocks each of which has pairwise distinct characteristic numbers.

Condition (ii) means that all characteristic numbers have the same sign.

As can easily be seen (cf. Remark 3 in 7.2.1), the assumption that A is a block matrix can be replaced by the condition of diagonalizability. However, in the multi-dimensional case the diagonalizability condition for A should be supplemented with a different type of condition. These questions will be discussed in Appendix 2 to the present section. The solvability theorem for the generalized homogeneous Cauchy problem for system (1) is proved following the plan of the previous subsection.

THEOREM. *Let system* (1) *be q-hyperbolic* (*i. e., let the above conditions* (i), (ii), *and* (iii) *hold*). *Then for every $S>0$ there are $\nu_0>0$, $R>0$, and R_1 such that for $|s|\leqslant M$, $\nu\geqslant\nu_0$, and $\gamma<-R\nu^q-R_1$ equation* (1) *has a unique solution $u=(u_1,\dots,u_M)$, $u_j\in H^{(s+t_j-1|q),q-1}_{[\gamma,\nu N]+}$, for any right-hand side $f=(f_1,\dots,f_M)$, $f_j\in H^{(s+s_j|q)}_{[\gamma,\nu N]+}$, and this solution satisfies the inequality*

$$\sum_{j=1}^{M}|[u_j]|_{(s+t_j-1|q),[\gamma,\nu N]}\leqslant C\sum\|(\mathcal{P}u)_j\|_{(s+s_j|q),[\gamma,\nu N]}, \tag{30}$$

where on the left-hand side we have the norm (6.4).

PROOF. 1) First of all, following the plan of 7.2.2, we derive inequality (30) from estimate (6.9) for scalar operators. As in the case of hyperbolic equations, we first consider an operator whose q-principal part $\mathcal{P}_q$ has diagonal form, i. e.,

$$\deg_q P_{ij}<t_j-s_i,\quad i\neq j.$$

Because the left- and right-hand sides of this inequality involve integers, we have

$$\deg_q P_{ij}\leqslant t_j-s_i-1.$$

Hence, by assumption,

$$\begin{aligned}|[u]|_{(s+t_j-s_j-1|q),[\gamma,\nu N]}&\leqslant C\|P_{jj}(y,D)u_j\|_{(s|q),[\gamma,\nu N]}\\&=C\Big\|f_j-\sum_{k\neq j}P_{jk}u_k\Big\|_{(s|q),[\gamma,\nu N]}\\&\leqslant C\|f_j\|_{(s|q),[\gamma,\nu N]}+C_1\sum_{k=1}^{M}\|u\|_{(s+t_k-s_j-1|q),[\gamma,\nu N]}.\end{aligned}$$

Replacing s by $s+s_j$ we obtain

$$|[u]|_{(s+t_j-1|q),[\gamma,\nu N]}\leqslant C\|f\|_{(s+s_j|q),[\gamma,\nu N]}+C_1\sum\|u_k\|_{(s+t_k-1|q),[\gamma,\nu N]}.$$

Adding together these estimates and noting that

$$\|u_k\|_{(s+t_k-1|q),[\gamma,\nu N]}\leqslant\nu^{-1}|[u]|_{(s+t_k-1|q),[\gamma,\nu N]}$$

we derive (30). The further derivation of this inequality is carried out with the aid of literal repetition of the arguments in 7.2.2. Furthermore, as in the scalar case, we note that if $\mathcal{P}(x,t,\xi,\tau)$ is a q-hyperbolic symbol, then the symbol $\mathcal{P}^*(x,-t,\xi,-\tau)$ also possesses the same property, and in this case the intersection of the cones of q-hyperbolicity of the symbols $\det \mathcal{P}^{\lambda *}(x,-t,\xi,-\tau)$ contains the ray $\{-\nu N, \nu \geqslant 0\}$. Hence, this operator satisfies inequality (30) with $t_1,\dots,t_M$ and $s_1,\dots,s_M$ replaced by $-s_1,\dots,-s_M$ and $-t_1,\dots,-t_M$, respectively. Performing change of variable $t \to -t$ we arrive at an estimate dual to (30):

$$(30') \qquad \sum |[v_j]|_{(-s-s_j-1|q),[-\gamma,-\nu N]} \leqslant C \sum \|(\mathcal{P}^* v)_j\|_{(-s-t_j|q),[-\gamma,-\nu N]}.$$

The solvability of system (1) in the space

$$\prod_{j=1}^{M} H_{[\gamma,\nu N]}^{(s+t_j-1|q),q-1}$$

for any right-hand side belonging to

$$\prod_{j=1}^{M} H_{[\gamma,\nu N]}^{(s+s_j|q)}$$

is proved by using inequalities (30) and (30′) and repeating the arguments in 7.2.4. To prove the solvability in spaces of functions that vanish for $t<0$ it is necessary to replace the weight $\exp(-\gamma t)$ in (30′) by $(\exp(\gamma' t)+\exp(\gamma'' t))^{-1}$, where $\gamma' \leqslant \gamma'' \leqslant \gamma$, following the scheme of §3 or, equivalently, to replace the norm of v with weight $\exp(-\gamma t)$ by the norms of χv and $(1-\chi)v$ with weights $\exp(-\gamma' t)$ and $\exp(-\gamma'' t)$, respectively. To this end the corresponding estimate must be proved for a scalar q-hyperbolic operator. In this case the derivation of the estimate for the weight $(\exp(-\gamma' t)+\exp(-\gamma'' t))^{-1}$ from the corresponding estimate for $\exp(\gamma t)$ is carried out by repeating literally the similar derivation in the case of hyperbolic operators, and here we will not dwell on this question.

7.4. q-parabolic systems. In this subsection q is an even positive integer. System (1) is said to be q-parabolic if its symbol $\mathcal{P}(y,\xi,\tau)$ satisfies the following conditions:
(i) the matrix $\mathcal{P}$ is q-nondegenerate in the sense of 7.1;
(ii) $\det \mathcal{P}(y,\xi,\tau)$ is a uniformly q-parabolic symbol.

THEOREM. *Let system* (1) *be q-parabolic. Then for each $S>0$ there is $\gamma_0=\gamma_0(S)$ such that for $\gamma \leqslant \gamma_0$ and any right-hand side $f=(f_1,\dots,f_M)$, $f_j \in H_{[\gamma]+}^{(s+s_j|q)}$, $j=1,\dots,M$, there exists a unique solution $u=(u_1,\dots,u_M)$, $u_j \in H_{[\gamma]+}^{(s+t_j-1|q),q}$, $j=1,\dots,M$, to system* (1), *and it satisfies the inequality*

$$(31) \qquad \begin{aligned} &|\gamma| \sum_{j=1}^{M} \|u_j\|_{(s+t_j-1|q),[\gamma]} + \sum_{j=1}^{M} \|u_j\|_{(s+t_j-1|q),q,[\gamma]} \\ &\qquad \leqslant \operatorname{const} \sum_{j=1}^{M} \|(\mathcal{P}u)_j\|_{(s+s_j|q),[\gamma]}, \quad u \in \prod H_{[\gamma]}^{(\infty)}. \end{aligned}$$

The proof of this theorem follows the plan of Theorem 7.3 with the only difference that here there is no need in considering the partition of the q-principal part into blocks. Inequality (31) can be somewhat refined in view of Remark 3 in 5.5.

Appendix 1. Hyperbolic systems with diagonalizable principal part

As we already said (see the remark in 7.2.1), it is natural to abandon the assumption of block structure of the principal part in the definition of hyperbolic systems in 7.2.1 and replace it by the condition of diagonalizability. In the case of first-order systems this generalization was suggested by Petrovskii **[P1]**. The present appendix is devoted to Petrovskii's definition in the context of somewhat more general systems considered in **[Le]**).

1. In his original work **[P2]** Petrovskii studied nondegenerate systems (7.1) (in the terminology of 7.1) under the following two additional assumptions.

1) If $t_1, \dots, t_M$ and $s_1, \dots, s_M$ are the numbers involved in the definition of the principal part, then the numbers $s_1, \dots, s_M$ coincide, and therefore it can be assumed (without loss of generality) that $s_1 = \dots = s_M = 0$, $t_j \geqslant 0$, $j = 1, \dots, M$.

2) (7.1) is a Kovalevskaya's system, i. e., the diagonal operators P_{jj} are resolved with respect to the highest derivative $D_t^{t_j}$, and the off-diagonal operators P_{ij}, $i \neq j$, do not involve derivatives of order t with respect to t_j.

Under these conditions the transformation

$$u_{kj} = D_t^k u_j, \qquad k = 0, \dots, t_j - 1, \quad j = 1, \dots, M,$$

of the unknown functions with their subsequent renumbering results in a system of the type

$$\mathcal{P}(y, D)u = f, \quad \mathcal{P}(y, D) = D_t E - Q(y, D_x), \tag{1}$$

where E is a unit $M \times M$ matrix and the matrix

$$Q(y, \xi) = \|Q_{ij}(y, \xi)\|, \qquad i, j = 1, \dots, M, \tag{2}$$

has the following structure (cf. **[Le]**):

$$\deg Q_{ij} \leqslant s_j - s_i + 1, \quad i, j = 1, \dots, M. \tag{2$'$}$$

The principal part of the matrix (2) is defined according to the scheme of §7, i. e.,

$$Q(y, \xi) = Q_0(y, \xi) + Q'(y, \xi), \tag{3}$$

where $Q_0 = \|Q_{0ij}\|$ and $\deg Q_{0ij} = s_j - s_i + 1$. Hence, $Q_0(y, \xi)$ is a homogeneous matrix with respect to ξ in the following sense:

$$Q_0(y, \xi) = |\xi| S^{-1}(|\xi|) Q_0(y, \xi/|\xi|) S(|\xi|), \tag{4}$$

where

$$S(c) = \|\delta_{ij} c^{s_i}\|, \quad i, j = 1, \dots, M.$$

The matrix Q' in (3) consists of lower terms, i. e.

$$Q'(y, \xi) = \|Q'_{ij}(y, \xi)\|, \quad \deg Q'_{ij} \leqslant s_j - s_i. \tag{5}$$

2. Following **[P1]**, we refer to system (7.1) with a symbol satisfying conditions (2)–(5) as a generalized hyperbolic system if the conditions below hold.

(a) For $|\xi| = 1$ and $y \in \mathbb{R}^{n+1}$ the matrix $Q_0(y, \xi)$ has elementary divisors of degree no higher than 1, and the number of coinciding elementary divisors does not depend on the choice of the point $(y, \xi) \in \mathbb{R}^{n+1} \times S^{n-1}$, where S^{n-1} is the unit sphere in $\mathbb{R}^n_\xi$.

(b) The characteristic numbers $\lambda_{0j}(y, \xi)$, $j = 1, \dots, M$, of the matrix $Q_0(y, \xi)$ are real.

As was shown by Petrovskii (see **[P1,** Chapter 3, §1]), in this case the following proposition holds.

PROPOSITION. *Let the matrix $Q_0(y, \xi)$ satisfy conditions* (a) *and* (b). *Then it possesses the following properties.*

(A) *For each point $(y^0, \xi^0) \in \mathbb{R}^{n+1} \times S^{n-1}$ there is a neighborhood $U(y^0, \xi^0) \in \mathbb{R}^{n+1} \times S^{n-1}$ where a matrix $C_U(y; \xi)$ is defined that possesses the following properties:*

1) *$C_U(y, \xi)$ is a smooth real function of the variables $(y, \xi) \in U$;*
2) *there is $\delta > 0$ such that*

$$|\det C_U(y, \xi)| > \delta, \quad (y, \xi) \in U; \tag{6}$$

3) *for $(y, \xi) \in U$ the matrix $Q_0(y, \xi)$ is represented in the form*

$$Q_0(y, \xi) = C_U^{-1}(y, \xi)\Lambda_0(y, \xi)C_U(y, \xi), \tag{7}$$

where $\Lambda_0(y, \xi)$ is a diagonal matrix.

(B) *The characteristic numbers $\lambda_0 j(y, \xi)$ of the matrix $Q_0(y, \xi)$, $|\xi| = 1$, are smooth real functions, and their multiplicities do not depend on y and ξ.*

The matrix $C_U(y, \xi)$ can be extended as a homogeneous function of degree 0 to a conic neighborhood of the set $U(y^0, \xi^0)$ by setting

$$C_U(y, \xi) = C_U(y, \xi|\xi|^{-1}), \qquad (y, \xi|\xi|^{-1}) \in U(y^0, \xi^0).$$

Then (7) turns into

$$Q_0(y, \xi) = \left(C(y, \xi)S(|\xi|)\right)^{-1} \Lambda_0(y, \xi) \left(C(y, \xi)S(|\xi|)\right). \tag{8}$$

3. Thus, we consider the system

$$\begin{gathered} D_t u - Q(y, D)u = f, \\ u = (u_1, \dots, u_M), \quad f = (f_1, \dots, f_M). \end{gathered} \tag{9}$$

THEOREM. *Let system* (9) *have the form described in subsection* 1 *of this appendix, and let it be a generalized hyperbolic system in the sense of subsection* 2. *Then for all $L > 0$ there is $\gamma_0 = \gamma_0(L)$ such that for $|s| \leqslant L$ and $\gamma \leqslant \gamma_0$ the inequality*

$$|\gamma| \sum_{j=1}^{M} \|u_j\|_{s+s_j,[\gamma]} \leqslant c \sum_{j=1}^{M} \|f_j\|_{s+s_j,[\gamma]} \tag{10}$$

holds.

Recall that, according to the accepted notation, for $s = 0$ we write $H^\lambda_{[\gamma]}$ and $\|\,.\,\|_{\lambda,[\gamma]}$ instead of $H^{(0|q),\lambda}_{[\gamma]}$ and $\|\,.\,\|_{(0|q),\lambda,[\gamma]}$, respectively.

REMARK. Using the expressions for the derivatives $D_t u_j$, $D_t^2 u_j$, etc. obtained from equations (9) we can replace (10) by a more general inequality

$$(10') \qquad |\gamma| \sum \|u_j\|_{(k),s+s_j,[\gamma]} \leqslant c \sum \|f_j\|_{(k),s+s_j,[\gamma]},$$

$k = 0, 1, 2, \ldots$. The extension of this inequality to arbitrary real values of k can be carried out using pseudodifferential operators, and we do not dwell on this purely technical question.

Based on inequality (10) (more precisely, on its generalization (10′) for each $k \in \mathbb{R}$) we can apply the scheme of §7 to derive the existence and uniqueness theorem for system (9). We do not dwell on this question either.

4. To emphasize the main idea of the derivation of (10), we start with a system having constant coefficients:

$$(11) \qquad P(D)u = D_t u - Q(D_x)u = f.$$

To simplify the notation, we first assume that the diagonalizing matrix $C(\xi)$ is defined on the unit sphere S^{n-1}, i. e.,

$$(12) \qquad Q_0(\xi) = S^{-1}(|\xi|)C^{-1}(\xi)\Lambda_0(\xi)C(\xi)S(|\xi|).$$

Let us denote

$$[[u, v]]_{s,[\gamma]} = \left(\sum_{j=1}^{M} [u_j, v_j]^2_{s,[\gamma]}\right)^{1/2},$$

$$\widetilde{S}(D_x) = S\left(\sqrt{1 + |D_x|^2}\right) = \left\|\delta_{ij}(1 + |D_x|^2)^{s_j/2}\right\|,$$

and consider the form

$$\begin{aligned} J &= -\operatorname{Im}\left[\left[C(D_x)\widetilde{S}(D_x)f, C(D_x)\widetilde{S}(D_x)u\right]\right]_{s,[\gamma]} \\ (13) \qquad &= -\operatorname{Im}\left[\left[C(D_x)\widetilde{S}(D_x)D_t u, C(D_x)\widetilde{S}(D_x)u\right]\right]_{s,[\gamma]} \\ &\quad + \operatorname{Im}\left[\left[C(D_x)\widetilde{S}(D_x)Q(D_x)u, C(D_x)\widetilde{S}(D_x)u\right]\right]_{s,[\gamma]} = J_1 + J_2. \end{aligned}$$

Applying the identity

$$-\operatorname{Im}[D_t z, z]_{[\gamma]} = \frac{1}{2}\int \exp(2\gamma t)\frac{\partial}{\partial t}|z|^2 = -\gamma\|z\|^2_{[\gamma]}$$

for $\gamma < 0$ we derive

$$\begin{aligned} J_1 &= |\gamma|\left[\left[C(D_x)\widetilde{S}(D_x)u, C(D_x)\widetilde{S}(D_x)u\right]\right]_{s^2,[\gamma]} \\ &= |\gamma|\int \exp(2\gamma t)(1 + |\xi|^2)^s \left\langle C(\xi)\widetilde{S}(\xi)\widehat{u}(\xi, t), C(\xi)\widetilde{S}(\xi)\widehat{u}(\xi, t)\right\rangle \, d\xi \, dt. \end{aligned}$$

Since the elements of the matrix $C(\xi)$ are homogeneous functions of degree 0 that are continuous at $\xi \neq 0$, there is δ_1 such that the absolute values of the elements of this matrix do not exceed δ_1. By virtue of (6), the norm of $C^{-1}(\xi)$ does not exceed $\kappa\delta_1^{M-1}\delta^{-1} = \delta_2$, where κ depends on M only. Using the inequality

$$\langle z, z\rangle = \langle C^{-1}Cz, C^{-1}Cz\rangle \leqslant \delta_2^2\langle C_z, C_z\rangle$$

we find a lower bound for J_1:

$$(14)\qquad \begin{aligned} J_1 &\geqslant \delta_2^{-2}|\gamma| \int \exp(2\gamma t)(1+|\xi|^2)^s \left\langle \widetilde{S}(\xi)\widehat{u}(\xi,t), \widetilde{S}(\xi)\widehat{u}(\xi,t)\right\rangle \, d\xi\, dt \\ &= \delta_2^{-2}|\gamma| \sum_{j=1}^{M} \|u_j\|^2_{s+s_j,[\gamma]}. \end{aligned}$$

We now estimate the second term on the right-hand side of (13). Setting

$$C(D_x)\widetilde{S}(D_x)u = v$$

we rewrite the form in question as

$$J_2 = \operatorname{Im}[[R(D_x)v, v]]_{s,[\gamma]},$$

where

$$(15)\qquad R(\xi) = C(\xi)\widetilde{S}(\xi)Q(\xi)\widetilde{S}^{-1}(\xi)C^{-1}(\xi).$$

We will show that there is $C > 0$ such that

$$(16)\qquad J_2 \geqslant -C \sum_{j=1}^{M} \|u_j\|^2_{s+s_j,[\gamma]}.$$

Adding together (15) and (16) and selecting γ from the condition $C|\gamma|^{-1} < \delta_2^{-2}/2$ we prove that

$$(17)\qquad \delta_3|\gamma| \sum_{j=1}^{M} \|u_j\|^2_{s+s_j,[\gamma]} \leqslant J.$$

Applying the Schwarz inequality to the form (13) we find an upper bound

$$\text{const}\left(\sum \|u_j\|^2_{s+s_j,[\gamma]}\right)^{1/2} \left(\sum \|f_j\|_{s+s_j,[\gamma]}\right)^{1/2}$$

for it. Comparing this with (17) we arrive at (10). To prove (16) we rewrite the matrix (15) in the form

$$(18)\qquad \begin{aligned} R(\xi) &= C(\xi)S(|\xi|)Q_0(\xi)S^{-1}(|\xi|)C^{-1}(\xi) \\ &\quad + C(\xi)\Big[\big(S(\sqrt{1+|\xi|^2}) - S(|\xi|)\big)Q_0(\xi)S^{-1}(|\xi|) \\ &\qquad + S(|\xi|)Q_0(\xi)\big(S^{-1}(\sqrt{1+|\xi|^2}) - S^{-1}(|\xi|)\big) \\ &\qquad + \widetilde{S}(\xi)Q'(\xi)\widetilde{S}^{-1}(\xi)\Big]C^{-1}(\xi) \\ &= R_0(\xi) + R'(\xi). \end{aligned}$$

By the definition of generalized hyperbolicity or, more precisely, according to inequality (8), we have $R_0(\xi) = \Lambda_0(\xi)$, whence, upon passing to the Fourier transform, we obtain

$$(19)\qquad \begin{aligned} \operatorname{Im}[[R_0(D)v, v]]_{s,[\gamma]} &= \operatorname{Im}[[\Lambda_0(D)v, v]]_{s,[\gamma]} \\ &= \int \exp(2\gamma t) \int (1+|\xi|^2)^s \sum_{j=1}^{M} \operatorname{Im}[\lambda_{0j}(\xi)|\widehat{u}_j(\xi,t)|^2]\, d\xi\, dt = 0 \end{aligned}$$

(here we have used the fact that the roots $\lambda_{0j}(\xi)$ are real).

Now, to prove (16) it suffices to check the uniform boundedness of the elements of the matrix $R'(\xi)$ in (18) for $\xi \in \mathbb{R}^n$. Since the elements of the matrices $C(\xi)$ and $C^{-1}(\xi)$ are uniformly bounded, is suffices to establish the uniform boundedness of the elements of the matrices in square brackets in (18). The boundedness of the elements of the third matrix follows directly from the expression for $Q'(\xi)$ (see (5)).

Consider an element with indices (i, j) in the first of the matrices. It is expressed as

$$\left((1+|\xi|^2)^{s_i/2} - |\xi|^{s_i}\right) Q_{0ij}(\xi)(1+|\xi|^2)^{-s_j/2} \overset{\text{def}}{=} G_{ij}(\xi). \tag{20}$$

The inequality

$$\sqrt{1+|\xi|^2} - |\xi| = \left(|\xi| + \sqrt{1+|\xi|^2}\right)^{-1} \leqslant (1+|\xi|^2)^{-1/2}$$

implies an estimate for (20):

$$|G_{ij}(\xi)| \leqslant \text{const}\, |\xi|^{s_j - s_i + 1}(1+|\xi|^2)^{(s_i - s_j - 1)/2}.$$

If $s_j - s_i + 1 \geqslant 0$, then the expression on the right-hand side does not exceed a constant, and in the case $s_j - s_i + 1 < 0$ the function (20) vanishes identically.

An element with indices (i, j) in the third matrix does not exceed

$$\begin{aligned} &\text{const}\, |\xi|^{s_j+1} \left|(1+|\xi|^2)^{-s_j/2} - |\xi|^{-s_j}\right| \\ &\qquad \leqslant \text{const}\, |\xi|(1+|\xi|^2)^{-s_j/2} \left|(1+|\xi|^2)^{s_j/2} - |\xi|^{s_j}\right| \leqslant \text{const}. \end{aligned}$$

In conclusion we consider the case when it is impossible to choose a unified diagonalizing matrix on the sphere $|\xi| = 1$. In this situation we cover the sphere by a system of neighborhoods U^j in each of which a diagonalizing matrix C^j is defined, take a partition of unity $\{\psi^j(D)\}$ subordinate to the covering $\bigcup U^j$, and consider the form

$$-\operatorname{Im} \sum_j [[C^j \widetilde{S} \psi^j f, C^j \widetilde{S} \psi^j u]]_{s,[\gamma]}.$$

For each term in this form we can repeat the above arguments and thus derive inequality (10) in which u and f should be replaced by $\psi^j(D)u$ and $\psi^j(D)f$, respectively. Adding together the resulting inequalities and recalling that $\{\psi^j(\xi)\}$ is a partition of unity we arrive at (10).

5. Now we give a sketch of the proof of inequality (10) in the case of variable coefficients assuming that the reader is familiar with the traditional technique of pseudodifferential operators, the sharp form of Gårding's inequality, and the theory of elliptic pseudodifferential operators on a manifold without boundary.

Since inequality (10) involves the large parameter $|\gamma|$, we can localize it (as was already done many times). In other words, when proving (10) we can assume, without loss of generality, that the support of the function $u(y)$ lies in a region of small diameter and that the matrix operator (3) coincides with its principal part.

To simplify the notation, we assume that the symbol of the diagonalizing matrix $C(y, \xi)$ is defined on the entire sphere $|\xi| = 1$. To this matrix symbol, there corresponds the pseudodifferential matrix operator $C(y, D_x)$, which acts on the variables x and depends on the parameter t.

As in the case of constant coefficients, consider the form

$$J = -\operatorname{Im}\big[\big[C(y, D_x)\widetilde{S}(D_x)f, C(y, D_x)\widetilde{S}(D_x)u\big]\big]_{s,[\gamma]}.$$

Replacing f by expression (1) and integrating by parts we represent the form in question as a sum of three forms:

$$\begin{aligned} J &= |\gamma|\big[\big[C(y, D_x)\widetilde{S}(D_x)u, C(y, D_x)\widetilde{S}(D_x)u\big]\big]_{s,[\gamma]} \\ &\quad + \operatorname{Im}\big[\big[(D_t C)(y, D_x)\widetilde{S}(D_x)u, C(y, D_x)\widetilde{S}(D_x)u\big]\big]_{s,[\gamma]} \\ &\quad + \big[\big[C(y, D_x)\widetilde{S}(D_x)Q_0(y, D_x)u, C(y, D_x)\widetilde{S}(D_x)u\big]\big]_{s,[\gamma]} = J_1 + J_1' + J_2. \end{aligned} \tag{21}$$

We will show that for every $L > 0$ there are positive constants K_1, K_2, and K_3 such that for $|s| \leqslant L$ we have

$$J_1 \geqslant K_1|\gamma| \sum \|u_j\|^2_{s+s_j,[\gamma]}, \tag{22}$$

$$J_1' \geqslant -K_2 \sum \|u_j\|^2_{s+s_j,[\gamma]}, \tag{23}$$

$$J_2 \geqslant -K_3 \sum \|u_j\|^2_{s+s_j,[\gamma]}. \tag{24}$$

Since $C(y, D_x)$ is a matrix of pseudodifferential operators that are bounded in any space $H^{\lambda}_{[\gamma]}$, the form J has an upper bound

$$\text{const}\Big(\sum \|u_j\|^2_{s+s_j,[\gamma]}\Big)^{1/2} \Big(\sum \|f_j\|^2_{s+s_j,[\gamma]}\Big)^{1/2}.$$

Taking a sufficiently large $|\gamma|$, we derive (10). Because $(D_t C)(y, D)$ is a matrix of pseudodifferential operators that are continuous in any space $H^{\lambda}_{[\gamma]}$, inequality (23) is obvious.

We begin with the derivation of (22). First of all, note that $C(y, D_x)\widetilde{S}(D_x)$ is an elliptic (in Petrovskii's sense) matrix pseudodifferential operator. The well-known results of the theory of elliptic systems imply the a priori estimate

$$\sum \|u_j\|^2_{s+s_j,[\gamma]} \leqslant C \sum \|h_j\|^2_{s,[\gamma]} + c_N \sum \|u_j\|^2_{s+s_j-N,[\gamma]}, \tag{25}$$

where $h = (h_1, \dots, h_M)$, $h = C(y, D)\widetilde{S}(D_x)u$, and N is an arbitrary positive number. If inequality (25) is regarded in the space of functions $u = (u_1, \dots, u_M)$, $u_j \in H^{s+s_j}_{[\gamma]}$, whose supports (with respect to x) lie in a region $\Omega \subset \mathbb{R}^n$, then the constant C_N depends on the diameter of Ω and tends to zero as $\operatorname{diam}\Omega \to 0$. More precisely, for each $L > 0$ there is $\kappa_0 = \kappa_0(L)$ such that under the condition $\operatorname{diam}\Omega \leqslant \kappa_0$ the constant C_N is less than $1/2$, i. e., (22) holds.

The above assertion is readily proved by contradiction. Indeed, if $C_N(\Omega)$ does not tend to zero as $\operatorname{diam}\Omega \to 0$, then there is a sequence of regions Ω^θ, $\operatorname{diam}\Omega^\theta \to 0$, and a sequence of vector-valued functions u^θ, $\operatorname{supp} u^\theta \subset \Omega^\theta$, such that

$$\sum \|u_j^\theta\|^2_{s+s_j-N,[\gamma]} = 1, \qquad \sum \|h_j^\theta\|_{s,[\gamma]} \to 0, \quad \theta \to \infty,$$

where $h^\theta = C\widetilde{S}u^\theta$. By (25), the sequence u^θ is uniformly bounded in $\prod H^{s+s_j}_{[\gamma]}$ and, therefore, weakly compact in this space and compact in $\prod H^{s+s_j-N}_{[\gamma]}$, $N > 0$. Therefore there is a subsequence (we retain the notation u^θ for it) such that

$$u^\theta \to u \quad \text{in} \quad \prod H^{s+s_j-N}_{[\gamma]}, \qquad C(y, D_x)\widetilde{S}(D_x)u = 0, \quad \operatorname{supp} u = \{0\} \times \mathbb{R}.$$

If $s + s_j - N > -n/2$, $j = 1\dots, M$, then there cannot exist a vector-valued function u that has a support at a point (with respect to x) and is not identically zero. If $s + s_j - N < -n/2$, then, generally, such a function exists and is a linear combination of the derivatives of $\delta(x)$. However, by virtue of the property of hypoellipticity of elliptic operators, such a function cannot be a solution to the homogeneous elliptic system

$$C(x, t, D_x)\widetilde{S}(D_x)u = 0.$$

We now proceed to the proof of (24). Set

$$(1 + |D|^2)^{s/2} C(y, D_x)\widetilde{S}(D_x)u = v. \tag{26}$$

The ellipticity of $C(y, D_x)$ immediately implies the existence of a parametrix, i. e., a pseudodifferential operator $R(y, D)$ such that

$$R(y, D)C(y, D) = E + T, \tag{27}$$

where T is an infinitely smoothing operator.

The method of constructing the parametrix implies that the operator R can be represented in the form

$$R(y, D) = C^{-1}(y, D) + R_{-1}(y, D), \tag{28}$$

where the first term on the right-hand side is a pseudodifferential operator with symbol $C^{-1}(y, \xi)$ and R_{-1} is an operator of order -1.

Applying the operator $R(y, D)(1 + |D|^2)^{-s/2}$ to both sides of (26) and using (27) we get

$$u = \widetilde{S}^{-1}(D_x)R(y, D_x)(1 + |D_x|^2)^{-s/2}v + \widetilde{S}^{-1}T\widetilde{S}u. \tag{29}$$

Substituting (29) into the form J_2 we rewrite the latter as follows:

$$J_2 = \operatorname{Im}[[Hv, v]]_{[\gamma]} + \operatorname{Im}[[Gu, C\widetilde{S}u]]_{s,[\gamma]}, \tag{30}$$

where

(31)
$$\begin{aligned} H &= (1 + |D|^2)^{s/2} C(y, D_x)\widetilde{S}(D_x)Q_0(y, D_x)\widetilde{S}^{-1}(D_x)R(y, D_x)(1 + |D_x|^2)^{-s/2}, \\ G &= C(y, D_x)\widetilde{S}(D_x)Q_0(y, D_x)\widetilde{S}^{-1}(D_x)T\widetilde{S}(D_x). \end{aligned}$$

Since T is an infinitely smoothing operator, the operator G also possesses this same property, and therefore the second form in (30) has a lower bound of the type of the right-hand side of (24).

When the symbol of the operator Q_0 and, consequently, the symbol of the operator R, do not depend on y_0, then, as was already shown in §4, the symbol of the operator H is a sum of a diagonal matrix $\Lambda_0(\xi)$ and a bounded matrix. Applying the standard technique of pseudodifferential operators we represent (31) in the form

$$H = \Lambda_0(y, D_x) + H_0,$$

where H_0 is a zero-order operator. Thus, the first form on the right-hand side of (30) is equal to

$$\operatorname{Im}\sum_{j=1}^{M}[\lambda_{0j}(y, D_x)v_j, v_j]_{[\gamma]} + \operatorname{Im}[[H_0 v, v]]_{[\gamma]}. \tag{32}$$

Since H_0 is a bounded operator, the second form on the right-hand side has a lower bound

$$-\operatorname{const}[[v,v]]_{[\gamma]} = -\operatorname{const}[[C\widetilde{S}u, C\widetilde{S}u]]_{s,[\gamma]} \geqslant -\operatorname{const}\sum \|u_j\|^2_{s+s,[\gamma]}.$$

As to the first form in (32), by the sharp form of Gårding's inequality, we have

$$\operatorname{Im}[\lambda_{0j}(y,D_x)v_j, v_j]_{[\gamma]} \geqslant -\operatorname{const}\|v_j\|_{[\gamma]},$$

and therefore the first form also has a lower bound of the type of the right-hand side of (24).

Appendix 2. q-hyperbolic systems with diagonalizable q-principal part

In Appendix 1 we extended the class of hyperbolic systems of §7 by replacing the assumption of block structure of the principal symbol in the definition of these systems by a weaker and more natural assumption of diagonalizability. In the present appendix we make an attempt to perform such a replacement for q-hyperbolic systems, $q > 1$. However, in this case we obtain a different (but not broader) class of systems. While weakening the conditions on the behavior of the principal symbol in the real domain we substantially strengthen the requirements for its behavior in the complex domain. The results of Appendix 2 were announced in **[V4]**.

1. Hence, as in Appendix 1, we will consider an $M \times M$ system

$$\mathcal{P}(y,D)u = f, \quad \mathcal{P}(y,D) = D_t E - \mathcal{Q}(y,D_x), \tag{1}$$

$$\mathcal{Q}(y,\xi) = \mathcal{Q}_0(y,\xi) + \mathcal{Q}'(y,\xi), \tag{2}$$

where $\mathcal{Q}_0 = \|\mathcal{Q}_{0ij}\|$, $\mathcal{Q}' = \|\mathcal{Q}'_{ij}\|$, and it is assumed that there are nonnegative integers $s_1, \dots, s_M$ and an odd number $q > 1$ such that

$$\deg \mathcal{Q}_{0ij} = s_j - s_i + q, \quad \deg \mathcal{Q}'_{ij} \leqslant s_j - s_i + q - 1. \tag{3}$$

Let the matrix $\mathcal{Q}_0(y,\xi)$ satisfy condition (a) in §1 of Appendix 1. Then, according to the assumption in this section, we can cover $\mathbb{R}^{n+1} \times S^{n-1}$ by a finite number of neighborhoods U in each of which there exists a matrix $C_U(y,\xi)$ that diagonalizes the principal symbol $\mathcal{Q}_0(y,\xi)$ for $(y,\xi) \in U$. We introduce the following notation:

$$\mathcal{Q}_0(y,\xi + ivN) = \mathcal{Q}_0(y,\xi) + iv\mathcal{Q}_{1N}(y,\xi) + \sum_{j\geqslant 2} v^j \mathcal{Q}_{jN}(y,\xi), \tag{4}$$

$$H_U(y,\xi) = C_U(y,\xi)\mathcal{Q}_{1N}(y,\xi)C_U^{-1}(y,\xi), \quad (y,\xi) \in U, \quad |\xi| = 1. \tag{5}$$

DEFINITION. System (1) will be called a generalized q-hyperbolic system if

(A) the principal symbol $\mathcal{Q}_0(y,\xi)$ satisfies conditions (a) and (c) in §2 of Appendix 1;

(B) there is a direction $N \in \mathbb{R}^n$ such that the symmetric matrix

$$H_U(y,\xi) + {}^t H_U(y,\xi) \tag{6}$$

is positive definite, i. e., for every $z \in \mathbb{R}^M$

$$\operatorname{Re}\langle H_U(y,\xi)z, z\rangle > \kappa |z|^2, \qquad (y,\xi) \in U, \quad |\xi| = 1. \tag{7}$$

REMARK. 1) If the matrix $Q_0(y,\xi)$ is symmetric, then $Q_{1N}(y,\xi)$ possesses the same property, and the matrix C_U is orthogonal. In this case condition (B) passes into

$$C_U(y,\xi)Q_{1N}(y,\xi)C_U^{-1}(y,\xi) > 0, \quad (y,\xi) \in U.$$

2) In case $\xi \in \mathbb{R}^1$ the matrix Q_0 has the form $Q_0(y,\xi) = A(y)\xi^q$. Therefore condition (A) means that the matrix $A(y)$ is similar to a diagonal matrix and possesses real characteristic numbers. Condition (B) means that all these numbers have the same sign. Hence, in this special case the above class of q-hyperbolic systems is an extension of the class in 7.3. We failed to investigate completely the relationship between these classes for $\xi \in \mathbb{R}^n$, $n > 1$. A partial answer to this question is given by the following result.

PROPOSITION. *Let the matrix $Q_0(\xi)$ in* (3) *satisfy conditions* (A) *and* (B) *of the definition; as a consequence, in this case its characteristic numbers $\lambda_{0j}(\xi)$, $j = 1,\dots,M$, are smooth functions of ξ, $|\xi| = 1$ (see the proposition in* §2 *of Appendix* 1). *Let $N \in \mathbb{R}^n$ be the vector entering condition* (B). *Then*

$$\langle \operatorname{grad} \lambda_{0j}(\xi), N\rangle > \kappa|\xi|^{q-1}, \quad j = 1,\dots,M.$$

This proposition implies that if all characteristic numbers $\lambda_{0j}(\xi)$ are pairwise distinct for $\xi \neq 0$, then q-hyperbolic systems in the sense of the above definition are q-hyperbolic in the sense of 7.3.

2. Proof of the proposition. To simplify the notation, assume that the diagonalizing matrix $C(\xi)$ is defined throughout the unit sphere. Consider the auxiliary matrix

$$A(\xi) = Q_0(\xi + ivN) - ikvE + iRv^qE, \qquad k > 0, \quad |\xi| = 1, \tag{8}$$

and the corresponding Hermitian form

$$J = \operatorname{Im}\langle C(\xi)A(\xi)z, C(\xi)z\rangle, \quad |\xi| = 1. \tag{9}$$

LEMMA. *Let the conditions of the proposition hold. Then there is a small number $k > 0$ and a sufficiently large $R > 0$ such that the form* (9) *is positive definite.*

Assuming that the lemma is true we complete the proof of the proposition. Consider the matrix iA instead of A. The positive definiteness of (9) implies that $\operatorname{Re}\langle CiAz, Cz\rangle = -\operatorname{Im}\langle CAz, Cz\rangle < 0$. By the well-known Lyapunov theorem (see, e. g., **[DK]**), the spectrum of the matrix iA lies in the left half-plane, i. e.,

$$\operatorname{Re}(i\lambda_{0j}(\xi + ivN) + kv - Rv^q) < 0, \qquad |\xi| = 1, \quad j = 1,\dots,M,$$

so that

$$\operatorname{Im}\lambda_{0j}(\xi + ivN) \geqslant kv - Rv^q, \qquad |\xi| = 1, \quad j = 1,\dots,M. \tag{10}$$

If ξ in (10) is replaced by $\xi/|\xi|$, where $\xi \in \mathbb{R}^n$, and v is replaced by $v/|\xi|$, then, since λ_{0j} is a homogeneous function of degree q, we arrive at the inequality

$$\operatorname{Im}\lambda_{0j}(\xi + ivN) \geqslant kv|\xi|^{q-1} - Rv^q.$$

Since $\lambda_{0j}(\xi + ivN)$ is a homogeneous function of ξ and v of degree q, and we have $\operatorname{Im}\lambda_{0j}(\xi) = 0$, the function $\operatorname{Im}\lambda_j(\xi + ivN)$ is divisible by v, i. e.,

$$\operatorname{Im}\lambda_j(\xi + ivN) = vP_j(\xi, v).$$

The substitution of this expression into (10) results in

$$P_j(\xi,\nu) \geqslant k|\xi|^{q-1} - R\nu^{q-1}.$$

Setting $\nu = 0$ in this inequality and noting that $P_j(\xi,0) = \langle N, \operatorname{grad} \lambda_{0j}(\xi)\rangle$ we complete the proof of the proposition.

PROOF OF THE LEMMA. We replace $Q_0(\xi + i\nu N)$ by expansion (4) and rewrite (9) in the form

$$\begin{aligned} J = {} & \operatorname{Im}\langle CQ_0 z, Cz\rangle + \nu \operatorname{Re}\langle CQ_{1N} z, Cz\rangle \\ & + \sum_{j\geqslant 2} \nu^j \langle CQ_j z, Cz\rangle + (-k\nu + R\nu^q)\langle Cz, Cz\rangle. \end{aligned} \tag{11}$$

Since $C(\xi)Q_0(\xi) = \Lambda_0(\xi)C(\xi)$, and the matrix $\Lambda_0(\xi)$ is real and diagonal, the first term on the right-hand side of (11) is equal to zero. Furthermore, according to (5) and (7),

$$\nu \operatorname{Re}\langle CQ_{1N} z, Cz\rangle = \nu \operatorname{Re}\langle HCz, Cz\rangle \geqslant \kappa |Cz|^2. \tag{12}$$

Taking $k < \kappa$ we obtain the inequality

$$J \geqslant (\kappa_1 \nu + R\nu^q)\langle Cz, Cz\rangle - \sum_{j\geqslant 2} \nu^j |\langle CQ_j z, Cz\rangle|. \tag{13}$$

Now it remains to note that for $j \geqslant 2$ we have

$$\begin{aligned} \nu^j |\langle CQ_j z, Cz\rangle| &= \nu^j \left|\langle CQ_j C^{-1} Cz, Cz\rangle\right| \leqslant \nu^j K_j |Cz|^2 \\ &\leqslant \left(K_j \varepsilon \nu + K_j \varepsilon^{-(q-1)/(q-j)} \nu^q\right). \end{aligned}$$

By choosing ε from the condition

$$\varepsilon \sum K_j \leqslant \kappa_1/2$$

and then taking R satisfying the condition

$$R \geqslant 2 \sum K_j \varepsilon^{-(q-1)/(q-j)}$$

we prove that

$$J \geqslant (\kappa_2 \nu + R\nu^q)\langle Cz, Cz\rangle > 0$$

because the matrix $C(\xi)$ is nonsingular. The lemma is proved.

3. We now state the main result.

THEOREM. *Let system (1) satisfy all conditions of the definition in §1. Then for each $L > 0$ there is $\nu_0 = \nu_0(L)$ such that under the conditions $|s| \leqslant L$, $\nu \geqslant \nu_0(s)$, and $\gamma < -R\nu^q$ (where R is sufficiently large) the inequality*

$$|\gamma| \sum \|u_j\|_{s+s_j,[\gamma,\nu N]} + \nu \sum \|u_j\|_{s+s_j+q-1,[\gamma,\nu N]} \leqslant C \sum \|f_j\|_{s+s_j,[\gamma,\nu N]} \tag{14}$$

holds.

PROOF. Inequality (14) is proved following the scheme in Appendix 1. However, the presence of the additional parameter ν and the related new terms makes the proof, which contains many technical details, much more cumbersome. The proof below is rather sketchy because we omit some technical details that are not very important.

3.1. The general plan of the derivation of (14) is the following. First of all, we apply the transformation

$$u = \exp(-\nu\langle N, x\rangle)v, \quad f = \exp(-\nu\langle N, x\rangle)g,$$

to pass to the system

$$D_t v - Q(y, D_x + i\nu N)v = g \tag{15}$$

involving the large parameter ν. We will prove the inequality

$$|\gamma| \sum \|v_j\|_{s+s_j,[\gamma]} + \nu \sum \|v_j\|_{s+s_j+q-1,[\gamma]} \leqslant C \sum \|g_j\|_{s+s_j,[\gamma]} \tag{16}$$

for (15), where (in contrast with Appendix 1) the norms are defined with the aid of the pseudodifferential operators

$$(1 + |D_x + i\nu N|^2)^{\lambda/2} \tag{17}$$

depending on the parameter ν. To simplify the presentation, we do not indicate this dependence in the notation of the norms. As in Appendix 1, we assume, for simplicity, that the symbol $C(y, \xi)$ is defined globally. Consider the quadratic form

$$J = -\operatorname{Im}\Big[\Big[C(y, D_x)\widetilde{S}(D_x)g, C(y, D_x)\widetilde{S}(D_x)v\Big]\Big]_{s,[\gamma]}, \tag{18}$$

where

$$\widetilde{S}(D_x) = \Big\|(1 + |D_x + i\nu N|^2)^{s_j/2}\delta_{ij}\Big\|.$$

Our aim is to prove the inequality

$$C_1|\gamma| \sum \|v_j\|^2_{s+s_j,[\gamma]} + \nu \sum \|v_j\|^2_{s+s_j+(q-1)/2,[\gamma]} \leqslant J \tag{19}$$

for $\nu \geqslant \nu_0$ and $\gamma < -R\nu^q$, where the numbers ν_0 and R are constructed in a special manner.

3.2. First of all we show that inequality (16) and, consequently, (14) follow from (19). Let $C(y, D_x) = \|C_{ij}(y, D_x)\|$. The operators $C_{ij}(y, D_x)$ are $H^\lambda_{[\gamma]} \to H^\lambda_{[\gamma]}$ bounded, where the norm in $H^\lambda_{[\gamma]}$ is defined via the pseudodifferential operator (17). The standard estimation shows that the norms of C_{ij} in these spaces have upper bounds not depending on ν. By this property, applying the Schwarz inequality we derive

$$\begin{aligned} J &\leqslant [[C\widetilde{S}g, C\widetilde{S}g]]^{1/2}_{s,[\gamma]}\,[[C\widetilde{S}v, C\widetilde{S}v]]^{1/2}_{s,[\gamma]} \\ &\leqslant C_3\left(\sum \|g_j\|^2_{s+s_j,[\gamma]}\right)\left(\sum \|v_j\|^2_{s+s_j,[\gamma]}\right)^{1/2}. \end{aligned} \tag{20}$$

Thus, the first term on the right-hand side of (19) is estimated via the right-hand side of (20), which implies the inequality

$$|\gamma| \sum \|v_j\|_{s+s_j,[\gamma]} \leqslant C_4 \sum \|g_j\|_{s+s_j,[\gamma]}. \tag{16'}$$

Similarly, applying the Schwarz inequality we find

$$(20')\qquad \begin{aligned} J &\leqslant [[C\widetilde{S}g, C\widetilde{S}g]]^{1/2}_{s-(q-1)/2,[\gamma]}\,[[C\widetilde{S}g, C\widetilde{S}g]]^{1/2}_{s+(q-1)/2,[\gamma]} \\ &\leqslant C_5 \Big(\sum \|g_j\|^2_{s+s_j-(q-1)/2,[\gamma]}\Big)\Big(\sum \|v_j\|^2_{s+s_j+(q-1)/2,[\gamma]}\Big)^{1/2}. \end{aligned}$$

Estimating the second term on the left-hand side of (19) by means of the right-hand side of (20′) we obtain

$$\nu \sum \|v_j\|_{s+s_j+(q-1)/2,[\gamma]} \leqslant C_6 \sum \|g_j\|_{s+s_j-(q-1)/2,[\gamma]}.$$

The replacement of s by $s + (q-1)/2$ in this inequality results in

$$(16'')\qquad \nu \sum \|v_j\|_{s+s_j+q-1,[\gamma]} \leqslant C_6 \sum \|g_j\|_{s+s_j,[\gamma]}.$$

Adding together (16′) and (16″) we arrive at (16).

3.3. To derive (19) we replace g by the left-hand side of (15) and integrate by parts the expression containing $D_t v$. Then the form (18) is written as

$$(21)\qquad \begin{aligned} J &= |\gamma|[[C\widetilde{S}v, C\widetilde{S}v]]_{s,[\gamma]} + \operatorname{Im}[[(D_tC)\widetilde{S}v, C\widetilde{S}v]]_{s,[\gamma]} \\ &\quad + \operatorname{Im}[[C\widetilde{S}Q(y, D_x + i\nu N)v, C\widetilde{S}v]]_{s,[\gamma]} = J' + J'' + I. \end{aligned}$$

Since $C(y, D_x)$ and $(D_tC)(y, D_x)$ are zero-order pseudodifferential operators and their $H^\lambda_{[\gamma]} \to H^\lambda_{[\gamma]}$ norms are uniformly bounded for $\nu > 0$, the form J'' admits a lower bound:

$$(22)\qquad J'' \geqslant -K_1 \sum \|v_j\|^2_{s+s_j,[\gamma]}.$$

The remaining part of Appendix 2 is devoted to finding a lower bound for the form I. We will show that there is $\delta > 0$ such that for each $\varepsilon > 0$ the inequality

$$(23)\qquad \begin{aligned} I &\geqslant \delta\nu[[C\widetilde{S}v, C\widetilde{S}v]]_{s+(q-1)/2,[\gamma]} \\ &\quad - \varepsilon\nu \sum \|v_j\|^2_{s+s_j+(q-1)/2,[\gamma]} - K(\varepsilon)\nu^q \sum \|v_j\|^2_{s+s_j,[\gamma]} \end{aligned}$$

holds.

The comparison of (21), (22), and (23) shows that

$$(24)\qquad \begin{aligned} J &\geqslant |\gamma|[[C\widetilde{S}v, C\widetilde{S}v]]_{s,[\gamma]} + \delta\nu[[C\widetilde{S}v, C\widetilde{S}v]]_{s+(q-1)/2,[\gamma]} \\ &\quad - \varepsilon\nu \sum \|v_j\|^2_{s+s_j+(q-1)/2,[\gamma]} - (K_1 + K(\varepsilon)\nu^q) \sum \|v_j\|^2_{s+s_j,[\gamma]}. \end{aligned}$$

Since the original inequality (14) involves the large parameters ν and γ, we can localize it. Thus, it can be assumed, without loss of generality, that Q does not contain lower terms, i. e., $Q \equiv Q_0$, and the vector-valued function v has a sufficiently small support. In this case the inequality

$$\sum \|v_j\|^2_{s+s_j+\lambda,[\gamma]} \leqslant K_3[[C\widetilde{S}v, C\widetilde{S}v]]_{s+\lambda,[\gamma]}$$

holds (cf. the derivation of (22) and (23) in §5 of Appendix 1). Setting $\lambda = 0$ and $(q-1)/2$ in this inequality, substituting it into (24), and choosing some appropriate ε, R, and $\gamma < -R\nu^q$ we arrive at (19).

3.4. Thus, the proof of the theorem has been reduced to the proof of (23). Since the further estimation involves operators acting on the variables x, and the variable t

is an unessential parameter with respect to which the integration with weight $\exp(\gamma t)$ is performed, below we omit the index $[\gamma]$. We also set $b = (q-1)/2$.

We expand the operator $Q_0(y, D_x + i\nu N)$ into a series in powers of ν and rewrite the form I in (21) as

$$\begin{aligned} I = & \operatorname{Im}[[C\widetilde{S}Q_0(y, D_x)v, C\widetilde{S}v]]_s + \nu \operatorname{Re}[[C\widetilde{S}Q_{1N}(y, D_x)v, C\widetilde{S}v]]_s \\ & + \sum_{k \geqslant 2} \nu^k \operatorname{Im}[[C\widetilde{S}Q_{kN}(y, D_x)v, C\widetilde{S}v]]_s = I_0 + I_1 + \sum_{k\geqslant 2} I_k \end{aligned} \tag{25}$$

(recall that $Q \equiv Q_0$). We will establish the following lower bounds for the forms in (25):

$$I_1 \geqslant \delta\nu[[C\widetilde{S}v, C\widetilde{S}v]]_{s+b} - \varepsilon \sum \|v_j\|^2_{s+s_j+b} - K(\varepsilon)\nu^q \sum \|v_j\|^2_{s+s_j}, \tag{26}$$

$$I_k \geqslant -\varepsilon\nu \sum \|v_j\|^2_{s+s_j+b} - K_1(\varepsilon)\nu^q \sum \|v_j\|^2_{s+s_j}, \tag{27_k}$$

where $k = 0, 2, 3, \dots$ and ε is an arbitrarily small number. By adding together (26) and (27_k) and choosing a sufficiently small ε we derive (23).

The estimation of the forms I_k, $k \geqslant 2$, is simple. Indeed, an element Q_{kNij} of the matrix Q_{kN} is a differential operator of order $s_j - s_i + q - k$. Because $l(k) = (q+1)/2 - k < (q-1)/2$, by means of integration by parts with respect to x we transfer $l(k)$ derivatives to the term $C\widetilde{S}v$. The application of the Schwarz inequality for $2 \leqslant k \leqslant b$ results in

$$\begin{aligned} I_k \geqslant & -K_6\nu^k \Big(\sum \|v_j\|^2_{s+s_j+b}\Big)^{1/2} \Big(\sum \|v_j\|^2_{s+s_j+(q+1)/2-k}\Big)^{1/2} \\ \geqslant & -\varepsilon_1\nu \sum \|v_j\|^2_{s+s_j+b} - K_6^2(4\varepsilon_1)^{-1}\nu \sum \nu^{2k-2}\|v_j\|^2_{s+s_j+(q+1)/2-k}. \end{aligned}$$

It now remains to note that, according to the interpolating inequalities, we have

$$\begin{aligned} \nu^{2k-2}\|v_j\|^2_{s+s_j-k+(q+1)/2} \leqslant & \, \varepsilon_2 \sum \|v_j\|^2_{s+s_j+(q-1)/2} \\ & + \varepsilon_2^{-(q-1)/(2k-2)}\nu^{q-1} \sum \|v_j\|^2_{s+s_j}. \end{aligned}$$

In the case $b+1 \leqslant k \leqslant q-1$ we have

$$\begin{aligned} I_k \geqslant & -K_7\nu^k \Big(\sum \|v_j\|^2_{s+s_j+q-k}\Big)^{1/2} \Big(\sum \|v_j\|^2_{s+s_j}\Big)^{1/2} \\ \geqslant & -\varepsilon\nu \sum \|v_j\|^2_{s+s_j+b} - K_7^2(4\varepsilon)^{-1}\nu^q \sum \|v_j\|^2_{s+s_j}. \end{aligned}$$

For $k \geqslant q$ the inequality (27_k) is obvious.

3.5. To derive (26) we use the transformation

$$w = \widetilde{\Lambda}^s(D_x)C(y, D_x)\widetilde{S}(D_x)v, \quad \widetilde{\Lambda}(\xi) = (1 + |\xi + i\nu N|^2)^{1/2}. \tag{28}$$

The elliptic operator $C(y, D_x)$ has a parametrix, i. e., a pseudodifferential operator $R(y, D_x)$ possessing the following properties:

$$R(y, D_x) = C^{-1}(y, D_x) + R_{-1}(y, D_x), \quad R(y, D_x)C(y, D_x) = E + T, \tag{29}$$

where R_{-1} is an operator of order -1 and T is an operator of order $-\infty$. Applying the operator $R\widetilde{\Lambda}^{-s}$ to both sides of (28) we obtain

$$v = \widetilde{S}^{-1}R\widetilde{\Lambda}^{-s}w - \widetilde{S}T\widetilde{S}v. \tag{30}$$

Therefore the form we are interested in can be written as

$$\begin{aligned} I_1 &= \nu \operatorname{Re}\big[\big[\widetilde{\Lambda}^s C\widetilde{S}Q_{1N}\widetilde{S}^{-1}R\widetilde{\Lambda}^{-s}w, w\big]\big] - \nu \operatorname{Re}\big[\big[\widetilde{\Lambda}^s C\widetilde{S}Q_{1N}\widetilde{S}^{-1}T\widetilde{S}v, w\big]\big] \\ &= \nu \operatorname{Re}\big[\big[H(y, D_x)w, w\big]\big] + \nu \operatorname{Re}[[Gw, w]] \\ &\quad - \nu \operatorname{Re}\big[\big[\widetilde{\Lambda}^1 C\widetilde{S}Q_{1N}\widetilde{S}^{-1}T\widetilde{S}v, w\big]\big] = I_{10} + I_{11} + I_{12}, \end{aligned}$$

where (cf. (5))

$$H(y,\xi) = C(y,\xi)S(|\xi|)Q_{1N}(y,\xi)S^{-1}(|\xi|)C^{-1}(y,\xi), \tag{32}$$

$$G(y,\xi) = \widetilde{\Lambda}^s(D_x)C(y,D_x)\widetilde{S}(D_x)Q_{1N}(y,D)\widetilde{S}^{-1}(D_x)R\widetilde{\Lambda}^{-s}(D_x) - H(y,D_x). \tag{33}$$

For the forms in (31) we will establish the following estimates:

$$I_{10} \geqslant \delta_1 \nu \sum \|w_j\|^2, \tag{34}$$

$$I_{1l} \geqslant -\varepsilon\nu \sum \|w_j\|_b^2 - K(\varepsilon)\nu^q \|w_j\|^2, \quad l = 1, 2, \tag{35_l}$$

which will imply (26).

We first derive (34). According to the definition in §1, the matrix $H(y,\xi) + {}^tH(y,\xi)$ is positive definite for $\xi \neq 0$, and its elements are homogeneous functions of ξ of degree $q - 1 = 2b$. By the well-known Gårding inequality, we have

$$\nu \operatorname{Re}[[H(y,D_x)w, w]] \geqslant \delta_0 \nu \left\|(1 + |D_x|^2)^{b/2} w\right\|^2 - K\nu\|w\|^2. \tag{36}$$

Furthermore,

$$\begin{aligned} \left\|(1 + |D_x|^2)^{b/2} w\right\| &\geqslant \left\|(1 + |D_x + i\nu N|^2)^{b/2} w\right\| \\ &\quad - \left\|\left((1 + |D_x + i\nu N|^2)^{b/2} - (b + |D_x|^2)^{b/2}\right) w\right\|. \end{aligned} \tag{37}$$

To estimate the second term on the right-hand side we employ an elementary inequality, namely if $b \geqslant 1$ and $c > d > 0$, then $c^b - d^b \leqslant b(c-d)c^{b-1}$. Replacing c and d in this inequality by $(1 + |\xi + i\nu N|^2)^{1/2}$ and $(1 + |\xi|^2)^{1/2}$, respectively, we find

$$\begin{aligned} &(1 + |\xi + i\nu N|^2)^{b/2} - (1 + |\xi|^2)^{b/2} \\ &\quad \leqslant b\big[(1 + |\xi + i\nu N|^2)^{1/2} - (1 + |\xi|^2)^{1/2}\big](1 + |\xi + i\nu N|^2)^{(b-1)/2} \\ &\quad \leqslant b|N|^2\nu^2(1 + |\xi + i\nu N|^2)^{(b-2)/2} \\ &\quad \leqslant \varepsilon_1(1 + |\xi + i\nu N|^2)^{b/2} - c(\varepsilon_1)\nu^{q-1}. \end{aligned} \tag{38}$$

In view of (36)–(38) we obtain (34). To prove (35_1) we note that G is a pseudodifferential operator with symbol

$$\begin{aligned} G(y,\xi) &= -H + \sum_{|\alpha+\beta+\gamma|\geqslant 0} (\alpha!\beta!\gamma!)^{-1} \\ &\qquad \times (\widetilde{\Lambda}^s)^{(\alpha)} D_x^\alpha (C\widetilde{S})^{(\beta)} D_x^\beta (Q_{1N}\widetilde{S}^{-1})^{(\gamma)} D_x^\gamma C^{-1}\widetilde{\Lambda}^{-s} \\ &= C\left(\widetilde{S}Q_{1N}\widetilde{S} - S(|\xi|)Q_{1N}S^{-1}(|\xi|)\right)C^{-1} + \sum_{|\alpha+\beta+\gamma|>0} \\ &= G_0(y,\xi) + G_1(y,\xi). \end{aligned} \tag{39}$$

The elements of the matrices entering the sum G_1 have an upper bound

$$\text{const}(1+|\xi+i\nu N|)^{-|\alpha|}(1+|\xi|)^{q-1-|\beta|-|\gamma|} \leqslant \text{const}(1+|\xi|)^{q-2}.$$

As has been shown, the elements of the diagonal matrix

$$\left[S((1+|\xi+i\nu N|^2)^{1/2})-S(|\xi|)\right]S^{-1}\left((1+|\xi+i\nu N|^2)^{1/2}\right)$$

have an upper bound

$$\text{const}\,\nu^2(1+|\xi+i\nu N|^2)^{1/2},$$

which implies that the elements of the matrix

$$G_0 = C\left[(\widetilde{S}-S)Q_{1N}\widetilde{S}+SQ_{1N}(\widetilde{S}-S)\right]C^{-1}$$

can be estimated in terms of

$$\text{const}\,\nu^2(1+|\xi+i\nu N|^2)^{(q-3)/2} \leqslant \varepsilon_2(1+|\xi+i\nu N|^2)^b - C(\varepsilon_2)\nu^{q-1}.$$

Hence, we have

$$|G(y,\xi)| \leqslant \varepsilon_3(1+|\xi+i\nu N|^2)^b - C(\varepsilon_3)\nu^{q-1}.$$

Similar inequalities hold for the derivatives of $G(y,\xi)$. Thus, if the operator G in the form I_{11} is replaced by the pseudodifferential operator with symbol (39), we obtain (35). As to the estimate for the norm of the difference $G - G(y,D_x)$, we do not dwell on it. To prove it one should pass to the Fourier transform, which results in a matrix integral operator with a kernel of the type $W(\xi-\xi',\xi')$. The estimation of this operator with respect to the second variable is carried out as in the case of symbol (39), and with respect to the first variable it decreases faster than any power function. The Schur inequality allows one to complete the proof.

To prove (35_2) we rewrite the form I_{12} as

$$I_{12} = -\nu\,\text{Re}[[z,\widetilde{\Lambda}^b w]] \geqslant -\varepsilon\nu\sum\|w_j\|_b^2 - (4\varepsilon)^{-1}\nu\sum\|z_j\|^2,$$

where

$$\begin{aligned} z &= \widetilde{\Lambda}^{s-b}C\widetilde{S}Q_{1N}\widetilde{S}^{-1}T\widetilde{S}v \\ &= \widetilde{\Lambda}^{s-b}C\widetilde{S}Q_{1N}\widetilde{S}^{-1}\widetilde{\Lambda}^{-s}(1+|D_x|^2)^N\left((1+|D_x|^2)^N T\widetilde{\Lambda}^s\widetilde{S}v\right). \end{aligned}$$

The norm of the expression in parentheses on the right-hand side is estimated in terms of const$\sum\|v_j\|_{s+s_j}^2$. Arguing as in the above estimation we find that the $L_2 \to L_2$ norm of the operator on the right-hand side has an upper bound const ν^{q-1}, which implies (35_2).

3.6. So, it remains to prove (27_0). Replacing v by (30) we rewrite the form in question as a sum of three forms:

$$\begin{aligned} I_0 &= \text{Im}\left[\left[\widetilde{\Lambda}^s C\widetilde{S}Q_0\widetilde{S}^{-1}R\widetilde{\Lambda}^{-s}w,w\right]\right] - \text{Im}\left[\left[\widetilde{\Lambda}^s C\widetilde{S}Q_0\widetilde{S}^{-1}T\widetilde{S}v,w\right]\right] \\ &= \text{Im}[[A(y,D_x)w,w]] + \text{Im}[[Bw,w]] - \text{Im}\left[\left[\widetilde{\Lambda}^s C\widetilde{S}Q_0\widetilde{S}^{-1}T\widetilde{S}v,w\right]\right] \\ &= I_{00}+I_{01}+I_{02}, \end{aligned}$$

where

$$A(y,\xi) = C(y,\xi)S(|\xi|)Q_0(y,\xi)S^{-1}(|\xi|)C^{-1}(y,\xi),$$
$$B = \widetilde{\Lambda}^s C\widetilde{S}Q_0\widetilde{S}^{-1}R\widetilde{\Lambda}^{-s} - A(y,D).$$

According to the imposed conditions, the matrix $A(y,\xi)$ is diagonal, and its elements are the eigenvalues of the matrix $Q_0(y,\xi)$, i. e., real homogeneous functions of degree $q = 2b+1$. By the sharp form of the Gårding inequality, we have

$$I_{00} \geqslant -\operatorname{const}\sum \left\|(1+|D_x|^2)^b w_j\right\|^2.$$

Arguing as in the derivation of (34) we conclude that

$$\begin{aligned} I_{00} &> (-\varepsilon_1 - C_1)\sum \|w_j\|_b^2 - C(\varepsilon_1)\nu^{q-1}\sum \|w_j\|^2 \\ &\geqslant -\left((\varepsilon_1 + C_1)\nu_0^{-1}\right)\nu \sum \|w_j\|_b^2 - C'(\varepsilon_1)\nu^q \sum \|w_j\|^2 \end{aligned}$$

for $\nu \geqslant \nu_0$. The first constant on the right-hand side can be made arbitrarily small by means of an appropriate choice of a sufficiently large constant ν_0.

The forms I_{01} and I_{02} are estimated using the arguments similar to those in the previous subsection, and we do not dwell on this question.

CHAPTER III

Mixed Problem for Hyperbolic Equations

Introduction

Let a differential operator $P(y; D)$ be defined in a cylindrical domain $\Omega \times \mathbb{R}_t$, where $\Omega \subset \mathbb{R}^n_x$, and let a set of differential operators $B_j(y'; D)$ be defined on the boundary $\partial\Omega \times \mathbb{R}_t$. By a mixed problem we mean the problem of determining a function $u(x, t)$, $x \in \Omega$, $t \geqslant 0$, from the equation

$$(1) \qquad P(y; D)u(x, t) = f(x, t)$$

with boundary conditions

$$(2) \qquad B_j(y', D)u(x', t) = g_k(x', t), \qquad x' \in \partial\Omega, \quad j = 1, \dots, \mu,$$

and initial conditions

$$D^k_t u(x, 0) = \varphi_k(x), \qquad k = 1, \dots, m - 1.$$

As in the case of the Cauchy problem, we will confine ourselves to zero Cauchy data (the general case is reduced to zero initial data by means of a transformation of the unknown function). As in Chapter II, we extend all functions entering (1) and (2) by zero to $t < 0$; then equation (1) will hold for all $t \in \mathbb{R}$, and the initial conditions will be replaced by

$$(3) \qquad u(x, t) = f(x, t) = 0, \qquad g_1(x', t) = \dots = g_\mu(x', t) = 0, \quad t < 0.$$

We will confine ourselves to the case when Ω is a half-space:

$$(4) \qquad \mathbb{R}^n_{N+} = \{x \in \mathbb{R}^n, \quad \langle N, x\rangle > 0\}.$$

In the end of this chapter we will make some remarks on the transition from the case of a half-space to a general type of domain.

The solvability of the mixed problem for second-order hyperbolic equations with classical (Dirichlet or Neumann type) boundary conditions was first proved in the mid-thirties by Kryzanski and Schauder **[KS]**; also see **[L]** and **[D]**. Contrary to the case of the Cauchy problem, attempts to construct a satisfactory theory for a mixed problem with general type boundary conditions have not been meeting with success for three decades. The results obtained in this field had a particular character and were related either to the case of two independent variables or to special classes of operators. Some comparatively general results were obtained only for the solvability of positive local problems for symmetric hyperbolic systems (Friedrichs **[F1, 2]** and some other authors) and symmetrizable systems (Friedrichs and Lax **[FL]** and Agranovich **[Ag1]**).

The first important result in studying general type mixed hyperbolic problems was obtained by Agmon **[A]** who considered the mixed problem in a half-space for a scalar strictly hyperbolic operator with constant coefficients in the principal part. Agmon introduced the notion of uniform Lopatinskii's condition and derived an a priori estimate under this condition and an additional assumption that the multiplicity of the roots of the principal symbol with respect to the variable dual to the normal to the boundary does not exceed two (Agmon's condition). In the late sixties a complete theory of mixed problems satisfying uniform Lopatinskii's condition was simultaneously constructed in two versions, namely for first-order systems by Kreiss **[Kr]** (also see **[R]**, **[Ra]**, **[Ag2, 3]**, and **[M]**, where some important additions to Kreiss' result are contained) and for a single high-order equation by Sakamoto **[S1]** (also see **[B]**, where some results under Agmon's condition were obtained, and **[GV4]**, where further references can be found). Sakamoto **[S1]** proved an energy estimate under some natural assumptions, but in the proof of the existence theorem it was required additionally that the boundary operators in (2) satisfy Schechter's normality conditions. To abandon this constraint, Gindikin and Volevich **[GV4]** introduced a dual problem and proved an energy estimate for it.

Many results on mixed problems that do not satisfy uniform Lopatinskii's condition stand outside the framework of the present book. For related references see **[GV4]**, **[VI]**, and also **[Es1]**.

In this chapter the presentation of the mixed problem for hyperbolic operators follows the paper **[GV4]**. The general plan is the following. Section 1 is devoted to conditions under which problem (1)–(3) is considered. Various equivalent forms of uniform Lopatinskii's condition are presented. In §2 the main result on the solvability of problem (1)–(3) is stated and the necessity of uniform Lopatinskii's condition and the other conditions of §1 is proved. Of great importance is §3 where the proof of the solvability of problem (1)–(3) is reduced to establishing two energy estimates, which we call the basic and dual estimates. In §4 some auxiliary material is compiled that relates to pseudodifferential operators and the Green formulas and is used in the derivation of energy estimates. Sections 5 and 6 are devoted to the proof of the basic estimate. In §5 localization and microlocalization of the main result are carried out and an estimate in a simpler domain, which we call the parabolic zone, is considered. A more difficult part of the proof, namely the estimation in the so-called hyperbolic zone is presented in §6. Finally, in §7 the results of §§5 and 6 are extended to problems with more general boundary conditions, which makes it possible to prove the dual estimate.

§1. Basic assumptions

1.1. Conditions on the symbol of the operator. In this chapter we will assume that the variables $y = (x, t)$ are chosen so that the vector N in (0.4) is directed along the axis x_1, i. e., $N = (1, 0, \dots, 0)$; the corresponding half-space is denoted $\mathbb{R}^n_+$. The selected variable x_1 will often be denoted z, i. e., we will write $y = (z, y')$, $y' = (x', t)$, and $x' = (x_2, \dots, x_n)$. The variables dual to y' and x' will be denoted as η' and ξ', respectively.

It will be assumed that the coefficients of the differential operators entering (0.1) and (0.2) belong to C^∞ and turn into constants for large values of the arguments. By $P_0(y, \eta)$ we will denote the principal homogeneous parts of the symbols of the operators in (0.1) and (0.2), i. e.,

$$P(y,\eta) = P_0(y,\eta) + P'(y,\eta), \qquad \deg P_0 = m, \quad \deg P' < m,$$
$$B_j(y,\eta) = B_{0j}(y,\eta) + B_j'(y,\eta), \quad \deg B_{0j} = \beta_j, \quad \deg B_j' < \beta_j, \quad j = 1, \dots, \mu.$$

In what follows we will always assume that the orders of the operators B_j are less than the order of P, i. e., $\beta_j < m$. All conditions below relate only to the principal parts.

Condition (A). $P_0(y, \xi, \tau)$ is a uniformly strictly hyperbolic symbol.

Condition (B). The direction $N = (1, 0, \dots, 0)$ is not characteristic at the points of the plane $x_1 = 0$, i. e., $P_0(0, y', 1, 0, \dots, 0) \neq 0$.

By condition (B), the polynomial $P_0(0, y', \xi, \tau)$ can be resolved with respect to the highest power ξ_1^m. We normalize the coefficients of the symbol so that in a neighborhood of the plane $x_1 = 0$ it is written in the form

$$(1) \qquad P_0(y, \eta) = \xi_1^m + \sum_{j=1}^{m} P_{0j}(y, \eta')\xi_1^{m-j},$$

where $P_{0j}(y, \eta')$ are real homogeneous polynomial symbols in η' of order j.

In the sequel we denote by X the set of parameters (y', ξ', τ), $y' \in \mathbb{R}^n$, $\xi' \in \mathbb{R}^{n-1}$, $\operatorname{Re}\tau \in \mathbb{R}$, $\operatorname{Im}\tau \leqslant 0$, and the symbol (1) will be denoted $P_0(X, \xi_1)$.

Consider the algebraic equation

$$(2) \qquad P_0(X, \lambda) = 0.$$

By condition (A), this equation has no real roots, and, in view of condition (B), the roots $\lambda_{0j}(X)$ are continuous functions of the multi-dimensional parameter X. Thus, for all values of the parameter $X = (y', \xi', \tau)$, $\operatorname{Im}\tau < 0$, the number of the roots of equation (2) in the half-planes $\operatorname{Im}\lambda > 0$ and $\operatorname{Im}\lambda < 0$ does not depend on X. We denote these numbers by m_+ and m_-, respectively. Since there are no real roots, we have

$$(3) \qquad m_+ + m_- = m.$$

REMARK. As follows from the further detailed analysis of the behavior of the roots of equation (2), in the case of even values of m the numbers m_+ and m_- coincide and are equal to $m/2$. For odd values of m one of these numbers is equal to $(m+1)/2$, whereas the other is equal to $(m-1)/2$, i. e., the modulus of their difference does not exceed 1.

Condition (C). $m_+ = \mu$, i. e., the number of the roots of equation (2) in the upper half-plane coincides with the number of boundary conditions in (0.2).

1.2. Condition on the boundary operators (uniform Lopatinskii's condition). Denote by $\mathfrak{M}(X)$ the space of solutions $V(z)$ to the homogeneous ordinary differential equation

$$(4) \qquad P_0(X, D_z)V(z) = 0 \quad (D_z = -id/dz).$$

This space is finite-dimensional, and its dimension is equal to the degree of the polynomial P_0:

$$\dim \mathfrak{M}(X) = m.$$

Let $\mathfrak{M}_+(X)$ denote the subspace of those solutions $V(z)$ for which $V(z) \to 0$, $z \to +\infty$. Since equation (2) has no real roots, the subspace $\mathfrak{M}_+$ coincides with the subspace of solutions to (4) that are uniformly bounded on the half-line $z \geqslant 0$, and we have

$$\dim \mathfrak{M}_+ = m_+ = \mu.$$

We say that a set of operators $\{P_0, B_{01}, \dots, B_{0\mu}\}$ satisfies Lopatinskii's condition if the boundary value problem

$$B_{0j}(X, D_z)V(0) = c_j \in \mathbb{C}, \qquad j = 1, \dots, \mu, \tag{5}$$

for equation (4) on the half-line possesses a unique solution $V(z) \in \mathfrak{M}_+(X)$.

Consider the operator $R\colon \mathbb{C}^\mu \to \mathbb{C}^\mu$ that maps a set of constants $c_1, \dots, c_\mu$ to the germ of order μ of the solution $V(z)$ to problem (4), (5):

$$R\colon \mathbb{C}^\mu \to \mathbb{C}^\mu \quad \Big(\{c_1, \dots, c_\mu\} \longmapsto \{V(0), \dots, (D_z^{\mu-1}V)(0)\}\Big). \tag{6}$$

This is a linear operator, and, according to Lopatinskii's condition, it has no kernel. Hence, R is determined by a nondegenerate $\mu \times \mu$ matrix, and (6) is an isomorphism. This implies the inequality

$$\sum_{j=1}^{\mu} \Big|(D_z^{j-1}V)(0)\Big| \leqslant C(X) \sum_{j=1}^{\mu} |(B_j(X, D_z)V)(0)| \tag{7}$$

for the solution to (4), (5).

It is known that in the case of elliptic or parabolic problems the parameter X runs over a closed set, the constants $C(X)$ in (7) are uniformly bounded, and Lopatinskii's condition itself guarantees Fredholm's character or unique solvability of the problem. As to the above hyperbolic problems, in this case the parameter X runs over an open set, and, generally, the constant in (7) may tend to $+\infty$ as $\operatorname{Im}\tau \to -0$. We will deal with the so-called uniform Lopatinskii condition, namely

Condition (D). For any value of the parameter

$$X = (y', \xi', \tau), \qquad y' \in \mathbb{R}^{n+1}, \quad |\xi'|^2 + |\tau|^2 = 1, \quad \operatorname{Im}\tau < 0, \tag{8}$$

problem (4), (5) possesses a unique solution, and in the domain (8) the constants $C(X)$ in (7) have a constant upper bound C.

1.3. Equivalent form of uniform Lopatinskii's condition. We stated uniform Lopatinskii's condition in the form that is suitable for a priori estimates but does not make it possible to answer the question whether a concrete system of operators satisfies this condition. Therefore here we state it in terms of the coefficients of the operators $B_{01}, \dots, B_{0\mu}$. To this end we need the following well-known assertion on factorization of polynomials.

PROPOSITION 1. *Let*

$$A(\omega, \lambda_0) = \sum_{j=0}^{m} a_j(\omega)\lambda^{m-j}, \tag{9}$$

where $a_j(\omega)$ are functions of class $C^l(\Omega)$, $l \geqslant 0$, depending on the parameter $\omega \in \Omega$, and let

$$|a_0(\omega)| > a > 0 \quad \forall \omega \in \Omega. \tag{10}$$

Assume that the roots of A (with respect to λ) split into J groups (counting multiplicities) such that for all $\omega \in \Omega$ the roots of the jth group belong to a set $U_j \subset \mathbb{C}$, and all *these sets are at positive distances from one another. Then the polynomial* (9) *admits factorization*:

$$A(\omega, \lambda) = \prod_{j=1}^{J} A_j(\omega, \lambda), \tag{9'}$$

where the roots of the polynomials $A_j(\omega, \lambda)$ *belong to* U_j, $j = 1, \dots, J$, *for all* $\omega \in \Omega$, *the coefficients of the polynomials* $A_j(\omega, \lambda)$ *possess the same degree of smoothness as the coefficients of* (9), *and the coefficients in the highest powers of* λ *in* $A_j(\omega, \lambda)$ *are uniformly separated from zero.*

The proof of the proposition can easily be derived from the standard proof of the Weierstrass preparation theorem (see **[GR,** Chapter II B]). In view of (10), we can divide the coefficients of polynomial (9) by $a_0(\omega)$ (with preservation of smoothness) and assume in what follows that $a_0(\omega) \equiv 1$. Denote by $\lambda_{jl}(\omega)$ the roots of (9) that belong to the jth group (counting multiplicities). Of course, these functions do not possess any smoothness properties. We set

$$A_j(\omega, \lambda) = \prod_{l=1}^{l_j} (\lambda - \lambda_{jl}(\omega)) = \lambda^{l_j} + a_{j1}(\omega)\lambda^{l_j - 1} + \dots + a_{jl}(\omega).$$

The coefficients $a_{jr}(\omega)$ are elementary symmetric functions of the roots $\lambda_{jl}(\omega)$ and are expressed in terms of the functions

$$\sum_{l=1}^{l_j} \lambda_{jl}(\omega)^r = \frac{1}{2\pi i} \int_{\gamma_j} \frac{\partial A(\omega, \lambda)}{\partial \lambda} \frac{\lambda^r}{A(\omega, \lambda)} \, d\lambda, \tag{11}$$

where the contour γ_j encircles the set U_j and lies at a positive distance from the sets U_k, $k \neq j$. By virtue of the imposed conditions, we can assume that

$$|A(\omega, \lambda)| > a_j > 0, \qquad \lambda \in \gamma_j, \quad \omega \in \Omega,$$

which implies that the left-hand sides in (11) are charactertized by the same degree of smoothness as the coefficients of A. The proposition is proved.

According to the proposition and conditions (A), (B), and (C), the symbol P_0 can be represented in the form

$$P_0(X, \lambda) = P_{0+}(X, \lambda) P_{0-}(X, \lambda). \tag{12}$$

The polynomials $P_{0\pm}$ are resolved with respect to the highest powers of λ, and their roots lie in the half-planes $\pm \operatorname{Im} \lambda > 0$. In case the coefficients of P do not depend on x and t for large $|x| + |t|$, the coefficients of $P_{0\pm}$ possess the same property.

The polynomials P_{0+} are suitable for describing the space $\mathfrak{M}_+(X)$ of decreasing solutions to equation (4).

PROPOSITION 2. *For a function* $V(z)$ *on the half-line* $z \geqslant 0$ *the following conditions are equivalent*:

(i) $V(z) \in \mathfrak{M}_+(X)$, *i. e.*, $V(z)$ *satisfies equation* (4), *and* $D^j V(z) \to 0$, $z \to +\infty$, $j = 0, 1, 2, \dots$;

(ii) *there are constants* $q_1, \dots, q_\mu$ (*in general, depending on* X) *such that*

$$V(z) = \frac{1}{2\pi i} \sum_{l=1}^{\mu} q_l \int_{\gamma_+} \lambda^{l-1} P_{0+}^{-1}(X, \lambda) e^{i\lambda z}\, dz, \tag{13}$$

where the contour γ_+ *lies in the half-plane* $\operatorname{Im} \lambda > 0$ *and encircles all the zeros of* $P_0(X, \lambda)$ *that have positive imaginary parts*;

(iii) $V(z)$ *is a solution to the equation*

$$P_{0+}(X, D_z) V(z) = 0. \tag{4_+}$$

PROOF. (i)⇒(ii). We extend $V(z) \in \mathfrak{M}_+$ as zero to $z < 0$, and let $\widehat{V}(\lambda)$ be the Fourier transform of this function:

$$\widehat{V}(\lambda) = \frac{1}{2\pi} \int_0^\infty e^{-i\lambda z} V(z)\, dz.$$

The function $\widehat{V}(\lambda)$ is holomorphic in the half-plane $\operatorname{Im} \lambda > 0$. By virtue of equation (4), we have

$$P_0(X, \lambda) \widehat{V}(\lambda) = c_0 \lambda^{m-1} + \dots + c_{m-1},$$

where the constants $c_0, \dots, c_{m-1}$ are expressed in terms of the germ at zero of the function $V(z)$. By the inversion formula, we can write

$$V(z) = \frac{1}{2\pi i} \int_{-\infty}^{\infty} e^{i\lambda z} Q(\lambda) P_0^{-1}(x, \lambda)\, d\lambda, \tag{13$'$}$$

where Q is a polynomial of degree no higher than $m-1$. Since $Q(\lambda) P_0^{-1}(\lambda) = O(|\lambda|^{-1})$ for $|\lambda| \to \infty$, the integration contour in (13′) can be deformed by replacing the real axis by a bounded contour γ_+ lying in the half-plane $\operatorname{Im} \lambda > 0$ and encircling all zeros of the polynomial P that belong to the upper half-plane. The fraction Q/P_0 can be written as

$$\frac{Q(\lambda)}{P_0(\lambda)} = \frac{Q_+(X, \lambda)}{P_{0+}(X, \lambda)} + \frac{Q_-(X, \lambda)}{P_{0-}(X, \lambda)},$$

where the degrees of the polynomials $Q_\pm$ are less than those of $P_{0\pm}$. According to the Cauchy theorem, the fraction Q_-/P_{0-} gives a zero contribution to integral (13′), and we thus arrive at (13).

Furthermore, we have the implications (ii)⇒(iii)⇒(ii) ⇒(i). The first and third of them are obvious, and the second implication is proved in the same way as the already established implication (i)⇒(ii).

We introduce the following notation. Let $B_{0j}'(X, \lambda)$ be the remainder on dividing $B_{0j}(X, \lambda)$ by $P_{0+}(X, \lambda)$, i. e.,

$$B_{0j}'(X, \lambda) \equiv B_{0j}(X, \lambda) \pmod{P_{0+}(X, \lambda)}, \qquad j = 1, \dots, \mu,$$

where B_{0j}' is a polynomial in λ of degree no higher than $\mu - 1$, i. e.,

$$B_{0j}'(X, \lambda) = \sum_{k=1}^{\mu} b_{jk}(X) \lambda^{\mu-k}.$$

We now set

$$\mathcal{B}(X) = \|b_{jk}(X)\|, \quad j,k = 1,\dots,\mu, \tag{14}$$

$$\mathrm{Lop}(P_0, B_{01},\dots,B_{0\mu})(X) = \det \mathcal{B}(X). \tag{15}$$

Note that, according to Proposition 2, any solution $V(z) \in \mathfrak{M}_+(X)$ to problem (4), (5) is a solution to the problem

$$P_+(x, D_z)V(z) = 0, \quad (B'_{0j}(x, D_z)V)(0) = c_j, \quad j = 1,\dots,\mu. \tag{16}$$

Hence, condition (D) is equivalent to the unique solvability of problem (16), and the germ of the solution satisfies the uniform estimate (7). We will restate this condition in terms of the matrix (14).

PROPOSITION 3. *For problem* (4), (5) *the conditions below are equivalent*:

(i) *Condition* (D) *from* 1.1 *holds.*

(ii) *The matrix* $\mathcal{B}(x)$ *is uniformly nonsingular, i. e., there is a constant* $K > 0$ *such that*

$$|\mathrm{Lop}(P_0, B_{01},\dots,B_{0\mu})(y', \xi', \tau)| \geqslant K, \tag{17}$$

$$\forall y' \in \mathbb{R}^n, \quad |\xi|^2 + |\tau|^2 = 1, \quad \mathrm{Im}\,\tau \leqslant 0. \tag{17'}$$

PROOF. (i)$\Rightarrow$(ii). Let condition (D) hold and let $V(z) \in \mathfrak{M}_+$ be a solution to problem (4), (5). By Proposition 2, there are $q_1,\dots,q_\mu$ such that $V(z)$ is representable in the form (13). The substitution of this expression into the boundary conditions (5) results in

$$c_j = (B_{0j}(x, D_z)V)(0) = \sum_k \sum_l b_{jk}(X) q_l \frac{1}{2\pi i} \int_{\gamma_+} \lambda^{\mu-k+l-1} P_{0+}^{-1}(X, \lambda)\, d\lambda. \tag{18}$$

We note that

$$\frac{1}{2\pi i} \int_{\gamma_+} \lambda^{\mu-k+l-1} P_{0+}^{-1}(X, \lambda)\, d\lambda = \delta_l^k. \tag{19}$$

Indeed, the contour γ_+ can be transformed into a circle of a large radius R, after which R can be made to tend to infinity. With account of (19), relations (18) go into the system

$$\mathcal{B}(X)q = c, \quad q = (q_1,\dots,q_\mu), \quad c = (c_1,\dots,c_\mu). \tag{20}$$

Note that relations of the type (19) and representation (13) imply that

$$D^k V(0) = \sum_{l=1}^{\mu} q_l \frac{1}{2\pi i} \int_{\gamma_+} \lambda^{\mu-k+l-1} P_{0+}^{-1}(X, \lambda)\, d\lambda = q_{\mu-k},$$

and therefore (7) goes into the inequality

$$|q| \leqslant C|c|, \tag{21}$$

which means that system (20) possesses a unique solution, and, consequently, the matrix $\mathcal{B}(X)$ is nonsingular. Inequality (21) can now be rewritten in the form

$$|\mathcal{B}^{-1}(X)c| \leqslant C|c|. \tag{21'}$$

Since c is an arbitrary vector, we obtain an estimate for the norm of the matrix $\mathcal{B}^{-1}$:

$$|\mathcal{B}^{-1}| \leqslant C. \tag{22}$$

Consider the well-known inequality (see **[K,** Chapter I, (4.12)])

$$|\det T| \leqslant \kappa |T|^{\mu-1} |T^{-1}|^{-1},$$

where T is a nonsingular $\mu \times \mu$ matrix and κ does not depend on T. If T is replaced in this inequality by $\mathcal{B}(X)$, we obtain

$$|\det \mathcal{B}(X)|^{-1} = |\det \mathcal{B}^{-1}(X)| \leqslant \kappa |\mathcal{B}^{-1}(X)|^{\mu-1} |\mathcal{B}(X)| \leqslant K^{-1},$$

i. e., (17) and (17′) are consequences of (D).

(ii)⇒(i). If (17) holds, then system (20) possesses a solution, and this solution satisfies (21). Therefore, the solution to problem (4), (5) can be looked for in the form (13), and (21) implies (7).

In conclusion we consider some simple examples of problems for hyperbolic equations that satisfy or do not satisfy condition (D).

Consider the Dirichlet problem in the half-space $x_1 \geqslant 0$ for the wave equation

$$P(D)u = (D_1^2 + \cdots + D_n^2 - D_{n+1}^2)u = f, \tag{23}$$

$$B(D)u\big|_{x_1=0} = u(0, y') = \varphi(y'). \tag{24}$$

In this case we have

$$P_+(\xi', \tau, \lambda) = \lambda - \left(\sqrt{\tau^2 - |\xi'|^2}\right)_+,$$

where $(\sqrt{\lambda})_+$ denotes the branch of the root that has a positive imaginary part, and we have $\operatorname{Lop}(P, B) \equiv 1$.

We now consider equation (23) with the Neumann boundary condition

$$B(D)u\big|_{x_1=0} = \frac{\partial u}{\partial x_1}\Big|_{x_1=0} = \varphi(y'). \tag{25}$$

In this case

$$\operatorname{Lop}(P, D_1) = \left(\sqrt{\tau^2 - |\xi'|^2}\right)_+,$$

i. e., Lopatinskii's condition is violated for $\operatorname{Im}\tau > 0$, $|\xi'| = |\operatorname{Re}\tau|$, and hence uniform Lopatinskii's condition does not hold for problem (23), (25).

§2. Statement of the main result. Necessary conditions for the solvability of the mixed problem

2.1. The space $H^{m,s}_{[\gamma]}(\mathbb{R}^{n+1}_+)$. Whereas in Chapter II, while studying the Cauchy problem, we singled out the variable $t = x_{n+1}$ playing the role of time, in the present section the role of the selected variable for the mixed problem in the half-space $x_1 \geqslant 0$ will be played by x_1.

Denote by $H^{m,s}_{[\gamma]}(\mathbb{R}^{n+1})$ the space of functions (distributions) with square of the norm given by the formula

$$\iint_{\operatorname{Im}\tau=\gamma} (1 + |\xi|^2 + |\tau|^2)^m (1 + |\xi'|^2 + |\tau|^2)^s |\widehat{u}(\xi, \tau)|^2 \, d\xi \, d\tau.$$

The elements of this space are characterized by the "degree of smoothness m" with respect to all variables and the additional "degree of smoothness s" with respect to the variables y'. In the case of natural numbers m the above norm is equivalent to

$$\|u\|^2_{m,s,[\gamma]} = \sum_{j=0}^{m} \iint\limits_{\operatorname{Im}\tau=\gamma} |\xi_1|^{2j}(1+|\xi'|^2+|\tau|^2)^{m+s-j}|\widehat{u}(\xi,\tau)|^2\,d\xi\,d\tau.$$

Denoting by $\widehat{u}'(x_1,\xi',\tau)$ the partial Fourier-Laplace transform of $u(y)$ with respect to the variables y' and passing to the Parseval relation with respect to x_1 we rewrite this norm in the form

$$(1)\qquad \|u\|^2_{m,s,[\gamma]} = \sum_{j=0}^{m} \int_{-\infty}^{\infty} \iint\limits_{\operatorname{Im}\tau=\gamma} (1+|\xi'|^2+|\tau|^2)^{m+s-j}|D_1^j\widehat{u}'(x_1,\xi',\tau)|^2\,d\xi\,d\tau\,dx_1.$$

We now denote by $H^{m,s}_{[\gamma]}(\mathbb{R}^{n+1}_+)$ the space of the restrictions to the half-space $x_1 \geqslant 0$ of the functions belonging to $H^{m,s}_{[\gamma]}(\mathbb{R}^{n+1}_+)$. The natural norm in this space is obtained from (1) by replacing integration along the line $-\infty < x_1 < \infty$ by integration over the half-line $0 \leqslant x_1 \leqslant \infty$. Thus, we set

$$(2)\qquad \begin{aligned}&\|u,\mathbb{R}^{n+1}_+\|^2_{m,s,[\gamma]}\\ &= \Big(\sum_{j=0}^{m} \int_{0}^{\infty} \iint\limits_{\operatorname{Im}\tau=\gamma} (1+|\xi'|^2+|\tau|^2)^{m+s-j}|D_1^j\widehat{u}'(x_1,\xi',\tau)|^2\,d\xi\,d\tau\,dx_1\Big)^{1/2}.\end{aligned}$$

We also note that (2) implies

$$(3)\qquad \|u,\mathbb{R}^{n+1}_+\|_{m,s,[\gamma]} = \Big(\sum \|D_1^j u\|^2_{0,m+s-j,[\gamma]}\Big)^{1/2}.$$

Let $H^{m,s}_{[\gamma]}(\mathbb{R}^{n+1})_+$ and $H^{m,s}_{[\gamma]}(\mathbb{R}^{n+1}_+)_+$ denote the subspaces of those functions belonging to the above spaces that vanish for $t < 0$.

We will need the space $H^{(s)}_{[\gamma]}(\mathbb{R}^n)$ of functions (distributions) supported at the plane $x_1 = 0$; the corresponding norm is

$$(4)\qquad \langle g\rangle_{(s),[\gamma]} = \Big(\iint\limits_{\operatorname{Im}\tau=\gamma} (1+|\tau|^2+|\xi'|^2)^s|\widehat{g}(\xi',\tau)|^2\,d\tau\,d\xi'\Big)^{1/2}.$$

Let $H^{(s)}_{[\gamma]}(\mathbb{R}^n)_+$ be the subspace of the elements vanishing for $t < 0$.

2.2. The mixed problem. The main part of the present chapter is devoted to the derivation of a priori estimates and the proof of the solvability of the mixed problem in $\mathbb{R}^{n+}_+$:

$$(5)\qquad P(y,D)u(y) = f(y), \qquad y \in \mathbb{R}^{n+1}_+,$$

$$(6)\qquad B_j(y',D)u(y)\big|_{x_1=0} = g_j(y'), \qquad j = 1,\dots,\mu,$$

$$(7)\qquad u(y) = f(y) = g_1(y') = \cdots = g_\mu(y') = 0, \qquad t < 0.$$

As was mentioned in the introduction, the coefficients of the operators entering (5) and (6) belong to C^∞ and are constant for large values of $|y|$ ($|y'|$). The main result of this chapter is the following

THEOREM. *For a set of operators $\{P, B_1, \dots, B_\mu\}$ the conditions below are equivalent.*

(I) *Conditions* (A), (B), (C), *and* (D) *in* §1 *hold.*

(II) *For each S there is $\gamma_0 = \gamma_0(S)$ such that for $|s| < S$ and $\gamma \leqslant \gamma_0$ the problem* (5)–(7) *possesses a unique solution $u \in H^{m,s}_{[\gamma]}(\mathbb{R}^{n+1}_+)_+$ for any right-hand sides $f \in H^{0,s+1}_{[\gamma]}(\mathbb{R}^{n+1}_+)_+$ and $g_j \in H^{m+s-\beta_j}_{[\gamma]}(\mathbb{R}^{n+1}_+)_+$, $j = 1, \dots, \mu$, and in this case we have*

$$D_1^j u(0, y') \in H^{(m+s-j)}_{[\gamma]}(\mathbb{R}^n), \quad j = 0, \dots, m-1, \tag{8}$$

and the a priori estimate

$$\begin{aligned} &|\gamma|\, \|u, \mathbb{R}^{n+1}_+\|^2_{m,s,[\gamma]} + \sum_{j=0}^{m-1} \langle D_1^j u(0, \cdot)\rangle^2_{m+s-j,[\gamma]} \\ &\leqslant C\left(|\gamma|^{-1}\|P(y, D)u\|^2_{0,s+1,[\gamma]} + \sum_{j=1}^{\mu}\langle B_j(y', D)u\rangle^2_{m+s-\beta_j,[\gamma]}\right) \\ &\qquad \forall u \in H^{(m,\infty)}_{[\gamma]}(\mathbb{R}^{n+1}_+) \end{aligned} \tag{9}$$

holds.

In the remaining part of the present section we will establish the simpler implication (II)⇒(I), i. e., we will prove the necessity of conditions (A), (B), (C), and (D). In §3 the proof of the existence and uniqueness theorem for problem (5)–(7) (i. e., the implication (I)⇒(II)) will be reduced to deriving (9) together with an estimate which, in a sense, is dual to (9). Sections 5 and 6 are devoted to the derivation of (9), and in §7 the dual estimate is derived and thus the proof of the theorem is completed.

2.3. Necessary conditions for the solvability of problem (5)–(7). As in the case of the Cauchy problem, we will prove that the solvability of the original problem implies the solvability of the problem for the equations and boundary conditions with constant coefficients that are obtained when the coefficients in the principal parts P_0, $B_{01}, \dots, B_{0\mu}$ are "frozen" at a point $y^0 \in \mathbb{R}^{n+1}$. If this point lies outside the plane $\{x_1 = 0\}$, we arrive at the Cauchy problem considered in Chapter II. In case y^0 lies in the plane $\{x_1 = 0\}$, we obtain a mixed problem. So, we have the following

PROPOSITION 1. *Let condition* (II) *of Theorem 2.2 hold and let a point $y^0 = (x_1^0, y'^0)$ belong to $\mathbb{R}^{n+1}$. Set $P(D) = P_0(y^0, D)$ and $B_j(D) = B_{0j}(y'^0, D)$. Then the following assertions hold.*

(i) *If $x_1^0 > 0$, then the equation*

$$P(D)u = f \tag{10}$$

possesses a unique solution $u \in H^{m,s}_{[\gamma]}(\mathbb{R}^{n+1})_+$ for any $f \in H^{0,s+1}_{[\gamma]}(\mathbb{R}^{n+1})_+$, $\gamma < 0$, and this solution satisfies the inequality

$$|\gamma|'\|u\|_{m,s,[\gamma]} \leqslant C'\|P(D)u\|_{0,s+1,[\gamma]} \quad \forall u \in H^{(\infty)}_{[\gamma]}(\mathbb{R}^{n+1}). \tag{11}$$

(ii) *Let $x_1^0 = 0$. Then the mixed problem in $\mathbb{R}^{n+1}_+$ for equation* (10) *with boundary conditions*

$$B_j(D)u\big|_{x_1=0} = g_j(y'), \quad j = 1, \dots, \mu, \tag{12}$$

possesses a unique solution $u \in H^{m,s}_{[\gamma]}(\mathbb{R}^{n+1}_+)_+$ *for any set of right-hand sides* $f \in H^{0,s+1}_{[\gamma]}(\mathbb{R}^{n+1}_+)_+$, $g_j \in (H^{s+m-\beta_j}_{[\gamma]})_+$, $j = 1, \dots, \mu$, $\gamma < 0$, *the inclusion relations* (8) *hold for this solution, and it satisfies the inequality*

$$
(13)\qquad
\begin{aligned}
&|\gamma|\,'\|u\|_{m,s,[\gamma]} + \sum_{j=0}^{m-1} {}'\langle D_1^j u\rangle_{m-s-j,[\gamma]} \\
&\qquad \leqslant C\Bigl(|\gamma|^{-1}\,'\|P(D)u\|_{0,s+1,[\gamma]} + \sum_{j=0}^{m-1} {}'\langle B_j(D)u\rangle_{m-s-\beta_j,[\gamma]}\Bigr),
\end{aligned}
$$

where $\cdot'\|\quad\|_{m,s,[\gamma]}$ *denotes the seminorms that are obtained from the norms* (2) *and* (4) *by replacing the factor* $1 + |\tau|^2 + |\xi'|^2$ *with* $|\tau|^2 + |\xi'|^2$. *The constants* C *entering* (11) *and* (13) *do not depend on the choice of* y^0.

Proposition 1 will be proved in 2.5.

Taking the Fourier-Laplace transform with respect to the variables y' for problem (10), (11), we arrive at the following problem on the half-line for the ordinary differential operator $P(D_z, \xi', \tau)$ with constant coefficients:

$$
(14)\qquad P(D_z, \xi', \tau)V(z) = F(z), \quad z > 0,
$$

$$
(15)\qquad (B_j(D_z, \xi', \tau)V)(0) = c_j, \quad j = 1, \dots, \mu.
$$

An immediate consequence of Proposition 1 is

PROPOSITION 2. *If Proposition* 1 *is true, then the following assertions hold*:

(i) *For every* $y^0 \in \mathbb{R}^{n+1}_+$ *and* $P(\xi, \tau) = P_0(y^0, \xi, \tau)$ *the inequality*

$$
(16)\qquad |\operatorname{Im}\tau|(|\xi_1| + |\xi'| + |\tau|)^m \leqslant C|P(\xi,\tau)|(|\xi'| + |\tau|), \quad \operatorname{Im}\tau \leqslant 0,
$$

holds;

(ii) *For each* $y'^0 \in \mathbb{R}^n$, $P(\xi, \tau) = P_0(0, y'^0, \xi, \tau)$, $B_j(\xi, \tau) = B_j(y'^0, \xi, \tau)$, $j = 1, \dots, \mu$, *problem* (14), (15) *has a unique solution* $V(z)$ *belonging to* $L_2(\mathbb{R}_+)$ *together with its derivatives up to order* m *for any* $F(z) \in L_2(\mathbb{R}_+)$ *and any set of constants* $(c_1, \dots, c_\mu) \in \mathbb{C}^\mu$, *and this solution satisfies the inequality*

$$
(17)\qquad
\begin{aligned}
&|\gamma| \sum_{j=0}^{m} (|\xi'|^2 + |\tau|^2)^{m-j} \int_0^\infty |D_z^j V(z)|^2\,dz + \sum_{j=0}^{m-1} (|\xi'|^2 + |\tau|^2)^{m-j} |(D_z^j V)(0)|^2 \\
&\qquad \leqslant C\Bigl(|\gamma|^{-1} \int_0^\infty |P(D_z, \xi', \tau)V(z)|^2 (|\xi'|^2 + |\tau|^2)\,dz \\
&\qquad\qquad + \sum_{j=0}^{m-1} (|\xi'|^2 + |\tau|^2)^{m-\beta_j} |c_j|^2\Bigr).
\end{aligned}
$$

Proposition 2 will be proved in the next subsection. We now complete the proof of the implication (II)⇒(I) under the assumption that this proposition is true.

Necessity of condition (A). The inequality

$$
|\operatorname{Im}\tau|(|\xi| + |\tau|)^{m-1} \leqslant C|P(\xi,\tau)|,
$$

which is equivalent to the condition of uniform strict hyperbolicity (see Chapter 2, §4) of the symbol $P(y, \xi, \tau)$, is a trivial consequence of (16).

Necessity of condition (B). We write

$$P_0(y,\xi,\tau) = a(y)\xi_1^m + O(|\xi_1|^{m-1}),$$

replace $(\xi_1, \xi', \sigma + i\gamma)$ in (16) by $(\rho\xi_1, \xi', \sigma + i\rho^\varepsilon\gamma)$, $0 < \varepsilon < 1$, divide both sides of (16) by $\rho^{m+\varepsilon}$, pass to the limit for $\rho \to +\infty$, and thus obtain

$$C^{-1} \leqslant |a(x_1^0, y'^0)| \quad \forall (x_1^0, y'^0),$$

i.e., (B) holds.

Necessity of conditions (C) and (D). It is proved by repeating literally the arguments used in the proof of Proposition 3 in 1.3. Indeed, by Proposition 2 (ii), problem (14), (15) is uniquely solvable for any right-hand side $F(z)$ and, in particular, for $F \equiv 0$. As was said in 2.1, the solution $V(z) \in \mathfrak{M}^+$ is representable in the form

$$V(z) = \frac{1}{2\pi i}\sum_{l=1}^{m_+} q_l \int \lambda^{l-1}(P^+(\lambda)^{-1}e^{i\lambda z})\,d\lambda, \tag{18}$$

where

$$q_l = (D^{m_+-l}V)(0).$$

The unique solvability of (14), (15) for $F \equiv 0$ is equivalent to the unique solvability of system (1.20) for any value of the parameter $X = (\xi', \sigma + i\gamma)$, $|\xi'|^2 + \sigma^2 + \gamma^2 = 1$, $\gamma \leqslant 0$, which implies that we have $\mu = m_+$ and that the Lopatinskii matrix $\mathcal{B}$ is nonsingular. Setting $|\xi'| + |\tau|^2 = 1$ in (17), assuming that $P(D_t, \xi', \tau)V = 0$, and discarding the first term on the left-hand side of (17) we derive inequality (1.7) with a constant not depending on X, i. e., condition (D) is proved.

2.4. Proof of Proposition 2 in 2.3. We confine ourselves to proving the second part of the assertion because the first part is simpler and was in fact established in Chapter II.

Thus, let problem (10), (12) have a unique solution $u \in H^{m,s}_{[\gamma]}(\mathbb{R}^{n+1}_+)_+$, where $s > 0$ is sufficiently large. We take right-hand sides of the type

$$f(y) = F(x_1)\varphi(y'), \qquad g_j(y') = \varphi(y')c_j, \quad j = 1, \dots, \mu,$$

where $F(x_1) \in L_2(\mathbb{R}_+)$ and $(c_1, \dots, c_\mu) \in \mathbb{C}^\mu$. Let us assume that the Fourier-Laplace transform $\widehat{\varphi}(\xi', \tau)$ is nonzero for almost all (ξ', τ), $\operatorname{Im}\tau < 0$. Then the partial Fourier-Laplace transform $\widehat{u}'(x_1, \xi', \tau)$ can be represented in the form

$$\widehat{u}'(x_1, \xi, \tau) = V(x_1, \xi', \tau)\widehat{\varphi}(\xi', \tau), \tag{19}$$

where V must be the solution to problem (14), (15). Furthermore, we substitute (19) into (13). Denoting the difference between the left-hand and right-hand sides of (17) as $G(\xi', \tau)$ and transferring the right-hand side of (9) to the left we obtain

$$\int |\widehat{\varphi}(\xi', \tau)|^2 G(\sigma, \xi')\,d\xi\,d\tau \leqslant 0.$$

Since the function $\widehat{\varphi}$ is nonzero almost everywhere and can assume arbitrary values, we conclude that $G(\tau, \xi') \leqslant 0$, i. e., (17) holds.

2.5. Proof of Proposition 1 in 2.2. The proof follows the plan of Proposition 4.2 in Chapter II. Select arbitrary functions $f \in H^{0,s+1}_{[\gamma]}(\mathbb{R}^{n+1})_+$ and $g_j \in H^{s+m-\beta_j}_{[\gamma]}$ and set

$$f_\varepsilon(x) = \varepsilon^{-m} f(x_1/\varepsilon, (y' - y'^0)/\varepsilon),$$
$$g_{j\varepsilon}(x) = \varepsilon^{-\beta_j} g_j((y' - y'^0)/\varepsilon), \quad j = 1, \dots, \mu.$$

By modifying the calculations in Proposition 4.2, Chapter II, we derive

$$\begin{aligned} f_\varepsilon{}'(x_1, \xi', \tau) &= \exp(-i\langle x'^0, \xi'\rangle - it^0\tau)\varepsilon^{n-m}\widehat{f}'(x_1/\varepsilon, \varepsilon\xi', \varepsilon\tau), \\ \widehat{g}_{j\varepsilon}(\xi', \tau) &= \exp(-i\langle x'^0, \xi'\rangle - it^0\tau)\varepsilon^{-\beta_j+n}\widehat{g}_j(\varepsilon\xi', \varepsilon\tau), \quad j = 1, \dots, \mu, \end{aligned}$$

whence

$$\begin{aligned} \|f_\varepsilon\|^2_{0,s+1,[\gamma/\varepsilon]} &= \varepsilon^{-2m-2s+n-1}\exp(2\gamma t^0/\varepsilon)\int_0^\infty \int_{\operatorname{Im}\tau=\gamma}\int_{R^{n-1}} \left|\widehat{f}'(x_1, \xi', \tau)\right|^2 \\ &\qquad \times (\varepsilon^2 + |\tau|^2 + |\xi'|^2)^{s+1}\, d\xi\, d\tau\, dx_1 \\ &= \varepsilon^{-2m-2s+n-1}\exp(2\gamma t^0/\varepsilon)('\|f\|_{0,s+1,[\gamma]} + O(\varepsilon)), \end{aligned}$$

where, as was mentioned above, $'\|f\|_{0,s+1,[\gamma]}$ and $'\langle g\rangle_{(\lambda),[\gamma]}$ denote the norms in which the factor $1 + |\tau|^2 + |\xi'|^2$ is replaced with $|\tau|^2 + |\xi'|^2$. Similarly, we obtain

$$\langle g_{j\varepsilon}\rangle^2_{(m+s-\beta_j),[\gamma/\varepsilon]} = \varepsilon^{-2m-2s+n}\exp(2\gamma t^0/\varepsilon)\left('\langle g_j\rangle_{(m+s-\beta_j),[\gamma]} + O(\varepsilon)\right).$$

According to condition (II), the mixed problem

$$P(y, D)v_\varepsilon = f_\varepsilon, \qquad B_j v_\varepsilon\big|_{x_1=0} = g_{j\varepsilon}, \quad j = 1, \dots \mu,$$

possesses a unique solution $v_\varepsilon \in H^{m,s}_{[\gamma/\varepsilon]}(\mathbb{R}^{n+1}_+)$, and this solution satisfies the inequality

$$\begin{aligned} &\frac{\gamma}{\varepsilon}\|v_\varepsilon\|^2_{m,s,[\gamma/\varepsilon]} + \sum_{j=0}^{m-1}\langle D_1^j v_\varepsilon\rangle^2_{(m+s-j),[\gamma/\varepsilon]} \\ &\qquad \leqslant C\left(\frac{\varepsilon}{\gamma}\|f_\varepsilon\|_{0,s+1,[\gamma/\varepsilon]} + \sum_{j=1}^{\mu}\langle g_{j\varepsilon}\rangle^2_{(m+s-\beta_j),[\gamma/\varepsilon]}\right) \qquad (20) \\ &\qquad \leqslant C\varepsilon^{2m-2s+n}\exp(2\gamma t^0/\varepsilon)\Big(\gamma^{-1\prime}\|f\|_{0,s+1,[\gamma]} \\ &\qquad\qquad + \sum{}'\langle g_j\rangle_{(m+s-\beta_j),[\gamma]} + O(\varepsilon)\Big). \end{aligned}$$

We now set

$$v_\varepsilon(x) = u_\varepsilon\left(\frac{x_1}{\varepsilon}, \frac{y' - y^{0\prime}}{\varepsilon}\right).$$

Then

$$\begin{aligned} \|v_\varepsilon\|^2_{m,s,[\gamma/\varepsilon]} &= \exp(2\gamma t^0/\varepsilon)\varepsilon^{-2m-2s+n+1}('\|u_\varepsilon\|^2_{m,s,[\gamma]} + O(\varepsilon)), \\ \langle D_1^j v_\varepsilon\rangle^2_{(m+s-j),[\gamma/\varepsilon]} &= \varepsilon^{-2m-2s+n}\left('\langle u_\varepsilon\rangle^2_{m+s-j,[\gamma]} + O(\varepsilon)\right). \end{aligned}$$

The substitution of this expression into (20) results in

$$
\begin{aligned}
|\gamma|' \|u_\varepsilon\|^2_{m,s,[\gamma]} &= \sum_{j=1}^{m-1}{}' \langle D_1^j u_\varepsilon \rangle_{(m+s-j),[\gamma]} + O(\varepsilon) \\
&\leqslant C \left(|\gamma|^{-1\prime} \|f\|_{0,s+1,[\gamma]} + \sum_{j=1}^{\mu}{}' \langle g_j \rangle^2_{(m+s-\beta_j),[\gamma]} \right).
\end{aligned} \tag{21}
$$

By (21), the sequence $\{u_\varepsilon\}$ is uniformly bounded in the space $H^{m,s}_{[\gamma]}(\mathbb{R}^{n+1}_+)$ for $\gamma < 0$, and consequently it is weakly compact in this space. We can take a subsequence that converges weakly to an element $u \in H^{m,s}_{[\gamma]}(\mathbb{R}^{n+1}_+)$, and in this case the Cauchy data up to order $m-1$ for the functions u_ε in the plane $x_1 = 0$ are weakly convergent to $D_1^j u(0, y')$, and the limit function satisfies (21) (where we should set $\varepsilon = 0$). It remains to show that u is a solution to problem (14), (15). We have

$$P(y, D) v_\varepsilon(y) = \varepsilon^{-m} f_\varepsilon(y),$$

i. e.,

$$\varepsilon^m P(y, D) u_\varepsilon \left(\frac{x_1}{\varepsilon}, \frac{y' - y^{0\prime}}{\varepsilon} \right) = f \left(\frac{x_1}{\varepsilon}, \frac{y' - y^{0\prime}}{\varepsilon} \right).$$

Performing the change of variables $x_1 \longmapsto x_1$, $y' \longmapsto y'^0 + \varepsilon y'$ we arrive at the equation

$$\varepsilon^m P \left(\varepsilon x_1, y'^0 + \varepsilon y', \frac{1}{\varepsilon} D \right) u_\varepsilon = f.$$

The differential operator on the left-hand side can be rewritten in the form

$$
\begin{aligned}
P(D) &+ \sum_{|\alpha| \leqslant m} \left(a_\alpha(\varepsilon x_1, y'^0 + \varepsilon y') - a_\alpha(0, y'^0) \right) D^\alpha \\
&+ \sum_{|\alpha| < m} \varepsilon^{m-|\alpha|} a_\alpha(\varepsilon x, y'^0 + \varepsilon y') D^\alpha \overset{\text{def}}{=} P(D) + P'_\varepsilon(y, D).
\end{aligned}
$$

It is clear that the sequence $P'_\varepsilon(y, D) u_\varepsilon$ is weakly convergent to zero, which implies that the limit function u satisfies (14). Formula (15) is verified in a similar way.

§3. Solvability of the mixed problem

In a sense, this section is similar to §3 of Chapter II, where the question of solvability of the Cauchy problem with zero initial data at $t = 0$ was reduced to the proof of two energy estimates for the original and formally adjoint operators. The solvability was proved in two stages. First, the solvability in $\mathbb{R}^{n+1}$ was derived in Sobolev's spaces with exponential weight $\exp(\gamma t)$ proceeding from some simple considerations of functional analysis. After that, at the second stage, the estimates with this weight were used for establishing estimates in the spaces $H^{(s)}_{[\gamma', \gamma'']}$. In turn, the latter estimates implied the solvability in spaces of functions vanishing for $t < 0$.

The mixed problem will be studied following a scheme similar in appearance to that of the Cauchy problem. At the first stage we will consider a boundary value problem in the half-space $x_1 > 0$ and prove its solvability in spaces with weight $\exp(\gamma t)$. At the second stage we will pass from estimates in $H^{(s)}_{[\gamma]}$ to estimates in $H^{(s)}_{[\gamma', \gamma'']}$ and prove the solvability of the boundary value problem in question for functions vanishing for

$t < 0$. This stage of the proof follows the scheme in §3 of Chapter II and does not require any essentially new techniques.

As to the first stage relating to the proof of the solvability of the boundary value problem in the half-space $x_1 \geqslant 0$, here there appear some fundamental difficulties. If the original problem is considered for smooth functions in the half-space $x_1 \geqslant 0$, then the adjoint problem involves distributions on the plane $x_1 = 0$ and distributions in $\mathbb{R}^{n+1}$ with support in $\mathbb{R}^{n+1}_+$. We do not know any method for deriving estimates for such a problem in the case of operators with variable coefficients. However, the estimation for the adjoint problem is not an end in itself, and our purpose is to prove the triviality of the null space of this problem. Since $\{x_1 = 0\}$ is not a characteristic plane of the operator under study, the theorems on partial hypoellipticity in the direction of x_1 hold for this operator, which implies that the null space of the adjoint problem consists of functions smooth in $x_1 \in [0, \infty)$ with range in spaces of distributions depending on the variable y'. It turns out that we can write down a boundary value problem (we will call it the transposed problem) whose solutions are elements of the null space of the adjoint problem, and this problem involves some distributions with support in the plane $\{x_1 = 0\}$. For the transposed problem we state an analog of inequality (2.9). In what follows (§7) we will present necessary and sufficient conditions for this inequality to hold. In the present section we use (2.9) and the estimate for the transposed problem for proving the solvability of the mixed problem.

3.1. Unbounded operator corresponding to the mixed problem. Our immediate task is to derive sufficient conditions for the existence of a solution $u \in H^{m,s}_{[\gamma]}(\mathbb{R}^{n+1}_+)$ to the problem

$$P(y, D)u(y) = f(y) \qquad y \in \mathbb{R}^{n+1}_+, \tag{1}$$

$$B_j(y', D)u(y)\big|_{x_1=0} = g_j(y'), \quad j = 1, \dots, \mu, \tag{2}$$

for $f \in H^{0,s+1}_{[\gamma]}(\mathbb{R}^{n+1}_+)$ ($g_j \in H^{(s+m-\beta_j)}_{[\gamma]}$, $j = 1, \dots, \mu$). We introduce the function spaces

$$E^s = H^{m,s}_{[\gamma]}(\mathbb{R}^{n+1}_+),$$
$$F^s = H^{0,s}_{[\gamma]}(\mathbb{R}^{n+1}_+),$$
$$G^s = \prod_{j=1}^{\mu} H^{(s+m-\beta_j-1/2)}_{[\gamma]}$$

and consider the unbounded operator

$$\mathfrak{A}\colon E^s \longrightarrow F^{s+1} \times G^{s+1/2} \quad (u \longmapsto \{Pu, Bu\}), \tag{3}$$

where $\{Pu, Bu\}$ denotes the tuple $P(y, D)u(y)$, $B_j(y', D)u(y)\big|_{x_1=0}$, $j = 1, \dots, \mu$.

Let us take E^{s+1} as the domain $D(\mathfrak{A})$. Then the graph of the operator $\mathfrak{A}$ will be contained in the graph of the continuous operator

$$E^s \longrightarrow F^s \times G^s \quad (u \longmapsto \{Pu, Bu\}), \tag{4}$$

which implies the closability of (3). Indeed, let $u_j \in E^{s+1}$, let $u_j \to 0$ in E^s, and let the sequence $\{Pu_j, Bu_j\}$ have a limit $\{f, g\}$ in $F^{s+1} \times G^{s+1/2}$. The continuity of the operator (4) implies that the sequence $\{Pu_j, Bu_j\}$ converges to zero in the norm of

$F^s \times G^s$, whence $\{f, g\} \equiv 0$. We will consider the closure of operator (3), for which the same notation will be retained.

Estimate (2.9) is extended by continuity to the space E^{s+1}, i. e., the inequality

(5)
$$|u, E^s| + \sum_{j=0}^{m-1} \langle D_1^j u(0, y)\rangle_{m+s-j,[\gamma]} \leqslant \text{const}\left(|Pu, F^{s+1}| + |Bu, G^{s+1/2}|\right), \quad u \in E^{s+1},$$

holds (at the moment the dependence of the constant on γ is inessential).

PROPOSITION. *Assume that inequality* (5) *holds for problem* (1), (2). *Then the unbounded operator* (3) *possesses the following properties*:

(i) *if* $u \in D(\mathfrak{A})$, *then* $D_1^j u(0, y') \in H_{[\gamma]}^{(m+s-j)}$, $j = 0, \dots, m-1$;

(ii) *for any function* $u \in D(\mathfrak{A})$ *inequality* (5) *holds*;

(iii) (5) *is an injective operator and has a closed range.*

PROOF. By the definition of an unbounded operator, the relation $u \in D(\mathfrak{A})$ means that there is a sequence $u_l \in E^{s+1}$ such that $u_l \to u$ in E^s and $\{Pu, Bu\} \to \mathfrak{A}u$ in $F^{s+1} \times G^{s+1/2}$. Inequality (5), which holds for all $u_j \in E^{s+1}$, implies that $D^j u_l(0, y')$ is a Cauchy sequence in $H_{[\gamma]}^{(s+m-j)}$, yielding (i). Furthermore, setting $u = u_j$ in (5) and passing to the limit as $j \to \infty$ we prove (ii). The validity of (5) for all $u \in D(\mathfrak{A})$ implies the injectivity of (3). Now let $\{f_j, g_j\}$ be a sequence of elements of the range of the operator (3), and let it be a Cauchy sequence in $F^{s+1} \times G^{s+1/2}$. In other words, let there exist a sequence $u_j \in E^{s+1}$ such that $Pu_j = f_j$, $Bu_j = g_j$. The fact that $\{f_j, g_j\}$ is a Cauchy sequence in $F^{s+1} \times G^{s+1/2}$ and inequality (5) imply that u_j is a Cauchy sequence in E^s. The limit $u \in E^s$ of this sequence belongs to the domain of operator (3), i. e., the sequence $\{f_j, g_j\}$ converges to an element of the image of operator (3). The proposition is proved.

3.2. Null space of the adjoint operator of (3). As was shown in the foregoing subsection, (3) is a closed injective operator with closed range. For the surjectivity of this operator (or, which is equivalent to surjectivity, for the existence of a bounded inverse operator) it is necessary and sufficient that the null space of the adjoint operator

(3′)
$$\mathfrak{A}^*\colon (F^s \times G^s)^* \longrightarrow (E^s)^*$$

be trivial:

(6)
$$\ker \mathfrak{A}^* = \varnothing.$$

Condition (6) is rather difficult to check because the description of the the domain of the operator (3′) may prove to be a complicated problem. However, we can derive a more efficient surjectivity condition for (3). To this end we consider the bounded operator

$$\mathfrak{A}_1\colon E^{s+1} \longmapsto F^{s+1} \times G^{s+1/2} \quad (u \longmapsto \{Pu, Bu\}).$$

It is clear that the graph of this operator is contained in that of (3) (with topology). Therefore the graph of the adjoint operator $\mathfrak{A}_1^*$ contains the graph of (3′), and hence condition (6) is sure to hold provided that

(6′)
$$\ker \mathfrak{A}_1^* = \varnothing.$$

To make condition (6′) more constructive we note that the form

$$[u, v] = \int_{R^{n+1}} u(y)\overline{v(y)}\, dy, \tag{7}$$

which was originally defined on $H_{[\gamma]}^{(\infty)}(\mathbb{R}_+^{n+1}) \times \overset{0}{H}{}_{[-\gamma]}^{(\infty)}(\mathbb{R}_+^{n+1})$ (where the index 0 means that we consider the subspace of elements vanishing for $x_1 < 0$), extends by continuity to

$$H_{[\gamma]}^{(m,s+1)}(\mathbb{R}_+^{n+1}) \times \overset{0}{H}{}_{[-\gamma]}^{-m,-s-1}(\mathbb{R}_+^{n+1})$$

and establishes an isomorphism of Banach spaces:

$$\left(H_{[\gamma]}^{m,s+1}(\mathbb{R}_+^{n+1})\right)' = \overset{0}{H}{}_{[-\gamma]}^{-m,-s-1}(\mathbb{R}_+^{n+1}).$$

Similarly, the duality

$$\int_{\mathbb{R}^{n+1}} u(y)\overline{v(y)}\, dy + \sum_{j=1}^{\mu} \int_{\mathbb{R}^n} g_j(y')\overline{h_j(y')}\, dy' \tag{8}$$

also establishes an isomorphism of Banach spaces:

$$\left(H_{[\gamma]}^{0,s+1}(\mathbb{R}_+^{n+1}) \times \prod_{j=1}^{\mu} H_{[\gamma]}^{s+m-\beta_j+1/2}(\mathbb{R}_+^{n+1})\right)'$$
$$= \overset{0}{H}{}_{[\gamma]}^{0,-s-1}(\mathbb{R}_+^{n+1}) \times \prod_{j=1}^{\mu} H_{[-\gamma]}^{(-s-m+\beta_j-1/2)}(\mathbb{R}_+^{n+1}).$$

Replacing u with $P(y, D)u$ and g_j with $B_j(y', D)u\big|_{x_1=0}$ in expression (8), we rewrite it in the form

$$\int (P(y, D)u)(y)\overline{v(y)}\, dy + \int \sum_{j=1}^{\mu} B_j(y', D)u(0, y')\overline{h_j}(y')\, dy'$$
$$= \left[u, P^*(y, D)v + \sum_{j=1}^{\mu} B_j^*(y', D)(\delta(x_1)h_j(y'))\right].$$

By what has been said, the abstract condition (6′) can be restated as a uniqueness condition for the problem in distributions:

$$P^*(y, D)v + \sum_{j=1}^{m} B_j^*(y', D)(\delta(x_1) \times h_j(y')) = h(y), \tag{9}$$

(9′) $v \in \overset{0}{H}{}_{[-\gamma]}^{0,-s-1}, \quad h \in \overset{0}{H}{}_{[-\gamma]}^{-m,-s-1}, \quad h_j \in H_{[-\gamma]}^{-m-s+\beta_j-1/2}(\mathbb{R}^n) \quad (j = 1, \dots, \mu).$

3.3. On smoothness of solutions to problem (9), (9′). The transposed boundary value problem. Let E be a space of distributions in $\mathbb{R}^{n+1}$, $\overset{0}{E}$ the subspace of elements with support in $x_1 \geqslant 0$, and $E(\mathbb{R}^{n+1}_+)$ the quotient space of E relative to the subspace of elements with support in $x_1 < 0$. Denote by π the canonical projector of E onto $E(\mathbb{R}^{n+1}_+)$, and let π_0 be the restriction of the projector to $\overset{0}{E}$:

$$\pi_0 : \overset{0}{E} \longrightarrow E(\mathbb{R}^{n+1}_+). \tag{10}$$

Operator (10) commuts with differential operators and maps to zero any distribution with support in the plane $\{x_1 = 0\}$. By what has been said, equation (9) implies that $\pi_0 v$ is a solution to the problem of the type

$$P^*(y, D)(\pi_0 v) = \pi_0 h, \tag{11}$$

$$\pi_0 v \in H^{0,-s-1}_{-[\gamma]}(\mathbb{R}^{n+1}_+), \quad \pi_0 h \in H^{-m,-s-1}_{[-\gamma]}(\mathbb{R}^{n+1}_+). \tag{11′}$$

We will be interested in the properties of the solution $\pi_0 v$ in the case when the right-hand side of (11) possesses an additional degree of smoothness with respect to the variables x_1. More precisely, let it belong to the subspace

$$H^{-m+l,-s-1-l}_{[-\gamma]}(\mathbb{R}^{n+1}_+), \quad l = 1, 2, \dots .$$

The index $[-\gamma]$ is inessential in this subsection, and we will omit it. Since the operator $P(y, D)$ (and, consequently, the operator $P^*(y, D)$ as well) is resolved with respect to the highest derivative D_1^m, the operator P^* is partially hypoelliptic up to the boundary in the subspace $\{x_1 \geqslant 0\}$. We have

LEMMA 1 ([**H2,** Theorem 4.3.1]). *Let conditions* (11) *and* (11′) *hold and let*

$$\pi_0 h \in H^{-m+l,-s-l}(\mathbb{R}^{n+1}_+) \tag{12}$$

for some positive integer l. Then

$$\pi_0 v \in H^{l,-s-l-1}(\mathbb{R}^{n+1}_+). \tag{12′}$$

Our primary goal is to study the null space of equation (9). To this end we will consider (9) in the case when the right-hand side possesses an additional degree of smoothness for $x_1 > 0$. We will need description of distributions from $\overset{0}{H}{}^{-m,-s-1}(\mathbb{R}^{n+1})$ and $\overset{0}{H}{}^{-s-1}(\mathbb{R}^{n+1})$ that belong to quotient spaces characterized by a higher degree of smoothness with respect to x_1 for $x_1 > 0$. The following lemma holds.

LEMMA 2. (a) *Let a distribution h_0 possess the properties*

$$h_0 \in \overset{0}{H}{}^{-m,-s-1}, \quad \pi_0 h_0 \in H^{0,-s-1-m}(\mathbb{R}^{n+1}_+).$$

Then h_0 is representable as

$$h_0 = \theta_0(x_1)h + \sum_{k=1}^{m} h_k(y') D_1^{k-1}\delta(x_1), \tag{13}$$

$$h \in H^{0,-s-1-m}(\mathbb{R}^{n+1}), \quad h_k \in H^{(-s-m+k-3/2)}(\mathbb{R}^{n+1}), \tag{13′}$$

where $\theta_0(x_1)$ is the characteristic function of the half-line $x_1 > 0$.

(b) *If* $v_0 \in \overset{0}{H}{}^{0,-s-1}$ *and* $\pi_0 v_0 \in H^{m,-s-1-m}(\mathbb{R}^{n+1})$, *then*

(14) $$v_0 = \theta_0(x_1)v, \quad v \in H^{m,-s-1-m}(\mathbb{R}^{n+1}).$$

PROOF. According to the definition of quotient space, there is an element $h \in H^{0,-s-1-m}(\mathbb{R}^{n+1})$ such that $\pi_0(h_0 - h) = 0$. Since the elements of $H^{0,-s-1-m}$ are functions belonging to $L_2(\mathbb{R}_{x_1})$, with range in the space of distributions, they can be multiplied by the characteristic function of the half-line $\{x \geqslant 0\}$, and $\theta_0(x_1)h$ also is a representative of the residue class h_0. Thus, we have $\pi_0(h_0 - \theta_0 h) = 0$, i. e., the distribution $h_0 - \theta_0 h$ has a support on the plane $x_1 = 0$. Let us prove that $r = h_0 - \theta_0 h$ has the form of the sum in (13).

Since $r(y)$ is a distribution with support at $x_1 = 0$, its Fourier-Laplace transform $\widehat{r}(z, \eta)$ is an entire function of z for almost all η'. Since $r \in H^{-m,-s-1}$, the integral

(15) $$\int_{\mathbb{R}^n}\int_{-\infty}^{\infty} |\widehat{r}(z,\eta')|^2(1+|z|^2+|\eta'|^2)^{-m}(1+|\eta'|^2)^{-s-1}\, d\eta' dz < \infty$$

is convergent. By the Liouville theorem, $\widehat{r}(z,\eta')$ is a polynomial in z of degree at most $m-1$, i. e., $\widehat{r}$ can be written as

$$\widehat{r}(z,\eta') = \sum_{k=1}^{m} \widehat{h}_k(\eta') z^{k-1}.$$

Substituting this expression into (15) we make change of variable $z = z'(1+|\eta'|^2)^{1/2}$ and perform integration with respect to z', which transforms (15) into

(15′) $$\sum^{n} c_{kl}\int_{\mathbb{R}} (1+|\eta'|^2)^{-m-s+(k+l-3)/2}\widehat{h}_k(\eta')\overline{\widehat{h}_l(\eta')}\, d\eta' < \infty,$$

where

$$c_{kl} = \int_{-\infty}^{\infty} z^{k+l-2}(1+z^2)^{-m}\, dz, \qquad k,l = 1,\dots,m-1.$$

It follows that the quadratic form

$$\sum c_{kl}\zeta_k\zeta_l = \int_{-\infty}^{\infty}\left|\sum_{k=1}^{m}\zeta_k z^{k-1}\right|^2(1+z^2)^{-m}\, dz$$

is positive definite, and consequently there is $\delta > 0$ such that

$$\sum c_{kl}\zeta_k\overline{\zeta}_l \geqslant \delta\sum|\zeta_k|^2.$$

The replacement of ζ_k by $\widehat{h}_k(\eta)(1+|\eta'|^2)^{-(m+s+3/2)/2+k/2}$, $k = 1,\dots,m$, in this inequality with subsequent integration of both sides with respect to η and application of (15′) leads to the conclusion that these functions are square integrable, and, consequently, they are the Fourier transforms of some distributions $h_k \in H^{(-m-s+k-3/2)}(\mathbb{R}^n)$. The lemma is proved.

We come back to problem (9), (9′) and assume that the first relation (13′) holds for the right-hand side h. Then h is representable in the form of the right-hand side of (13), and, by Lemma 1, we have $\pi_0 v \in H^{m,-s-1-m}(\mathbb{R}^{n+1})$, i. e., v satisfies the conditions of assertion (b) in Lemma 2, and v can be represented in the form (14). We substitute the resulting expressions into equation (9) and rewrite it as

$$\begin{aligned}\theta_0(P^*(y,D)v-h) = &-[P^*(y,D),\theta_0]v - \sum_{j=1}^{\mu} B_j^*(y',D)(\rho_j(y'))\delta(x_1)\\ &+\sum_{l=1}^{m} h_l(y')D_1^{l-1}\delta(x_1)=0.\end{aligned} \tag{16}$$

The left-hand side of (16) as a function of x_1 belongs to $L_2(\mathbb{R}_+)$ and its range lies in the space of distributions, whereas the right-hand side is a distribution with support in the plane $\{x_1=0\}$. It follows that the two sides vanish independently. Since the left-hand side of this relation is zero, the equation

$$P^*(x,D)v = h(x), \quad x_1>0, \tag{17}$$

$$v\in H^{m,-s-1}_{[-\gamma]}(\mathbb{R}^{n+1}), \quad h\in H^{0,-s-m-1}_{[-\gamma]}(\mathbb{R}^{n+1}), \tag{17'}$$

holds. To analyze the right-hand side we need the well-known lemma below (e.g., see **[GV1**, p. 250]).

LEMMA 3. *Let a differential operator $Q(y,D)$ have the form*

$$Q(y,D)=\sum_{j=0}^{m} Q_j(y,D^*_{y'})D_1^j.$$

Then

$$Q(y,D)(\theta_0(x_1)v)-\theta_0(x_1)Q(y,D)v=\sum_{j=1}^{m} h_j(y')D_1^{j-1}\delta(x_1),$$

$$h_j(y') = -i\sum_{k=0}^{m-j}((D_{y'}D_{x_1}^{m-j-k}u)(0,y')).$$

In view of Lemma 3, the right-hand side of (16) can be rewritten in the form

$$\sum_{l=1}^{m}\left[-E_l(y',D)v\big|_{x_1=0}-\sum_{j=1}^{\mu} H_{lj}(y',D')\rho_j(y')-h_l(y')\right]D_1^{l-1}\delta(x_1).$$

Since the distributions $D_1^{l-1}\delta(x_1)$, $l=1,\dots,m$, are linearly independent, the coefficients in the derivatives of $\delta(x_1)$ vanish, which implies the system of boundary conditions

$$E_l(y',D)v\big|_{x_1=0}+\sum H_{lj}(y',D_y)\rho_j(y')=h_l(y'), \quad l=1,\dots,m. \tag{18}$$

We have shown that the null space of problem (9), (9′) coincides with that of problem (16), (18); the latter problem will be called the transposed problem in relation to the original problem (1), (2).

As follows from the definition of the operators E_l and H_{lj}, we have

$$\deg E_l = m-l, \quad \deg H_{lj}=\beta_j-l+1. \tag{19}$$

In §7 we will establish analogs of conditions (A), (B), (C), and (D) of §1 that guarantee the fulfillment of the a priori inequality

$$
\begin{aligned}
&|\gamma|\,\|v\|^2_{m,h,[-\gamma]} + \sum^{m-1} \langle D_1^j v\rangle_{(m+h-j),[-\gamma]} + \sum_{k=1}^{\mu} \langle \rho_k\rangle_{(h+\beta_k+1),[-\gamma]} \\
&\qquad \leqslant C\Big(|\gamma|^{-1}\|P^*(y,D)v\|^2_{0,h+1,[-\gamma]} \\
&\qquad\qquad + \sum_{j=1}^{m} \langle E_j(y',D)v + \sum H_{jk}(y',D)\rho_k\rangle^2_{(j+h),[-\gamma]}\Big) \\
&\qquad v \in H^{m,h+1}_{[-\gamma]}, \quad (\rho_1,\dots,\rho_\mu) \in \prod_{j=1}^{\mu} H^{(h+\beta_k)}_{[-\gamma]}
\end{aligned}
\tag{20}
$$

for the solutions to problem (16), (18). Thus, we have proved the following

THEOREM. *Let inequality* (2.9) *hold for the problem* (1), (2) *and let inequality* (20) *with* $h = -s-1$ *hold for the transposed problem* (16), (18). *Then for all* $f \in H^{0,s+1}_{[\gamma]}(\mathbb{R}^{n+1}_+)$ *and* $g_j \in H^{(s+m-\beta_j)}_{[\gamma]}$, $j = 1,\dots,\mu$, *the problem* (1), (2) *possesses a unique solution* $u \in H^{m,s}_{[\gamma]}(\mathbb{R}^{n+1}_+)$.

3.4. On the solvability of the mixed problem. Now we prove the main implication (I)$\Rightarrow$(II) in Theorem 2.2 under the assumption that inequalities (2.9) and (20) are true. A trivial modification of the arguments in 3.1.4 of Chapter II shows that, on condition that (2.9) holds, it suffices to prove that problem (1), (2) with right-hand sides $f \in H^{0,s+1}_{[\gamma',\gamma'']}(\mathbb{R}^{n+1}_+)$ and $g_j \in H^{m+s-\beta_j}_{[\gamma',\gamma'']}$, $j = 1,\dots,\mu$, has a solution $u \in H^{m,s}_{[\gamma',\gamma'']}(\mathbb{R}^{n+1}_+)$. Here $\gamma' < \gamma'' < \gamma_0 < 0$, and the spaces $\Phi_{[\gamma',\gamma'']}$, $\Phi = H^{m,s}$, $H^{0,s+1}$, $H^{(s)}(\mathbb{R}^n)$, etc. are defined according to the scheme in §1 of Chapter 2. To prove the solvability in these spaces it suffices to show that the null space of the adjoint operator is trivial. Repeating the arguments of 3.2 we reduce this question to the problem of uniqueness for equation (9) with

$$
v \in H^{0,-s-1}_{[-\gamma',-\gamma'']}, \quad h \in H^{-m,-s-1}_{[-\gamma',-\gamma'']}, \quad h_j \in H^{-m-s+\beta_j-1/2}_{[-\gamma',-\gamma'']}. \tag{9''}
$$

Now we can repeat literally the arguments in 3.3 applied to spaces of distributions with weight $(\exp(\gamma' t) + \exp(\gamma'' t))^{-1}$. As a result, it turns out to be necessary to prove the uniqueness for problem (16), (18) in the spaces with that weight.

Recall (see §1 of Chapter 2) that if E is a Banach space with the norm $|\cdot, E|$, then $E_{[-\gamma',-\gamma'']}$, $\gamma' \leqslant \gamma''$, is defined as the linear span of $E_{[-\gamma']} + E_{[-\gamma'']}$ with Banach norm

$$
|u, E_{[-\gamma',-\gamma'']}| = |\chi(t)u, E_{[-\gamma'']}| + |(1-\chi)uE_{[-\gamma']}|,
$$

where $\chi(t) \in C^\infty(\mathbb{R})$, $\chi(t) = 1$ for $t \geqslant 1$, and $\chi(t) = 0$ for $t \leqslant -1$. The uniqueness of the solution to the problem under consideration is a direct consequence of the a priori estimate for problem (16), (18):

$$\|v\|^2_{m,h,[-\gamma',-\gamma'']} + \sum_{k=1}^{\mu} \langle \rho_k \rangle^2_{(h+\beta_k+1),[-\gamma',-\gamma'']} + \sum_{j=0}^{m-1} \langle D_1^j v \rangle_{(m+h-j),[-\gamma',-\gamma'']}$$
$$\leqslant C(\gamma',\gamma'')\Big(\|P^* v\|_{(0,h+1),[-\gamma',-\gamma'']} + \sum_{j=1}^{m} \langle E_j v + \sum_{k=1}^{\mu} H_{jk}\rho_k \rangle_{(j+h),[-\gamma',-\gamma'']}\Big) \tag{21}$$
$$\forall u \in H^\infty_{[-\gamma',-\gamma'']}.$$

The assertion below is proved following the plan of 3.2 of Chapter II.

PROPOSITION. *Let inequality* (20) *hold for problem* (16), (18). *Then for* $\gamma' \leqslant \gamma'' \leqslant \gamma_0$, *where* $\gamma_0 = \gamma_0(L)$, $|h| \leqslant L$, *and* γ_0 *is sufficiently large, inequality* (21) *holds.*

PROOF (cf. 3.2 of Chapter II). We set $\gamma = \gamma', \gamma''$ in (20), where $\gamma < \gamma' < \gamma'' < \gamma/2 < 0$ and γ is a large parameter, and add together the corresponding inequalities. The left-hand side of the resulting inequality will have upper and lower estimates expressed in terms of

$$|\gamma|\, \|v\|^2_{m,h,[-\gamma',-\gamma'']} + \sum_{k=1}^{\mu} \langle \rho_k \rangle^2_{(h+\beta_k+1),[-\gamma',-\gamma'']} + \sum_{j=1}^{m} \langle D_1^j v \rangle^2_{(m+h-j),[-\gamma',-\gamma'']} \tag{22}$$

with constants 4 and 1/4, respectively. As to the right-hand side, it is estimated (cf. 3.2 of Chapter II) as follows

$$\begin{aligned} &|\gamma|^{-1}\|P^* v\|^2_{0,h+1,[-\gamma',-\gamma'']} + \sum_{j=1}^{m} \Big\langle E_j v + \sum_{k=1}^{\mu} H_{jk}\rho_k \Big\rangle^2_{(j+h),[-\gamma',-\gamma'']} \\ &+ |\gamma|^{-1}\|[P^*,\chi]v\|^2_{0,h+1,[-\gamma']} + |\gamma|^{-1}\|[P^*,1-\chi]v]\|^2_{0,h+1,[-\gamma'']} \\ &+ \sum_{j=1}^{m}\Big\{ \langle [E_j,\chi]v\rangle^2_{(j+h),[-\gamma']} + \langle [E_j,(1-\chi)]v\rangle^2_{(j+h),[-\gamma'']} \\ &+ \sum_{k=1}^{\mu}\Big(\langle [H_{jk},\chi]\rho_k\rangle^2_{(j+h),[-\gamma']} + \langle [H_{jk},1-\chi]\rho_k\rangle^2_{(j+h),[-\gamma'']}\Big)\Big\} \end{aligned} \tag{23}$$

with some constant (not depending on γ). By the Leibniz formula, we have

$$[P^*,\chi] = \sum \chi_l P^{*(l)}(y,D), \quad \chi_l = D_t^l \chi / l!,$$

where the symbol $P^{*(l)}$ is the lth derivative of P^* with respect to τ. The other commutators have a similar form. Repeating literally the derivation of (16) and (17) in 3.2 of Chapter II we estimate the sum of the third and fourth terms in (23) via

$$\begin{aligned} &\text{const}\,|\gamma|^{-1}\sum \|P^{*(l)}v\|^2_{0,h+1,[-\gamma',-\gamma'']} \\ &\leqslant \text{const}\,|\gamma|^{-1}\sum_{l=1}^{m} \|v\|^2_{m,h+m-l+1,[-\gamma',-\gamma'']} \\ &\leqslant \text{const}\,|\gamma|^{-1}\|v\|^2_{m,h,[-\gamma',-\gamma'']}. \end{aligned}$$

Similarly (in view of (19)), for $\mu = \gamma', \gamma''$ and $\psi = \chi, 1 - \chi$ we derive

$$\begin{aligned}\langle [E_j, \psi] v\rangle_{(j+h),[-\mu]} &\leqslant \text{const} \sum_{\alpha>0} \langle E_j^{(\alpha)} v\rangle_{(j+h),[-\gamma',-\gamma'']}\\ &\leqslant \text{const} \sum_{|\alpha|\geqslant 1} \langle v\rangle_{(h+m-|\alpha|),[-\gamma',-\gamma'']}\\ &\leqslant \text{const}\, |\gamma|^{-1} \langle v\rangle_{(m+h),[-\gamma',-\gamma'']}.\end{aligned}$$

Furthermore, for $\mu = \gamma', \gamma''$ and $\psi = \chi, 1 - \chi$ we have

$$\begin{aligned}\langle [H_{jk}, \psi] \rho_k\rangle_{(j+h),[-\mu]} &\leqslant \text{const} \sum_{\alpha>0} \langle H_{jk}^{(\alpha)} \rho_k\rangle_{(j+h),[-\gamma',-\gamma'']}\\ &\leqslant \text{const} \sum_{|\alpha|\geqslant 1} \langle \rho_k\rangle_{(\beta_k+h+1-|\alpha|),[-\gamma',-\gamma'']}\\ &\leqslant \text{const}\, |\gamma|^{-1} \langle \rho_k\rangle_{(\beta_k+h+1),[-\gamma',-\gamma'']}.\end{aligned}$$

Combining these estimates and taking a sufficiently large $|\gamma|$ we arrive at the desired inequality (21).

§4. Some auxiliary assertions (pseudodifferential operators and estimates for quadratic forms)

When deriving basic inequalities in §§4 and 5 of Chapter II we used various estimates for quadratic forms of differential operators (see 4.1, 4.3, 4.4, and 4.5). In the present chapter we will also deal with this kind of estimates; however instead of differential operators with respect to all variables we will consider differential operators with respect to the variable x_1 whose coefficients will be pseudodifferential operators acting on the other variables and depending on x_1 as a parameter.

4.1. Denote by S^j the class of symbols $b(y, \eta', \gamma)$ belonging to C^∞ as functions of y and (η', γ) for $|\eta'| + |\gamma| \neq 0$ and satisfying the inequalities

(1) $$|D_y^\alpha D_{\eta'}^\beta D_\gamma^p b(y, \eta', \gamma)| \leqslant C_{\alpha\beta p} \Lambda(\eta', \gamma)^{j-|\beta|-p},$$

where

(2) $$\Lambda(\eta', \gamma) = (|\eta'|^2 + \gamma^2)^{1/2}.$$

With each function $b \in S^j$ we can associate the pseudodifferential operator

(3) $$_{[\gamma]}b(x, D', \gamma) = \exp(-\gamma t) b(y, D', \gamma)(\exp(\gamma t) g)$$

acting on the variable y' and depending on the parameter x_1. This operator transforms the spaces $H_{[\gamma]}^{(\pm\infty)}(\mathbb{R}^n)$ into themselves.

If $b(y, \eta', \gamma) = B(y, \xi', \sigma + i\gamma)$, where B is a holomorphic function of $\sigma + i\gamma$, then, as was already noted in §1 of Chapter I, the operators (3) are compatible for $\gamma = \gamma'$ and $\gamma = \gamma''$ (i. e., they coincide on the intersection of $H_{[\gamma']}^{(\infty)}$) and $H_{[\gamma'']}^{(\infty)}$), and the index $[\gamma]$ in the notation of the operators can be dropped.

The norm $\langle \cdot \rangle_{(s),[\gamma]}$ can be expressed in terms of an operator with symbol (2):

(4) $$\langle g\rangle_{(s),[\gamma]} = \langle {}_{[\gamma]}\Lambda^s(D', \gamma) g\rangle_{[\gamma]} = \langle \Lambda^s(D', \gamma) \exp(\gamma t) g\rangle.$$

We also set

(5)
$$\begin{aligned}\langle g,h\rangle_{(s),[\gamma]} &=\langle {}_{[\gamma]}\Lambda^s(D',\gamma)g, {}_{[\gamma]}\Lambda^s(D',\gamma)h\rangle\\ &=\langle \Lambda^s(D,\gamma)(\exp(\gamma t)g), \Lambda^s(D,\gamma)\exp(\gamma t)g\rangle,\end{aligned}$$

where

(5′)
$$\langle g,h\rangle = \int_{\mathbb{R}^n} g(y')\overline{h(y')}\,dy'.$$

The theory of conventional pseudodifferential operators can easily be extended to operators (3). We will need the following properties of these operators.

1) If $b(y,\eta',\gamma)\in S^j$, then

(6)
$$\langle b(y,D',\gamma)g\rangle_{(s),[\gamma]} \leqslant c_s(b)\langle g\rangle_{(s+j),[\gamma]}.$$

2) If $b_1(y,\eta',\gamma)\in S^{j_1}$ and $b_2(y,\eta',\gamma)\in S^{j_2}$, then

(7)
$$\big\langle (b_1(y,D',\gamma)\cdot b_2(y,D',\gamma)-(b_1b_2)(y,D',\gamma))g\big\rangle_{(s),[\gamma]} \leqslant c_s(b_1,b_2)\langle g\rangle_{(s+j_1+j_2-1),[\gamma]}.$$

3) If $b(x,\eta',\gamma)\in S^j$, then

(8)
$$\langle(\overline{b}-b^*)g\rangle_{(s),[\gamma]} \leqslant d_s(b)\langle g\rangle_{(s+j-1),[\gamma]},$$

where b^* denotes the adjoint operator of b relative to the form (5′).

4) If $b(x,\eta',\gamma)\in S^0$ and the symbol b is real and nonnegative, then

(9)
$$\operatorname{Re}\langle b(x,D',\gamma)g,g\rangle_{[\gamma]} > -c(b)\langle g\rangle^2_{(-1/2),[\gamma]}.$$

For the proof of assertions 1), 2), and 3) see, for instance, **[GV1,** Chapter 5]. Property 4) is an analog of the sharp form of Gårding's inequality for operators depending on a parameter. Its proof was outlined in **[Ag2]**. We note that the constants in (6)–(9) do not depend on γ. For our further aims refinement of the constant $c_s(b)$ in (6) will be needed. We have the following

LEMMA. *Let $b(y,\eta',\gamma)\in S^j$ and let*

$$|b(y,\eta',\gamma)| \leqslant B\Lambda(\eta',\gamma)^j, \qquad y\in\mathbb{R}^n,\quad \eta'\in\mathbb{R}^n,\quad \gamma\leqslant\gamma_0.$$

Then

(10)
$$\langle bg\rangle_{(s),[\gamma]} \leqslant (B^2+c|\gamma|^{-1})^{1/2}\langle g\rangle_{(s+j),[\gamma]}.$$

PROOF. 1) First we establish (10) for the case $s=j=0$. According to (4), we have

(11)
$$\begin{aligned}\langle bg\rangle^2_{[\gamma]} &=\langle bg,bg\rangle_{[\gamma]} = \langle b^*\cdot bg,g\rangle_{[\gamma]}\\ &=B^2\langle g\rangle^2_{[\gamma]} + \operatorname{Re}\langle (c-B^2)g,g\rangle_{[\gamma]} + \operatorname{Re}\langle (b^*\cdot b-c),g\rangle_{[\gamma]},\end{aligned}$$

where $c(y,\eta',\gamma)=|b(y,\eta',\gamma)|^2$. By property 4), the second term on the right-hand side has an upper bound

$$\operatorname{const}\langle g\rangle^2_{(-1/2),[\gamma]} \leqslant \operatorname{const}|\gamma|^{-1}\langle g\rangle^2_{[\gamma]}.$$

Note that

$$b^*\cdot b-c = (\overline{b}\cdot b-c)+(b^*-\overline{b})\cdot b.$$

Using properties 2) and 3) we find an upper bound for the second term on the right-hand side of (11):

$$\begin{aligned}(\langle(\overline{b}\cdot b-c)g\rangle_{[\gamma]}+\langle(b^*-\overline{b})\cdot bg\rangle_{[\gamma]})\langle g\rangle_{[\gamma]}\\ \leqslant \text{const}\langle g\rangle_{(-1),[\gamma]}\langle g\rangle_{[\gamma]}\leqslant \text{const}\,|\gamma|^{-1}\langle g\rangle^2_{[\gamma]}.\end{aligned}$$

2) In the general case we set $g={}_{[\gamma]}\Lambda^{-s-j}h$. Then the left-hand side of (10) takes the form

$$\langle bg\rangle_{(s),[\gamma]}=\langle\Lambda^s b\Lambda^{-s-j}h\rangle_{[\gamma]}\leqslant\langle b\Lambda^{-j}h\rangle_{[\gamma]}+\langle(\Lambda^s b\Lambda^{-s-j}-b\Lambda^{-j})h\rangle_{[\gamma]}.$$

By what was proved, the first term on the right-hand side does not exceed

$$(B^2+\text{const}\,|\gamma|^{-1})^{1/2}\langle h\rangle_{[\gamma]}=(B^2+\text{const}\,|\gamma|^{-1})\langle g\rangle_{(s+j),[\gamma]}.$$

According to property 2), the second term has an upper bound

$$\text{const}\,\langle h\rangle_{(-1),[\gamma]}\leqslant\text{const}\,|\gamma|^{-1}\langle g\rangle_{(s+j),[\gamma]}.$$

The lemma is proved.

4.2. We denote by $S[\xi_1]^{m,r}$ the class of symbols of the type

$$L(y,\eta,\gamma)=\sum_{j=0}^{m}b_j(y,\eta',\gamma)\xi_1^{m-j},\qquad b_j(y,\eta',\gamma)\in S^{j+r}. \tag{12}$$

For $r=0$ we will write $S[\xi_1]^m$. As in 4.4.3 of Chapter II, we define the set of norms

$$[L]_k=\max_{0\leqslant j\leqslant m,|\alpha|\leqslant k}\ \sup_{y,\eta',\gamma}|D_y^\alpha,b_j(y,\eta',\gamma)|\Lambda^{-j-r}(\eta',\gamma).$$

With every $L\in S[\xi_1]^{m,r}$ a continuous operator

$$H^{l+m,s+r}_{[\gamma]}(\mathbb{R}^{n+1}_+)\longrightarrow H^{l,s}(\mathbb{R}^{n+1}_+)\quad(u\longmapsto L(x,D,\gamma)u)$$

can naturally be associated. Lemma 4.1 implies an estimate for the norm of this operator.

LEMMA. *If $L(x,\eta',\gamma)\in S[\xi_1]^{m,r}$, then*

$$\|L(y,D,\gamma)u\|_{l,s,[\gamma]}\leqslant(\kappa[L]_0+c|\gamma|^{-1/2})\|u\|_{l+m,s+r,[\gamma]}, \tag{13}$$

where κ does not depend on L and γ, and c depends on the finite set of norms $[L]_k$ and does not depend on γ.

We will consider quadratic forms of functions on $\mathbb{R}^{n+1}_+$, for which the notation $[\cdot,\cdot]$ will be retained:

$$[u,v]=\int_{\mathbb{R}^{n+1}_+}u(y)\overline{v(y)}\,dy.$$

Accordingly, we set

$$\begin{aligned}[u,v]_{[\gamma]}&=[\exp(\gamma t)u,\exp(\gamma t)v],\\ [u,v]_{0,s,[\gamma]}&=[{}_{[\gamma]}\Lambda^s(D',\gamma)u,{}_{[\gamma]}\Lambda^s(D',\gamma)v]_{[\gamma]},\\ [u,v]_{l,s,[\gamma]}&=\sum_{k=0}^{l}[D_1^k u,D_1^k v]_{0,s,[\gamma]}.\end{aligned}$$

Let $S_0[\xi_1]^m$ denote the set of those symbols $L \in S[\xi_1]^m$ whose coefficients at ξ_1^m are real constants.

PROPOSITION 1. *If* $L \in S_0[\xi_1]^k$ *and* $M \in S_0[\xi_1]^{k-1}$, *then*

$$
\begin{aligned}
&\left|\operatorname{Im}[L(y,D,\gamma)u, M(y,D,\gamma)v]_{0,s,[\gamma]}\right| \leqslant (\kappa[L]_0[M]_0 + c|\gamma|^{-1/2}) \\
&\qquad \times \left(\|u\|^2_{k-1,s+1/2,[\gamma]} + \sum_{j=0}^{k-1} \langle D_1^j u(0,\cdot)\rangle^2_{k-1-j+s,[\gamma]} \right).
\end{aligned} \tag{14}
$$

PROOF. According to the definition of $S_0[\xi_1]^m$, we can write

$$L = a\xi_1^k + L', \quad M = b\xi_1^{k-1} + M', \quad L' \in S[\xi_1]^{k-1,1}, \quad M' \in S[\xi_1]^{k-2,1},$$

where a and b are real numbers. It follows that

$$
\begin{aligned}
\operatorname{Im}[Lu, Mu]_{0,s,[\gamma]} &= ab \operatorname{Im}[D_1^k u, D_1^{k-1} u]_{0,s,[\gamma]} \\
&\quad + \operatorname{Im}[L'u, Mu]_{0,s,[\gamma]} + a \operatorname{Im}[D_1^k u, M'u]_{0,s,[\gamma]} = J_1 + J_2 + J_3.
\end{aligned}
$$

The integration by parts of the form J_1 results in

$$J_1 \leqslant |ab| \langle D_1^{k-1} u\rangle^2_{(s),[\gamma]},$$

where the right-hand side can be estimated in terms of the right-hand side of (14). Furthermore, by the Schwarz inequality, we have

$$|J_2| \leqslant \|L'u\|_{0,s-1/2,[\gamma]} \, \|Mu\|_{0,s+1/2,[\gamma]}.$$

According to the lemma, the first and second factors on the right-hand side are estimated in terms of

$$(\kappa_1[L]_0 + c_1|\gamma|^{-1/2})\|u\|_{k-1,s+1/2,[\gamma]}$$

and

$$(\kappa_2[M_0] + c_2|\gamma|^{-1/2})\|u\|_{k-2,s+3/2,[\gamma]},$$

respectively. Noting that

$$\|u\|_{k-2,s+3/2,[\gamma]} \leqslant \|u\|_{k-1,s+1/2,[\gamma]}$$

we can estimate $|J_2|$ in terms of the right-hand side of (14). To estimate J_3 we use the identity

$$
\begin{aligned}
&\operatorname{Im}(D_1^k u \overline{M'u}) = -\frac{1}{2}\frac{\partial}{\partial x_1} \operatorname{Re}(D_1^{k-1} \overline{M'u}) + \frac{1}{2} \operatorname{Re} D_1^{k-1} u \overline{M''u} \\
&\left(M''(x,\eta',\gamma) = \frac{\partial}{\partial x_1} M'(x,\eta,\gamma) \right).
\end{aligned}
$$

Performing the integration we find

$$J_3 = \frac{a}{2} \operatorname{Re}\langle D_1^{k-1} u, M'u\rangle_{(s),[\gamma]} + \frac{a}{2} \operatorname{Re}[D_1^{k-1} u, M''u]_{0,s,[\gamma]}.$$

Since $M' \in S[\xi_1]^{k-2,1}$, we can apply Proposition 4.1 to estimate the first term on the right-hand side via the right-hand side of (14). As to the second term, it has an upper bound

$$\begin{aligned}\text{const}\,\|D_1^{k-1}u\|_{0,s,[\gamma]}\|u\|_{k-2,s+1,[\gamma]} &\leqslant \text{const}\,\|u\|^2_{k-1,s+1,[\gamma]}\\ &\leqslant \text{const}\,|\gamma|^{-1}\|u\|^2_{k-1,s+1/2,[\gamma]}.\end{aligned}$$

The proposition is proved.

PROPOSITION 2. *If* $L \in S[\xi_1]^{k-\theta}$, $M \in S[\xi_1]^{k-2+\theta}$, *and* $\theta = 0, 1$, *then*

$$|[Lu, Mu]_{0,s,[\gamma]}| \leqslant c\left(\|u\|^2_{k-1,s,[\gamma]} + \langle D_1^{k-1}u\rangle_{(s),[\gamma]}\sum_{j=0}^{k-2}\langle D^j u\rangle_{(s+k-2-j),[\gamma]}\right). \tag{15}$$

PROOF. If $\theta = 1$, then the left-hand side of (15) can be estimated trivially via the first term on the right-hand side. In the case $\theta = 0$ one should carry out integration by parts following Proposition 1.

4.3. We now single out a subclass of symbols belonging to $S[\xi_1]^k$ such that in application to the operators corresponding to them Proposition 1 in 4.2 admits an important improvement, namely the index $s + 1/2$ in the first norm on the right-hand side can be replaced by s.

Let O^j denote the set of symbols $b(y, \eta', \gamma) \in S^j$ possessing the following properties.

(i) $b(y, \eta', \gamma)$ is a homogeneous function of degree j in the variables η' and γ;

(ii) $\operatorname{Im} b(y, \eta', 0) = 0$.

Property (ii) implies that if $b \in O^j$, then $\operatorname{Re} b \in O^j$ and $\gamma^{-1}\operatorname{Im} b \in O^{j-1}$, i. e.,

$$\begin{gathered} b(y, \eta', \gamma) = b^0(y, \eta', \gamma) + ib^1(y, \eta, \gamma),\\ b^0 \in O^j, \quad b^1 \in O^{j-1}, \quad \operatorname{Im} b^0 = \operatorname{Im} b^1 \equiv 0. \end{gathered} \tag{16}$$

Now we denote by $O[\xi_1]^k$ the set of symbols of the type

$$L(x, \eta, \gamma) = b_0\xi_1^k + \sum_{j=1}^{k} b_j(x, \eta', \gamma)\xi_1^{m-j}, \quad b_j \in O^j, \tag{17}$$

where b_0 is a real constant. Note that

$$O[\xi_1]^k \subset S_0[\xi_1]^k.$$

A refined version of Proposition 1 of 4.2 is the following

PROPOSITION. *Let* $L \in O[\xi_1]^k$ *and* $M \in O[\xi_1]^{k-1}$. *Then*

$$\begin{aligned} &\left|\operatorname{Im}\,[L(y, D, \gamma)u, M(y, D, \gamma)u]_{s,[\gamma]}\right|\\ &\qquad\leqslant (\kappa[L]_0[M]_0 + c\gamma^{-1/2})\left(|\gamma|\,\|u\|^2_{k-1,s,[\gamma]} + \sum_{j=0}^{k-1}\langle D_1^j u\rangle_{(s+k-1-j),[\gamma]}\right), \end{aligned} \tag{18}$$

where κ does not depend on the symbols L and M, and c depends on the finite set of norms $[L]_k$, $[M]_k$, where $|s| < S$ and $k \leqslant k_0(S)$.

The proof of (18) is based on an important version of the Green formula for operators with symbols of class $O[\xi_1]^k$.

4.4. PROPOSITION. *Let $L \in O[\xi_1]^m$ and $M \in O[\xi_1]^{m-1}$. Then*

$$\begin{aligned}(19)\qquad &[L(y,D,\gamma)u, M(y,D,\gamma)u]_{s,[\gamma]} - [M(y,D,\gamma)u, L(y,D,\gamma)u]_{s,[\gamma]}\\ &\qquad = 2i\gamma G_{s\gamma}[u,u] + iH_{s\gamma}\langle u,u\rangle + R_{s\gamma}[u,u],\end{aligned}$$

where $G_{s\gamma}$ and $H_{s\gamma}$ are quadratic forms of the type

$$(20)\qquad G_{s,\gamma}[u,u] = \sum_{j,k=0}^{m-1} [g_{jk}(y,D',\gamma)D_1^{m-1-j}u, D_1^{m-k-1}u]_{s,[\gamma]},$$

$$(21)\qquad H_{s,\gamma}\langle u,u\rangle = \sum_{j,k=0}^{m-1} \langle h_{jk}(y',D',\gamma)D_1^{m-1-j}u, D_1^{m-k-1}u\rangle_{s,[\gamma]},$$

in which the symbols $g_{jk}(y,\eta',\gamma)$ belong to S^{j+k} and the symbols $h_{jk}(y',\eta',\gamma)$ are the values assumed by the symbols $h_{jk}(y,\eta',\gamma) \in S^{j+k}$ for $x_1 = 0$. The symbols g_{jk} and h_{jk} are Hermitian quadratic forms with constant coefficients with respect to the coefficients (in different powers of ξ_1^j) of the symbols L and M. As to the form $R_{s\gamma}$, it satisfies the inequality

$$(22)\qquad |R_{s,\gamma}[u,u]| \leqslant \text{const}\, \|u\|^2_{m-1,s[\gamma]}.$$

PROOF. We write

$$L = \sum_{j=0}^{n} b_j \xi_1^{m-j}, \quad M = \sum_{k=0}^{n} d_k \xi_1^{m-1-k}$$

and represent the left-hand side of (19) in the form

$$(23)\qquad [[Lu, Mu]]_{s,[\gamma]} - [Mu, Lu]_{s,[\gamma]} = \sum_{j=0}^{m}\sum_{k=0}^{m-1} J_{jk},$$

where

$$(24)\qquad J_{jk} = [b_j D_1^{m-j}u, d_k D_1^{m-k-1}u]_{s,[\gamma]} - [d_k D_1^{m-k-1}u, b_j D_1^{m-j}u]_{s,[\gamma]}.$$

We first assume that $j > 0$ and (for definiteness) $j \leqslant k$. Integrating by parts in the second form in (24) we transpose $k - j + 1$ derivatives from the left-hand side to the right-hand side, which results in

$$J_{jk} = [b_j D_1^{m-j}u, d_k D_1^{m-k-1}u]_{s,[\gamma]} - [d_k D_1^{m-j}u, b_j D_1^{m-k-1}u]_{s,[\gamma]} + \cdots = I_{jk} + \cdots,$$

where the dots symbolize the forms entering $H_{s\gamma}$ and $R_{s\gamma}$. The form I_{jk} can be rewritten as

$$I_{jk} = \left[(d_k^* b_j - b_j^* d_k)D_1^{m-j}u, D_1^{m-k-1}u\right]_{s,[\gamma]}.$$

Furthermore, we have

$$d_k^* b_j - b_j^* d_k = r_{jk}(y,D',\gamma) + r'_{jk},$$

where

$$r_{jk}(y,\eta',\gamma) = b_j(y,\eta',\gamma)\overline{d_k(y,\eta',\gamma)} - d_k(y,\eta',\gamma)\overline{b_j(y,\eta',\gamma)}. \tag{25}$$

The properties of pseudodifferential operators with symbols of class S^j (see 4.1) imply that the operator r'_{jk} can be represented in the form

$$r'_{jk} = r^0_{jk}\Lambda(D',\gamma)^{j+k-1},$$

by virtue of which the form

$$[r'_{jk}D_1^{m-j}u, D_1^{m-k-1}u]_{s,[\gamma]} = \left[(\Lambda^{-k+1}r'_{jk}\Lambda^{-j})\Lambda^j D_1^{m-j}u, \Lambda^{k-1}D_1^{m-k-1}u\right]_{s,[\gamma]}$$

can be included in $R_{s,\gamma}[u,u]$.

Replacing the operators b_j and d_k in (25) by their expressions of the type (16), i.e., setting $b_j = b_j^0 + i\gamma b_j^1$ and $d_j = d_j^0 + i\gamma d_j^1$, we obtain

$$r_{jk} = 2i\gamma(b_j^1 d_k^0 - b_j^0 d_k^1) \overset{\text{def}}{=} 2i\gamma r_{1jk}, \quad r_{1jk} \in S^{j+k-1}.$$

Now we note that

$$\begin{aligned}\gamma[r_{1jk}D_1^{m-j}u, D_1^{m-1-k}u]_{s,[\gamma]} =& \gamma[r_{1jk}\Lambda^{-k}D_1^{m-j}u, \Lambda^k D_1^{m-1-k}u]_{s,[\gamma]} \\ &+ \gamma[(r_{1jk} - \Lambda^k r_{1jk}\Lambda^{-k})D_1^{m-j}u, D_1^{m-1-k}u]_{s,[\gamma]}.\end{aligned}$$

Here the first form is one of the terms in the form $G_{s\gamma}$, and the second form can be immediately included in $H_{s\gamma}$.

Finally, consider the case $j = 0$. If $k = 0$, then

$$\begin{aligned}J_{00} =& - b_0 d_0 \left([D_1^m u, D_1^{m-1}u]_{s,[\gamma]} - [D_1^m u, D_1^{m-1}u]_{s,[\gamma]}\right) \\ =& 2ib_0 d_0 \langle D_1^{m-1}u, D_1^{m-1}u\rangle_{s,[\gamma]}.\end{aligned}$$

In the case $k > 0$ one should perform one-fold integration by parts. The proposition is proved.

4.5. In connection with Proposition 4.4, we make several important remarks.

1) If the symbols L and M do not depend on y, then the form $R_{s\gamma}[u,u]$ is identically zero. In this case we can pass to the partial Fourier transform in (19), which gives

(26)

$$\begin{aligned}- \operatorname{Im} \int_0^\infty \int L(D_1,\eta',\gamma)\widehat{u}'(x_1,\xi',\tau)\overline{M(D_1,\eta',\gamma)\widehat{u}'(x_1,\xi',\tau)}\,d\eta' dx_1 \\ = |\gamma| \operatorname{Re} \sum \int_0^\infty \int g_{jk}(\eta',\gamma) D_1^{m-1-j}\widehat{u}'(x_1,\xi',\tau)\overline{D_1^{m-1-k}\widehat{u}'(x_1,\xi',\tau)}\,d\eta' dx_1 \\ - \frac{1}{2}\sum \operatorname{Re} \int h_{jk}(\eta',\gamma)D_1^{m-1-j}\widehat{u}'(0,\xi',\tau)\overline{D_1^{m-1-k}\widehat{u}'(0,\xi',\tau)}\,d\eta'.\end{aligned}$$

2) If the support of $u(y)$ lies in the half-space $x_1 > \varepsilon > 0$, then the second term on the right-hand side should be dropped. Passing to the Fourier transform with respect

to x_1 we obtain for $s = 0$ the relation

$$\tag{27} \begin{aligned} \iint -\operatorname{Im}\left(L(\xi_1,\eta',\gamma)\overline{M(\xi_1,\eta',\gamma)}\right)|\widehat{u}(\xi,\tau)|^2\,d\xi\,d\tau \\ = |\gamma|\iint \operatorname{Re}\sum g_{jk}(\eta',\gamma)\xi_1^{2m-2-j-k}|\widehat{u}(\xi,\tau)|^2\,d\xi\,d\tau. \end{aligned}$$

It follows from (27) that

$$\tag{28} |\gamma|\operatorname{Re}\sum g_{jk}\xi_1^{2m-2-j-k} = -\operatorname{Im} L(\xi_1,\eta',\gamma)\overline{M(\xi_1,\eta',\gamma)}.$$

3) Applying the arguments in the proposition to the ordinary differential operators $L(D_1,\eta',\gamma)$ and $M(D_1,\eta',\gamma)$ we conclude that the expressions in the integrals with respect to η' in (26) are equal. In other words,

$$\tag{29} \begin{aligned} -\operatorname{Im}\int_0^\infty L(D_1,\eta',\gamma)V(x_1)\overline{M(D_1,\eta',\gamma)V(x_1)}\,dx_1 \\ =|\gamma|\operatorname{Re}\sum_{j,k}\int_0^\infty g_{jk}(\eta',\gamma)D_1^{m-1-j}V(x_1)\overline{D_1^{m-1-k}V(x_1)}\,dx_1 \\ -\frac{1}{2}\sum_{j,k}h_{jk}(\eta',\gamma')D_1^{m-1-j}V(0)\overline{D_1^{m-1-k}V(0)}. \end{aligned}$$

4) With the quadratic forms in (29) we will associate expressions in which the operators of differentiation acting on the complex conjugate function $\overline{V(x)}$ will be denoted as $\overline{D_1}$. Then (29) can be represented in the form

$$\tag{30} \begin{aligned} -\operatorname{Im}\int_0^\infty L(D_1,\eta',\gamma)M(\overline{D_1},\eta',\gamma)V(x_1)\overline{V(x_1)}\,dx_1 \\ =|\gamma|\operatorname{Re}\sum_{j,k}\int_0^\infty g_{jk}(\eta',\gamma)D_1^{m-1-j}\overline{D_1^{m-1-k}}V(x_1)\overline{V(x_1)}\,dx_1 \\ -\frac{1}{2}\sum_{j,k}h_{jk}(\eta',\gamma')D_1^{m-1-j}\overline{D_1^{m-1-k}}V(x_1)\overline{V}(x_1)\big|_{x_1=0}. \end{aligned}$$

If we set

$$\tag{31} \begin{aligned} G(\eta',\gamma,\Xi_1,\overline{\Xi_1}) = \sum g_{jk}(\eta',\gamma)\Xi_1^{m-j-1}\overline{\Xi}_1^{m-1-k}, \\ H(\eta',\gamma,\Xi_1,\overline{\Xi_1}) = \sum h_{jk}(\eta',\gamma)\Xi_1^{m-j-1}\overline{\Xi}_1^{m-1-k}, \end{aligned}$$

then (30) can be rewritten as

$$
\begin{aligned}
-\operatorname{Im}\int_0^\infty L(D_1,\eta',\gamma)M(\overline{D_1},\eta',\gamma)V(x_1)\overline{V(x_1)}\,dx_1 \\
=|\gamma|\operatorname{Re}\sum_{j,k}\int_0^\infty G(\eta',\gamma,D_1,\overline{D}_1)V(x_1)\overline{V(x_1)}\,dx_1 \\
-\frac{1}{2}H(\eta',\gamma',D_1,\overline{D_1})V(x_1)\overline{V(x_1)}\big|_{x_1=0}.
\end{aligned}
$$

§5. Basic estimate. Preliminary results

We showed in §4 that the proof of the main assertion of Theorem 2.2 (the implication (I)⇒(II)) is a consequence of inequalities (2.9) and (3.20). The present section is devoted to the proof of (2.9).

5.1. Statement of the energy estimate. Localization.

THEOREM. *Let conditions* (A), (B), (C), *and* (D) *of* §1 *hold. Then for every* $S > 0$ *there is* $\gamma_0 = \gamma_0(S)$ *such that for* $|s| \leqslant S$ *and* $\gamma \leqslant \gamma_0$ *the inequality*

$$
\begin{aligned}
&|\gamma|\,\|u\|^2_{m-1,s,[\gamma]} + \sum_{j=0}^{m-1}\langle D_1^j u\rangle^2_{(m-1+s-j),[\gamma]} \\
&\qquad\leqslant C\Big(|\gamma|^{-1}\|P(y,D)u\|^2_{0,s,[\gamma]} + \sum_{j=1}^{\mu}\langle B_j(y',D)u\rangle^2_{(m-1+s-\beta_j),[\gamma]}\Big) \qquad (1)\\
&\forall u \in H^{m,s}_{[\gamma]}(\mathbb{R}^{n+1}_+)
\end{aligned}
$$

holds.

Inequality (2.9) is a consequence of (1) and the condition that $\{x_1 = 0\}$ is not a characteristic plane. Indeed, according to the definition of the norm (2.3), we have

$$
|\gamma|\,\|u\|^2_{m,s,[\gamma]} = |\gamma|\,\|D_1^m u\|^2_{0,s,[\gamma]} + |\gamma|\,\|u\|^2_{m-1,s+1,[\gamma]} \tag{2}
$$

so that to derive (2.9) from (1) we must estimate the first term on the right-hand side via the right-hand side of (1) (where s is replaced by $s+1$). Since the boundary is not characteristic, the operator P can be written in the form (1.4), whence

$$
|\gamma|\,\|D_1^m u\|^2_{0,s,[\gamma]} \leqslant 2|\gamma|\|P(y,D)u\|^2_{0,s,[\gamma]} + \text{const}\,\|u\|^2_{m-1,s+1,[\gamma]}.
$$

Noting that

$$
|\gamma|\,\|Pu\|^2_{0,s,[\gamma]} \leqslant |\gamma|^{-1}\|Pu\|^2_{0,s+1,[\gamma]}
$$

we arrive at (2.9).

The norms entering inequality (1) contain the large parameter $|\gamma|$. Threfore, the following obvious assertion holds.

LEMMA. *Let inequality* (1) *hold for a given set of operators* $\{P, B_1, \dots, B_\mu\}$. *Then this inequality holds (possibly for larger values of* $|\gamma|$*) for the set*

$$\{P + P', B_1 + B_1' + B_1'', \dots, B_\mu + B_\mu' + B_\mu''\},$$

where P', B_j', *and* B_j'' *are operators with smooth coefficients such that* $\deg P' < \deg P$, $\deg B_j' < \deg B_j = \beta_j$, *and* $\deg B_j'' \leqslant \beta_j$, $j = 1, \dots, \mu$, *and the moduli of the coefficients of the operators* B_j'' *are sufficiently small.*

PROPOSITION. *Let there exist a finite covering*

$$\bigcup_\lambda U_\lambda \supset \overline{\mathbb{R}_+^{n+1}}$$

such that (1) *holds for those* $u \in H_{[\gamma]}^{m,s}(\mathbb{R}_+^{n+1})$ *whose supports belong to* U_λ. *Then* (1) *holds (possibly for larger values of* γ*) for all* $u \in H_{[\gamma]}^{m,s}(\mathbb{R}_+^{n+1})$.

PROOF (cf. 4.4.2 in Chapter II). Take a partition of unity subordinate to the covering, i. e., $\psi_j(y) \in \mathcal{D}(U_j)$, $\psi_j \geqslant 0$, and

$$\sum \psi_j^2(y) \equiv 1.$$

We write (1) for the functions $\psi_j(y)u(y)$ and then add together the resulting inequalities for all indices j. The right-hand side of the sum of these inequalities is equal to

$$\begin{aligned}
C|\gamma|^{-1} \sum_{j=1}^{J} \|P(\psi_j u)\|_{0,s,[\gamma]}^2 &+ C \sum_{j=1}^{J} \sum_{k=1}^{\mu} \langle B_k(\psi_j u)\rangle_{(m-1+s-\beta_k),[\gamma]}^2 \\
&= C|\gamma|^{-1} \sum_j \Big\| \psi_j P u + \sum D^\alpha \psi_j P^{(\alpha)} u/\alpha! \Big\|_{0,s,[\gamma]}^2 \\
&\quad + C \sum_j \sum_k \Big\langle \psi_j B_j u + \sum D^\alpha \psi_j B_j^{(\alpha)} u/\alpha! \Big\rangle_{(m-1+s-\beta_k),[\gamma]}.
\end{aligned}$$

Using the results of §4 (see 4.1 and 4.2) we find an upper bound for the above expression:

$$\begin{aligned}
C_1|\gamma|^{-1}\|Pu\|_{0,s,[\gamma]}^2 + C_2 \sum \langle B_k u\rangle_{(m-2+s-\beta_k),[\gamma]}^2 &+ C_3|\gamma|^{-1}\|u\|_{m-1,s,[\gamma]}^2 \\
&+ C_4 \sum_{k=0}^{m-2} \langle D_1^k u\rangle_{(m-2+s-\beta_k),[\gamma]}.
\end{aligned}$$

The last term is estimated by

$$C_4|\gamma|^{-1} \sum_{k=0}^{m-1} \langle D_1^k u\rangle_{(m-1+s-\beta_k),[\gamma]}.$$

The desired assertion will be proved if we show that the left-hand side

$$\sum_{j=1}^{J} \left(|\gamma|\, \|\psi_j u\|_{m-1,s,[\gamma]}^2 + \sum_{k=0}^{m-1} \langle D_1^k(\psi_j u)\rangle_{(m-1+s-k),[\gamma]}^2 \right) \tag{3}$$

is not less than

$$C_5\Big(|\gamma|\,\|u\|^2_{m-1,s,[\gamma]}+\sum\langle D_1^k u\rangle^2_{(m-1+s-k),[\gamma]}\Big)-C_6\|u\|^2_{m-1,s,[\gamma]}$$
$$-C_7\sum\langle D_1^k u\rangle^2_{(m-2+s-k),[\gamma]}.$$

To derive the required inequality it suffices in fact to repeat the derivation of the above estimates. For instance, let us estimate the first term in (3). We have

$$\begin{aligned}2|\gamma|\sum\|\psi_j u\|^2_{m-1,s,[\gamma]} &=2|\gamma|\sum_j\sum_{k=0}^{m-1}\|D_1^k(\psi_j u)\|^2_{0,m-1+s-k,[\gamma]}\\ &\geqslant|\gamma|\sum_j\sum_{k=0}^{m-1}\|\psi_j D_1^k u\|^2_{0,m-1+s-k,[\gamma]}\\ &\quad-4|\gamma|\sum_j\sum_{k=0}^{m-1}\sum_{l=1}^{k}\|D_1^l\psi_j D^{k-l}u)\|^2_{0,m-1+s-k,[\gamma]}/l!.\end{aligned}$$

The second term on the right-hand side has a lower bound

$$-C_8|\gamma|\,\|u\|^2_{m-2,s,[\gamma]}\geqslant-C_8\|u\|^2_{m-2,s+1,[\gamma]}\geqslant-C_8\|u\|^2_{m-1,s,[\gamma]}.$$

As to the first term, it is equal to

$$\begin{aligned}|\gamma|\sum_j\sum_k\|\Lambda^{m-1+s-k}(\psi_j D_1^k u)\|^2_{[\gamma]} &\geqslant\frac{1}{2}|\gamma|\sum_j\sum_k\|\psi_j\Lambda^{m-1+s-k}D_1^k u\|^2_{[\gamma]}\\ &\quad-2|\gamma|\sum_j\sum_k\|[\Lambda^{m-1+s-k},\psi_j]D_1^k u\|^2_{[\gamma]}.\end{aligned}$$

Since we have $\sum\psi_j^2\equiv 1$, the first term on the right-hand side is equal to

$$\frac{1}{2}|\gamma|\,\|u\|^2_{m-1,s,[\gamma]}.$$

Using the results of §4 we can show that the second term on the right-hand side has a lower bound

$$-C_9\|u\|^2_{m-1,s,[\gamma]}.$$

What has been proved permits us to make some important remarks that will be used below.

1) When proving inequality (1) we can assume, without loss of generality, that the support of $u(y)$ lies in a sufficiently small neighborhood of a point $y^0\in\mathbb{R}^{n+1}_+$. If y^0 does not belong to the boundary plane $\{x_1=0\}$, then (1) goes into the estimate for the Cauchy problem that was proved in §4 of Chapter II. Therefore, below we will assume that the support of $u(y)$ belongs to a neighborhood of the point $y^0=(0,y'^0)$.

2) If the support belongs to a small neighborhood of y^0, then the coefficients of the operators P and B_j can be varied arbitrarily outside this neighborhood (cf. 4.4.3 of Chapter II). Thus, in what follows we will assume that the operators in question have the form

(4) $$P(y,\eta)=P(\eta)+P(y',\eta),\quad B_j(y,\eta)=B_j(\eta)+B_j(y',\eta),\quad j=1,\dots,\mu,$$

where the norms $[P']_0$, $[B_1']_0, \dots, [B_\mu']_0$ are assumed to be arbitrarily small.

3) In view of the lemma, we can assume, without loss of generality, that the operators P, B_1, and B_μ contain no lower terms.

4) By virtue of the lemma, we can assume, without loss of generality, that the boundary operators have constant coefficients.

5.2. Microlocalization of inequality (1). Denote by $\Sigma^\pm$ a hemisphere in the space of the variables (η', γ), i. e.,

$$\Sigma^\pm = \{(\eta', \gamma) \in \mathbb{R}^{n+1},\ |\eta'|^2 + \gamma^2 = 1,\ \pm\gamma \leqslant 0\}.$$

If U is a subregion of $\Sigma^\pm$, then we denote by KU the corresponding cone

$$KU = \{\eta' = \rho\overline{\eta}',\ \gamma = \rho\overline{\gamma},\ (\overline{\eta}', \overline{\gamma}) \in U,\ \rho > 0\}.$$

For brevity, we will say that an inequality holds for a function $u(x_1, x', t)$ in U if it holds for each function u such that the support of its partial Fourier transform $\widehat{u}'(x_1, \xi', \sigma + i\gamma)$ belongs to KU for all x_1.

PROPOSITION. *Let there exist a finite covering $\bigcup_j U_j \supset \Sigma^+$ such that inequality (1) holds in every neighborhood U_j. Then (1) holds (possibly for larger values of $-\gamma$) for all $u \in H^{m,s}_{[\gamma]}(\mathbb{R}^{n+1}_+)$.*

PROOF. We take a partition of unity on the sphere Σ^+ subordinate to the covering $\bigcup U_\lambda$, i. e., nonnegative functions $\mu_j(\eta', \gamma) \in C^\infty$ such that $\operatorname{supp} \mu_j \in U_j$ and

$$\sum \mu_j^2(\eta', \gamma) \equiv 1. \tag{5}$$

The function $\mu_j(\eta', \gamma)$ can be extended to all values of $\eta' \in \mathbb{R}^n$ and $\gamma \leqslant 0$ as a homogeneous function of degree 0, i. e.,

$$\mu_j(\eta', \gamma) \overset{\text{def}}{=} \mu_j(\eta'\Lambda^{-1}(\eta', \gamma), \gamma\Lambda^{-1}(\eta', \gamma)),$$

where

$$\Lambda(\eta', \gamma) = (|\eta'|^2 + \gamma^2)^{1/2}.$$

According to the conditions of the proposition, if $u(y)$ is an arbitrary function belonging to $H^{m,s}_{[\gamma]}(\mathbb{R}^{n+1}_+)$, then $\mu_j(D', \gamma)u$ satisfies (1). Performing the summation of the resulting inequalities with respect to j and denoting by $|||u|||_{s,[\gamma]}$ the left-hand side of (1) we derive (by virtue of (5)) the inequality

$$\begin{aligned}\sum_j |||\mu_j(D'\gamma)u|||^2_{s,[\gamma]} = |||u|||^2_{s,[\gamma]} &\leqslant C|\gamma|^{-1} \sum \|P(y, D)(\mu_j u)\|^2_{0,s,[\gamma]} \\ &\quad + C \sum \langle B_j u\rangle^2_{m-1+s-\beta_j,[\gamma]}\end{aligned} \tag{6}$$

(recall that, by the hypothesis, the operators B_j have constant coefficients). Since the operator P is resolved with respect to the highest derivative with respect to x_1 (cf. (1.1)), we have

$$P(\mu_j u) = \mu_j P u + \sum_{k=1}^{m} (P_k(y, D')\mu_j - \mu_j P_k(y, D')) D_1^{m-k} u,$$

whence, by virtue of the results of §4, it follows that

$$\begin{aligned}|\gamma|^{-1}\|P(\mu_j u)\|^2_{0,s,[\gamma]} \leqslant & 2|\gamma|^{-1}\|\mu_j Pu\|^2_{0,s,[\gamma]} \\ & + \operatorname{const}|\gamma|^{-1}\sum_{k=1}^{m}\|D_1^{m-k}\|^2_{0,s+k-1,[\gamma]} \\ \leqslant & 2|\gamma|^{-1}\|Pu\|^2_{0,s,[\gamma]} + C|\gamma|^{-1}\|u\|^2_{m-1,s,[\gamma]}.\end{aligned}$$

The summation of these estimates with respect to j and the substitution into (6) yields (1).

Let us fix a sufficiently small number $\kappa > 0$ and set

$$\Sigma^+_{\text{parab}} = \{(\eta',\gamma) \in \Sigma^+, \quad \gamma \leqslant -\kappa\}, \tag{7}$$

$$\Sigma^+_{\text{hyperb}} = \{(\eta',\gamma) \in \Sigma^+, \quad -\kappa \leqslant \gamma \leqslant 0\}. \tag{8}$$

The regions (7) and (8) are called the parabolic and hyperbolic zones, respectively.

5.3. Proof of inequality (1) in the parabolic zone. The definition of the parabolic zone (7) implies that

$$\{(\eta',\gamma) \in \Sigma^+_{\text{parab}}\} \Longrightarrow \{\gamma \leqslant -\kappa\Lambda(\eta',\gamma)\}. \tag{7'}$$

In turn, (7′) implies that

$$|\gamma|^j\|u\|^2_{m-1,s,[\gamma]} \approx \|u\|^2_{m-1,s+j/2,[\gamma]}$$

in this zone, where the equivalence constant does not depend on γ. Hence, in the parabolic zone (7) inequality (1) assumes the form

$$\begin{aligned}&\|u\|^2_{m-1,s+1/2,[\gamma]} + \sum_{j=0}^{m-1}\langle D_1^j u\rangle^2_{(m-1+s-j),[\gamma]} \\ &\qquad \leqslant C\left(\|Pu\|^2_{0,s-1/2} + \sum_{j=1}^{\mu}\langle B_j u\rangle^2_{(m-1+s-\beta_j),[\gamma]}\right), \\ &u \in H^{m,s}_{[\gamma]}(\mathbb{R}^{n+1}_+), \quad \operatorname{supp}\widehat{u}'(x_1,\xi',\tau) \subset K\Sigma^+_{\text{parab}}.\end{aligned} \tag{1p}$$

Following §§4 and 5 of Chapter II, we derive the desired estimate as a consequence of a more precise estimate. The following assertion holds.

THEOREM. *Let conditions* (A), (B), (C), *and* (D) *of* §1 *hold, let operators* P, $B_1,\dots,B_\mu$ *have the form* (4), *and let the norms* $[P]_0$ *and* $[B_j]_0$, $j = 1,\dots,\mu$, *not exceed a sufficiently small number* δ. *Then there is a symbol* $Q(y,\eta,\gamma) \in S_0[\xi_1]^{m-1}$:

$$Q(y,\eta) = Q(\eta) + Q'(y,\eta), \qquad [Q']_0 \leqslant c\delta, \tag{9}$$

such that the inequality

$$\begin{aligned}&\|u\|^2_{m-1,s+1/2,[\gamma]} + \sum_{j=1}^{m-1}\langle D_1^j u\rangle^2_{(m-1+s-j),[\gamma]} \\ &\qquad \leqslant C\left(\operatorname{Im}[Pu,Qu]_{0,s,[\gamma]} + \sum_{j=1}^{\mu}\langle B_j u\rangle^2_{(m-1+s-\beta_j),[\gamma]}\right)\end{aligned} \tag{10}$$

holds in the region (7).

COROLLARY. *Inequality* (1p) *holds.*

PROOF. By the Schwarz inequality, the first term on the right-hand side of (10) can be estimated via

$$\|Pu\|_{0,s-1/2,[\gamma]}\|Qu\|_{0,s+1/2,[\gamma]} \leqslant \varepsilon^{-1}\|Pu\|^2_{0,s-1/2,[\gamma]} + C_1\varepsilon\|u\|^2_{m-1,s+1/2,[\gamma]}.$$

Selecting ε from the condition $CC_1\varepsilon < 1/2$ we arrive at (1p).

REMARK. Note that, in view of Proposition 1 in 4.2, the right-hand side can be estimated in terms of the left-hand side.

PROOF. The theorem is proved in several stages.

1) *Choice of the symbol.* We note that the imaginary parts of the roots $\lambda_j^{\pm}(y',\eta',\gamma)$ of the symbol $P_0(y',\lambda,\xi',\sigma+i\gamma)$ do not vanish on the compact set Σ^+_{parab}, and therefore there is a constant $\kappa_1 > 0$ such that

$$\pm\operatorname{Im}\lambda_j^{\pm}(y',\eta',\gamma) > \kappa_1\Lambda(\eta',\gamma), \qquad j=1,\dots,m_{\pm}. \tag{11}$$

Hence, according to Proposition 1 in 1.3, the symbols $P_{\pm}(y',\xi_1,\eta',\gamma)$ appearing in the factorization of P (see 1.3) belong to $S_0[\xi_1]^{m_\pm}$. We now set

$$Q(y',\xi_1,\eta',\gamma) = P_+(y,\xi_1,\eta',\gamma)\partial P_-(y,\xi_1,\eta',\gamma)/\partial\xi_1. \tag{12}$$

It is easy to see that the symbol (12) belongs to $Q \in S_0[\xi_1]^{m-1}$ and is representable in the form (9).

2) *Reduction of inequality* (10) *to the case of constant symbols.* This follows immediately from Proposition 1 in 4.2. Indeed, replacing the operators P and Q by their expansions (4) and (9) we rewrite the first term on the right-hand side of (10) in the form

$$\begin{aligned}\operatorname{Im}[P(D)u, Q(D)u]_{0,s,[\gamma]} &+ \operatorname{Im}[P'(y,D)u, Q(D)u]_{0,s,[\gamma]}\\ &+ \operatorname{Im}[P'(y,D)u, Q'(y,D)u]_{0,s,[\gamma]}.\end{aligned}$$

By Proposition 1 of 4.2, the sum of the second and third terms has an upper bound

$$(C_1\delta + C_2(\delta)|\gamma|^{-1/2})\left(\|u\|^2_{m-1,s+1/2,[\gamma]} + \sum_{j=0}^{k-1}\langle D_1^j u\rangle^2_{m-1+s-j,[\gamma]}\right).$$

Thus, if we manage to prove (10) with some constant C for the operators $P(D)$ and $Q(D)$, then, taking δ such that $CC_1\delta < 1/4$ and γ such that $CC_2(\delta)|\gamma|^{-1/2} \leqslant 1/4$, we arrive at (10).

3) *Reduction to an estimate on the half-line.* We have the following

LEMMA. *Symbols $P(\eta)$ and $Q(\eta)$ not depending on y satisfy inequality* (10) *if and only if for any function $V(z)$ belonging to $L_2(\mathbb{R}_+)$ together with the derivatives up to order m the inequality*

$$C_0\Big(\sum_{j=0}^{m}\Lambda^{2m-1-2j}\int_0^\infty |D^jV(z)|^2\,dz+\sum_{j=0}^{m-1}\Lambda^{2m-2-2j}|(D^jV)(0)|^2\Big)$$

$$\leqslant \operatorname{Im}\int_0^\infty P(D,\eta',\gamma)V(z)\overline{Q(D,\eta',\gamma)V(z)}\,dz \tag{13}$$

$$+\sum_{j=0}^{m}\Lambda^{2(m-1-\beta_j)}|(B_j(D,\eta',\gamma)V)(0)|^2$$

holds.

PROOF. The necessity of (13) for the validity of (10) in the case of constant symbols P and Q is proved by repeating literally the arguments of Proposition 2 in 2.3. To prove the sufficiency one should replace $V(x_1)$ in (13) by $\widehat{u}'(x_1,\xi',\sigma+i\gamma)$, $u\in H^{m,\infty}_{[\gamma]}(\mathbb{R}^{n+1}_+)$ and perform integration with respect to ξ' and σ.

Hence, the proof of (10) reduces to proving (13) on the half-line for ordinary differential operators depending on the parameter Λ.

4) *Estimate on the half-line* (*an inequality for a quadratic form*). We set

$$P_l(y,\eta,\gamma)=P(y,\eta,\gamma)/(\xi_1-\lambda_l^-(y,\eta',\gamma)).$$

Then, by the definition of the symbol Q, we have

$$Q(y,\eta,\gamma)=\sum_{l=1}^{m_-}P_l(y,\eta,\gamma).$$

The substitution of the expression for Q into the quadratic form results in

$$\operatorname{Im}\int_0^\infty P(D,\eta',\gamma)V(z)\overline{Q(D,\eta',\gamma)V(z)}\,dz$$

$$=\sum_{l=1}^{m_-}\int_0^\infty \operatorname{Im}\big[(D-\lambda_l^-(\eta',\gamma))(P_l(\eta',\gamma)V(z))\overline{P_l(D,\eta',\gamma)V(z)}\big]\,dz \tag{14}$$

$$=\sum_{l=1}^{m_-}\int_0^\infty(-\operatorname{Im}\lambda_l^-(\eta',\gamma))|P_lV|^2\,dz+\frac{1}{2}|(P_lV)(0)|^2.$$

By virtue of condition (7′), we have

$$-\operatorname{Im}\lambda_l^-(\eta',\gamma)\geqslant\omega_0\Lambda(\eta',\gamma)$$

in the parabolic zone, whence

$$\operatorname{Im}\int_0^\infty PV\overline{QV}\,dz\geqslant\omega_0\Lambda\sum_{l=1}^{m_-}\int_0^\infty|P_lV|^2\,dz+\frac{1}{2}|(P_lV)(0)|^2. \tag{15}$$

5) *Auxiliary assertions.* The derivation of (13) from (15) will be based on two assertions of a more general character, and we state them as the propositions below.

PROPOSITION 1. *Let a function $V(z)$ belong to $L_2(\mathbb{R}_+)$ together with the derivatives up to order $m-1$. Then the inequality*

$$(16)\qquad \sum_{j=0}^{m-2} \Lambda^{2(m-2-j)+1}|D^j V(0)|^2 \leqslant C_1 \int_0^\infty \sum_{j=0}^{m-1} \Lambda^{2(m-1-j)+1}|D^j V(z)|^2\,dz$$

holds.

PROPOSITION 2. *If operators P and $B_1, \dots, B_\mu$ satisfy Lopatinskii's condition, then the inequality*

$$(17)\qquad \begin{aligned} C_2 \sum_{j=0}^{m-2} \Lambda^{2(m-2-j)} \int_0^\infty |D^j V(z)|^2\,dz &\leqslant \int_0^\infty |P_l(D,\eta',\gamma)V(z)|^2\,dz \\ &+ \sum_{j=1}^{\mu} |B_j(D,\eta',\gamma)V(0)|^2 \Lambda^{2(m-1-\beta_j)-1} \end{aligned}$$

holds.

Proposition 1 is one of the simple versions of the theorem on smoothness of restrictions of functions that belong to Sobolev spaces, to manifolds of lower dimension. For the sake of completeness, we will present its proof in the next subsection.

Proposition 2 follows from the results on elliptic operators depending on a parameter that are presented in Appendix to this section.

First of all, we note that the definition of the symbols P_l implies that

$$P_l(\lambda,\eta',\gamma) = P_+(\lambda,\eta',\gamma)(P_-(\lambda,\eta',\gamma)/(\lambda - \lambda_l^-(\eta',\gamma)))$$

because the operators P_l and $B_1, \dots, B_\mu$ satisfy Lopatinskii's condition (which involves only the symbol P_+). Furthermore, we set $\eta' = \Lambda\overline{\eta}'$ and $\gamma = \Lambda\overline{\gamma}$, $\overline{\eta}'^2 + \overline{\gamma}^2 = 1$ and represent $P_l(\lambda,\eta',\gamma)$ as an elliptic symbol depending on one variable λ and, generally, on a large parameter Λ (the dependence on $\overline{\eta}'$ and $\overline{\gamma}$ is inessential). The theory presented in Appendix leads to inequality (17).

6) *Completion of the proof of the theorem.* Now we use (16) and (17) to derive (13) from (15). Adding $\sum \Lambda^{2m-1-2\beta_j}|(B_jV)(0)|^2$ to both sides of (15) and applying inequality (17) we find a lower bound

$$(18)\qquad C_3\Lambda \sum_{j=0}^{m-1} \Lambda^{2(m-1-j)} \int_0^\infty |D^j V(z)|^2\,dz + \frac{1}{2}\sum |(P_lV)(0)|^2$$

for the right-hand side of (13). Since the symbol P_l is resolved with respect to λ^{m-1}, we choose a small number $\varepsilon > 0$ and obtain the inequality

$$\frac{1}{2}\sum |(P_lV)(0)|^2 \geqslant |D^{m-1}V(0)|^2 - C_4\varepsilon \sum_{j=0}^{m-2} \Lambda^{2(m-1-j)}|D^j V(0)|^2.$$

According to Proposition 1, the second term on the right-hand side has a lower bound

$$-C_5\varepsilon\Lambda\sum_{j=0}^{m-1}\Lambda^{2(m-1-j)}\int_0^\infty |D^jV|^2\,dz.$$

Selecting a sufficiently small ε we derive the lower bound

$$C_6\Lambda\sum_{j=0}^{m-1}\Lambda^{2(m-1-j)}\int_0^\infty |D^jV|^2\,dz+C_7\sum_{j=0}^{m-1}\Lambda^{2(m-1-j)}|D^jV(0)|^2$$

for (18). This completes the proof of the theorem.

5.4. Proof of Proposition 1 in 5.3. We first assume that the function $V(z)$ and its derivatives up to order $m-1$ belong to L_2 throughout the entire line. Then, by the inversion formula for the Fourier transform, we obtain

$$D^jV(0)=(2\pi)^{-1/2}\int_{-\infty}^{\infty}\lambda^j\widehat{V}(\lambda)\,d\lambda,\quad j=0,\dots,m-2.$$

Multiplying and dividing the integrand by

$$\Big(\sum_{l=0}^{m-1}\Lambda^{2(m-1-l)}\lambda^{2l}\Big)^{1/2}$$

and applying the Schwarz inequality we derive

$$\begin{aligned}|D^jV(0)|^2\leqslant&(2\pi)^{-1}\int_{-\infty}^{\infty}\sum\Lambda^{2(m-1-l)}|\lambda^l\widehat{V}(\lambda)|^2\,d\lambda\\ &\times\int_{-\infty}^{\infty}\sum\lambda^{2j}\Big(\sum\Lambda^{2m-2-2l}\lambda^{2l}\Big)^{-1}\,d\lambda.\end{aligned}\tag{19}$$

By virtue of the Parseval relation, the first integral on the right-hand side is equal to

$$\sum\Lambda^{2(m-1-l)}\int_{-\infty}^{\infty}|D^lV(z)|^2\,dz.\tag{19'}$$

Since $j\leqslant m-2$, the second integral on the right-hand side of (10) is convergent. We now make change of variable $\lambda=\Lambda\theta$ and find that the integral is equal to

$$\text{const}\,\Lambda^{2j-2(m-1)+1}.\tag{19''}$$

Substituting (19′) and (19″) into (19), multiplying both sides of the inequality by $\Lambda^{2(m-1-j)}$, and performing summation with respect to $j=0,\dots,m-2$ we arrive at inequality (16) with integrals over the entire line on the right-hand side. To pass to integrals over the half-line we construct a linear extension operator. In other words, the proof of (16) is completed with the aid of the following

LEMMA. *Let a function $V(z)$ belong to $L_2(\mathbb{R}_+)$ together with its derivatives up to order k. Then there is a function $V^*(z)$ possessing the following properties*:

(i) $D^j V^*(z) \in L_2(\mathbb{R})$, $j = 0, \dots, k$;

(ii) $V^*(z) = V(z)$ *for* $z \geqslant 0$;

(iii) *there is a constant c depending on k only, such that*

$$\int_{-\infty}^{\infty} |D^j V^*(z)|^2 \leqslant c \int_{-\infty}^{\infty} |D^j V(z)|^2 \, dz, \qquad j = 0, \dots, k. \tag{20}$$

PROOF. For negative values of z we define $V(z)$ using Hestens' formula:

$$V^*(z) = \sum_{l=0}^{k-1} a_l V(-lz), \qquad z < 0, \tag{21}$$

where the numbers a_l are found from the condition of continuity at zero for the derivatives $D^j V^*(z)$, $j = 0, \dots, k$, i. e., from the linear system of equations

$$\sum_{l=0}^{k} (-l)^j a_l = 1, \qquad j = 1, \dots, k.$$

It is clear that the function $V^*(z)$ coinciding with $V(z)$ for $z > 0$ and defined for $z < 0$ by means of (21) satisfies all conditions of the lemma.

Appendix to §5

1. We consider estimates[1] for the boundary value problem for ordinary operators with constant coefficients depending on a parameter Λ:

$$Q(D, \Lambda) V(z) = F(z), \tag{1}$$

$$B_j(D, \Lambda) V(0) = g_j, \quad j = 1, \dots, \mu. \tag{2}$$

Here $Q(\lambda, \Lambda)$ and $B_j(\lambda, \Lambda)$, $j = 1, \dots, \mu$, are homogeneous polynomials of degrees m and $\beta_j < m$, respectively. Assume that the polynomial Q satisfies an ellipticity condition with parameter:

$$Q(\lambda, \Lambda) \neq 0, \quad \lambda^2 + \Lambda^2 \neq 0, \quad \Lambda \geqslant 0. \tag{3}$$

By virtue of (3), the equation

$$Q(\lambda, \Lambda) = 0$$

has no real roots, and the polynomial Q can be represented in the form

$$Q(\lambda, \Lambda) = Q_+(\lambda, \Lambda) Q_-(\lambda, \Lambda), \qquad Q_\pm(\lambda, \Lambda) \neq 0, \quad \pm \operatorname{Im} \lambda > 0.$$

We set $\mu = \deg Q_+$; let

$$B_j'(\lambda, \Lambda) = \sum_{k=1}^{\mu} b_{jk}(\Lambda) \lambda^{\mu - k} \equiv B_j(\lambda, \Lambda) \pmod{Q_+(\lambda, \Lambda)}. \tag{4}$$

[1] The results presented below are the simplest version of the results of Agranovich and Vishik [**AV**].

The coefficients $b_{jk}(\Lambda)$ are homogeneous functions of Λ of order $\beta_j - \mu + k$, i. e., $b_{jk}(\Lambda) = b_{jk}\Lambda^{\beta_j-\mu+k}$. We will say that the set of operators $\{Q, B_1, \dots, B_\mu\}$ satisfies uniform Lopatinskii's condition if

$$\det \|b_{jk}\| \neq 0. \tag{5}$$

This condition ensures the unique solvability of the homogeneous equation

$$Q(D,\Lambda)V(z) = 0 \tag{1'}$$

in the space of functions $V(z)$ belonging to $L_2(\mathbb{R}_+)$ together with their derivatives up to order m and satisfying conditions (2) with arbitrary constant $(g_1, \dots, g_\mu) \in \mathbb{C}^\mu$.

2. Consider the so-called canonical basis of problem (1), (1′), i.e., a set of functions $V_j(z,\Lambda) \in L_2(\mathbb{R}_+)$, $j = 1, \dots, \mu$, satisfying the conditions

$$Q(D,\Lambda)V_j(z,\Lambda) = 0, \tag{6}$$

$$B_k(D,\Lambda)V_j(z,\Lambda)\big|_{z=0} = \delta_{kj}, \quad k = 1, \dots, \mu. \tag{7}$$

By homogeneity, we have

$$V_j(z,\Lambda) = \Lambda^{-\beta_j} V_j(\Lambda z, 1).$$

Note that, by virtue of (3), the solution to problem (6), (7) belonging to $L_2(\mathbb{R}_+)$ must be an infinitely differentiable exponentially decreasing function for $z \to +\infty$ (cf. Proposition 2 in 1.3). Therefore,

$$\int_0^\infty |D^k V_j(z,\Lambda)|^2\, dz = \Lambda^{-2\beta_j+2k-1} \int_0^\infty |D^k V_j(z,1)|^2\, dz = C_k \Lambda^{-2\beta_j+2k+1}. \tag{8}$$

3. Our aim is to prove inequalities of the type (5.17) for solutions to problem (1), (2):

$$\begin{aligned} c\sum_{j=0}^{m-1} \Lambda^{2(m-1-j)} \int_0^\infty |D^j V(z)|^2\, dz &\leqslant \int_0^\infty |Q(D,\Lambda)V(z)|^2\, d\Lambda \\ &\quad + \sum_{j=1}^{\mu} |B_j(D,\Lambda)V(0)|^2 \Lambda^{2(m-1-\beta_j)-1}. \end{aligned} \tag{9}$$

First we prove (9) under the assumption that the homogeneous equation (1′) holds. In this case the solution $V(z)$ can be represented in the form

$$V(z) = \sum_{j=1}^{\mu} g_j V_j(z,\Lambda).$$

With account of (8), we obtain

$$\begin{aligned} \sum \Lambda^{2(m-1-k)} \int_0^\infty |D^k V(z)|^2\, dz &\leqslant \mu \sum_j |g_j|^2 \sum_k \Lambda^{2(m-1-k)} \int_0^\infty |D^k V(z)|^2\, dz \\ &= \text{const} \sum \Lambda^{2(m-1-\beta_j)-1} |g_j|^2. \end{aligned} \tag{10}$$

4. We now consider the general case. Denote by F^* the extension of the function F in (1) to the entire line by Hestens' formula (see 5.4); let $V^* \in L_2(\mathbb{R})$ be a solution to the equation

$$Q(D,\Lambda)V^*(z) = F^*(z), \qquad -\infty < z < \infty.$$

The solution to problem (1), (2) will be sought in the form $V = V^* + W$, where W is the solution to the problem

$$Q(D,\Lambda)W(z) = 0, \quad z > 0, \quad B_j(D,\Lambda)W\big|_{z=0} = g_j - h_j, \quad j = 1,\dots,\mu,$$

with

$$h_j = B_j(D,\Lambda)V^*(z)\big|_{z=0}.$$

According to the results of the foregoing subsection, we have

$$\sum_k \Lambda^{2(m-1-k)} \int_0^\infty |D^k W|^2\, dz \leqslant \text{const} \sum_j \Lambda^{2(m-1-\beta_j)-1}(|g_j|^2 + |h_j|^2). \tag{11}$$

Furthermore, passing to the Fourier transform in (10) we obtain

$$\sum \Lambda^{2(m-1-k)} \int_0^\infty |D^k V^*(z)|^2\, dz \leqslant \sum \Lambda^{2(m-1-k)} \int_{-\infty}^\infty |D^k V^*(z)|^2\, dz$$
$$= \int_{-\infty}^\infty \sum \Lambda^{2(m-1-k)}\lambda^{2k} |\widehat{V}^*(\lambda)|^2\, d\lambda \leqslant \text{const} \int_{-\infty}^\infty |Q(\lambda,\Lambda)\widehat{V}^*(\lambda)|^2\, d\lambda$$
$$= \text{const} \int_{-\infty}^\infty |F^*(z)|^2\, dz \leqslant \text{const} \int_0^\infty |F(z)|^2\, dz.$$

We thus derive an estimate for the left-hand side of (9) in terms of the right-hand side and the additional term

$$\sum_{j=1}^{\mu} \Lambda^{2(m-1-\beta_j)-1} |(B_j(D,\Lambda)V^*)(0)|^2. \tag{12}$$

To estimate (12), we repeat the arguments of 5.4. Namely we have

$$(B_j(D,\Lambda)V^*)(0) = (2\pi)^{-1/2} \int_{-\infty}^\infty B_j(\lambda,\Lambda)\widehat{V}^*(\lambda)\, d\lambda = \int_{-\infty}^\infty \frac{B_j(\lambda,\Lambda)}{Q(\lambda,\Lambda)} \widehat{F}^*(\lambda)\, d\lambda.$$

Since the degree of B_j is less than that of Q, the application of the Schwarz inequality results in

$$|B_j(D,\Lambda)V^*(0)|^2 \leqslant (2\pi)^{-1} \int |\widehat{F}^*(\lambda)|^2\, d\lambda \int_{-\infty}^\infty |B_j(\lambda,\Lambda)|^2 |Q(\lambda,\Lambda)|^{-2}\, d\lambda$$
$$\leqslant \text{const}\, \Lambda^{-2m+2\beta_j+1} \int_0^\infty |F(z)|^2\, dz.$$

Substituting this inequality into (12) we estimate the result by const $\int_0^\infty |F(z)|^2\,dz$. Inequality (9) is proved completely.

§6. Completion of the proof of the basic estimate

6.1. Preliminary remarks. In §5 we consecutively carried out localization and microlocalization of the basic estimate (5.1). In other words, we showed that in deriving this estimate we could assume that

1) the operators P and $B_1, \dots, B_\mu$ contain no lower terms;

2) the symbol of the operator P has the form

$$P(y,\eta) = P(\eta) + P'(y,\eta), \qquad [P']_0 \leqslant \delta,$$

where δ is a sufficiently small constant;

3) the operators $B_1, \dots, B_\mu$ have constant symbols;

4) the estimate must be proved for the functions $u(y)$ whose partial Fourier transforms $\widehat{u}'(x_1, \xi', \sigma + i\gamma)$ are supported in a conic neighborhood of a point, i. e., the function $\widehat{u}'(x_1, \xi', \sigma + i\gamma)$ is nonzero only if $(\eta', \gamma) \in KU$, where

$$U = \{\overline{\eta}', \overline{\gamma}, \quad |\overline{\eta}'|^2 + |\overline{\gamma}|^2 = 1,\ |\overline{\eta}' - \eta'^0| + |\overline{\gamma} - \gamma_0| < \varepsilon\}, \tag{1}$$

and ε is sufficiently small;

5) by the results of §5 (the estimates in the parabolic zone), it remains to consider the neighborhood (1) for the case $\gamma_0 = 0$.

The technique of the proof of (5.1) for this zone is rather complicated, and the present section is entirely devoted to it.

Roughly speaking, the idea is to find a symbol $Q(y,\eta) \in S[\xi_1]^{m-1}$ such that the inequality

$$\begin{aligned} &C\left(|\gamma|\,\|u\|^2_{m-1,s,[\gamma]} + \sum_{j=0}^{m-1} \langle D_1^j u\rangle^2_{m-1+s-j,[\gamma]}\right) \\ &\qquad \leqslant \operatorname{Im}[P(y,D)u, Q(y,D)u]_{(s),[\gamma]} + \sum_{j=1}^{\mu} \langle B_j u\rangle^2_{m-1-\beta_j+s,[\gamma]} \end{aligned} \tag{2}$$

holds.

In the case of functions whose support does not intersect the plane $\{x_1 = 0\}$ we can take $-P^{(1)} = -\partial P/\partial\tau$ as Q. As to the functions with supports in the half-space, the corresponding form can be estimated from below by the norm $|\gamma|\,\|u\|^2$ only if a certain norm of the Cauchy data $u(y)$ at $x_1 = 0$ is added. Hence, to derive (2) we should "correct" the symbol $P^{(1)}$ in a special way by adding a symbol Q^b to it such that the right-hand side of (2) has a lower bound

$$C_0 \sum_{j=1}^{m-1} \langle D_1^j u\rangle^2_{m-1+s-j,[\gamma]} - \lambda C_1 |\gamma|\,\|u\|^2_{m-1,s,[\gamma]},$$

where the constant λ can be made arbitrarily small.

A rather subtle construction of the symbol Q^b was suggested by Sakamoto [**1**].

To formulate precisely the asserions proved in this section we have to study the behavior of the symbol $P(y, \xi_1, \xi', \sigma + i\gamma)$ in the neighborhood of a (hyperbolic) point $(0, y^{0\prime}, \eta^{0\prime}, 0)$.

6.2. Structure of a hyperbolic symbol in the hyperbolic zone. Let $P(y, \eta)$ be a homogeneous strictly hyperbolic symbol. For brevity, we set

$$X = (x_1, y', \eta', \gamma), \quad X_0 = (0, y'^0, \eta'^0, 0)$$

and consider the symbol $P(X, \lambda)$ in the neighborhood of a point X_0. Since P is a real symbol, the polynomial $P(X_0, \lambda_0)$ can have a number, say l, of real roots λ_j^0 of multiplicity m_j $(j = 1, \dots, l)$ and a number, say r, of pairs of conjugate nonreal roots, which of course can be multiple. Thus, for $X = X_0$ we have

$$P(X_0, \lambda) = P_0(\lambda) P_1(\lambda) \cdots P_l(\lambda), \quad P_j(\lambda) = (\lambda - \lambda_j^0)^{m_j}, \ j > 0. \tag{3}$$

Now we consider the polynomial $P(X, \lambda)$ for X sufficiently close to X_0. In this case the roots of the polynomial $P(X, \lambda)$ split into $(l+1)$ nonintersecting groups. According to Proposition 1 in 1.3, to the division of the roots into the groups there corresponds the factorization

$$P(X, \lambda) = P_0(X, \lambda) P_1(X, \lambda) \cdots P_l(X, \lambda) \tag{4}$$

of the symbol possessing the following properties:

(a) $P_j(X_0, \lambda) = P_j(\lambda)$, $j = 0, \dots, l$;

(b) $P_j(y, \eta', \gamma, \lambda)$ is a smooth function of all arguments and a homogeneous function of (η', γ, λ) of degree m_j, $m_0 = \deg P_0(\lambda)$;

(c) the coefficients of $P_j(X, \lambda)$ assume real values at $\gamma = 0$.

Only the last of these properties needs an explanation. Let a point $X = (y, \eta', 0)$ be sufficiently close to $X_0 = (y'^0, \eta'^0, 0)$ and let $P_j(X, \lambda)$, $j > 0$, have a complex root $\lambda_j(X)$. Since $P(X, \lambda)$ is a real symbol, among the roots of the (generally, different) polynomial $P_k(X, \lambda)$ there is a complex conjugate root $\overline{\lambda_j(X)}$. In view of conditions (a) and (b), the root $\lambda_j(X)$ is sufficiently close to the real number λ_j^0, i. e., $\overline{\lambda_j(X)}$ is in fact a root of the symbol $P_j(X, \lambda)$. Hence, $P_j(X, \lambda)$ is a real polynomial. The coefficients of the polynomials P_j defined in a neighborhood of X_0 can easily be extended to $\mathbb{R}_+^{n+1} \times \mathbb{R}^{n+1}$ so that

$$P_j(y, \lambda, \eta', \gamma) \in O[\xi_1]^{m_j}, \qquad j = 0, \dots, l. \tag{5}$$

The polynomial $P(X, \lambda)$ and, consequently, all the polynomials $P_j(X, \lambda)$ have no roots for $\gamma < 0$. Therefore, the number $m_j^{\pm}$ of the roots of $P_j(X, \lambda)$ in the half-plane $\pm \operatorname{Im} \lambda > 0$, $m_j^+ + m_j^- = m_j$, does not depend on X. Consequently, the factorization

$$P_j(X, \lambda) = P_j^+(X, \lambda) P_j^-(X, \lambda), \quad P_\pm(X, \lambda) = \prod_{j=0}^{l} P_j^{\pm}(X, \lambda) \tag{6}$$

holds.

To illustrate factorization (3) we consider the simplest hyperbolic symbol corresponding to the wave operator:

$$P(X, \lambda) = \lambda^2 + |\xi|^2 - (\sigma + i\gamma)^2, \qquad X = (\xi', \sigma, \gamma),$$

in the neigborhood of a point $X_0 = (\xi', \sigma_0, 0)$, where $|\xi'|^2 + \sigma_0^2 = 1$. Depending on the value of $|\xi'^0|$ we have to consider the following three cases.

(a) $|\xi'^0| > |\sigma_0|$. In this case the polynomial $P(X_0, \lambda) = \lambda^2 + (|\xi'^0|^2 - \sigma_0^2)$ has a pair of complex conjugate pure imaginary roots, i. e., $P_0(X, \lambda)$ coincides with $P(X_0, \lambda)$ in a small neighbrhood of X_0.

(b) $|\xi'^0| < |\sigma_0|$. In this case $P(X, \lambda)$ has two distinct real roots for $X = X_0$, i. e.,

$$P(X, \lambda) = P_1 P_2 = (\lambda - ((\sigma + i\gamma)^2 - |\xi'|^2)^{1/2})(\lambda + ((\sigma + i\gamma)^2 - |\xi'|^2)^{1/2}).$$

For small $\gamma < 0$ one of the roots falls in the upper complex half-plane, whereas the other lies in the lower half-plane.

(c) $|\xi'^0| = |\sigma_0|$. Here there is one real root of multiplicity 2 for $X = X_0$, i. e.,

$$P(X, \lambda) = P_1(X, \lambda) = \lambda^2 + |\xi'|^2 - (\sigma + i\gamma)^2.$$

Following Sakamoto [**S1**], we set

$$F_{ji}(\lambda) = P(X_0, \lambda)(\lambda - \lambda_j^0)^{-i}, \qquad i = 1, \dots, m_j^-, \quad j = 1, 2, \dots, l. \tag{7}$$

If $m_j^- = 0$, we assume that $F_{ji} = 0$. Setting

$$\widehat{P}_j(X, \lambda) = \prod_{k \neq j} P_k(X, \lambda), \quad \widehat{P}_j^{\pm}(X, \lambda) = \prod_{k \neq j} P_k^{\pm}(X, \lambda),$$

we can rewrite the polynomial (7) in the form

$$F_{ji}(\lambda) = \widehat{P}_j(X_0, \lambda)(\lambda - \lambda_j^0)^{m-i} = P^+(X_0, \lambda)\widehat{P}_j^-(X_0, \lambda)(\lambda - \lambda_j^0)^{m_j^- - i}. \tag{7'}$$

For $j = 0$ we set

$$F_{0i}(\lambda) = \lambda^{m_0^- - i} P(X_0, \lambda)/P_0^-(X_0, \lambda) = P^+(X_0, \lambda)\widehat{P}_0^-(X_0, \lambda)\lambda^{m_0^- - i}. \tag{7''}$$

If we formally set $\lambda_0^0 = 0$, then the right-hand side of (7″) takes the form of the right-hand side of (7′).

A significant point in the derivation of the desired estimate in the hyperbolic zone is the following simple assertion.

PROPOSITION. *Let uniform Lopatinskii's condition hold. Then the system of polynomials* $\{F_{ji}(\lambda)\}$, $i = 1, \dots, m_j^-$, $j = 0, \dots, l$, *and* $\{B_j(\lambda) = B_j(X_0, \lambda)\}$, $j = 1, \dots, \mu$, *forms a basis in the space of polynomials of degree no higher than* $m - 1$.

PROOF. We have to show that these polynomials are linearly independent, i.e., there is no nontrivial system of constants q_{ji}, $i = 1, \dots, m_i^-$, $j = 0, \dots, l$, $h_1, \dots, h_\mu$, such that

$$H(\lambda) \overset{\text{def}}{=} \sum_{k=0}^{l} \sum_{i=1}^{m_k^-} q_{ki}(\lambda - \lambda_k^0)^{i-1}\widehat{P}_k(\lambda)P_k^+ + \sum h_j B_j(\lambda) \equiv 0. \tag{8}$$

Indeed, if the polynomial $H(\lambda)$ is identically zero, then the remainder on dividing this polynomial by $P^+(\lambda)$ is also identically zero, i. e.,

$$\sum h_j B_j'(\lambda) \equiv 0, \quad B_j'(\lambda) \equiv B_j(\lambda) \pmod{P^+(\lambda)}.$$

By uniform Lopatinskii's condition, the polynomials $\{B_j'(\lambda)\}$ are linearly independent, and therefore $h_1 = \dots = h_\mu = 0$. Similarly, the remainder on dividing $H(\lambda)$ by $P_k(\lambda)$ is equal to

$$\left[\sum_{j=1}^{m_k^-} q_{kj}(\lambda - \lambda_k^0)^{j-1}\right] P_k^+(\lambda)\widehat{P}_k(\lambda).$$

Since the polynomials $P_k^+(\lambda)$ and $\widehat{P}_k(\lambda)$ are not identically zero, the polynomial in square brackets must be identically zero, whence $q_{kj} = 0$, $j = 1, \dots, m_k^-$. The proposition is proved.

We extend the polynomials (7) as homogeneous functions to all values of the parameters:

(9) $$F_{ji}(\xi_1, \eta', \gamma) = \Lambda^{a_{ji}}(\eta', \gamma) F_{ji}(\xi_1 \Lambda^{-1}(\eta', \gamma)), \quad a_{ji} = \deg F_{ji}.$$

6.3. General plan for the derivation of the estimate in the hyperbolic zone. We will construct the symbol Q in (2) as the sum of three symbols:

$$Q(y, \eta, \gamma) = -P^{(1)}(y, \xi, \sigma + i\gamma) + Q^{(b)}(y, \eta, \gamma) + Q^{(h)}(y, \eta, \gamma)$$

each admitting its own estimate of "frozen" coefficients. So, the following assertions take place.

THEOREM 1. *Let $P(y, \eta)$ be a uniformly hyperbolic symbol (i. e., let condition (A) hold) and let*

$$P^{(1)}(y, \xi, \tau) = \partial P(y, \xi, \tau)/\partial \tau.$$

Then for each $S > 0$ there is $\delta_0 = \delta_0(S)$ such that for $\delta < \delta_0$ (where δ is the constant in condition 2) of 6.1) and $\gamma \leqslant \gamma_0(S, \delta_0)$ the inequality (cf. §4 of Chapter II)

(10)
$$-\operatorname{Im}[P(y, D)u, P^{(1)}(y, D)u]_{(s),[\gamma]} \geqslant C_1 |\gamma| \, \|u\|^2_{m-1,s,[\gamma]} - C_2 \sum_{j=0}^{m-1} \langle D_1^j u \rangle^2_{(m-1+s-j),[\gamma]}$$

holds.

THEOREM 2. *Let conditions (A) and (B) of §1 hold and let*

(11) $$B(y, \eta, \gamma) = P_0^+(y, \eta, \gamma) \widehat{P}_0(y, \eta, \gamma),$$

(11′) $$C(y, \eta, \gamma) = \partial P_0^-(y, \eta, \gamma)/\partial \xi_1,$$

(12) $$Q^{(P)}(y, \eta, \gamma) = B(y, \eta, \gamma) C(y, \eta, \gamma).$$

Then for all $S > 0$ there are $\delta_0 = \delta_0(S)$ and $\gamma_0 = \gamma_0(S, \delta_0)$ such that for $\delta \leqslant \delta_0$ and $\gamma \leqslant \gamma_0$ the inequality

(13)
$$\begin{aligned} &\operatorname{Im}[P(y, D)u, Q^{(P)}(y, D)u]_{(s),[\gamma]} \\ &\quad \geqslant \sum_{i=0}^{m_0^-} \langle F_{0i}(D, \gamma)u \rangle^2_{(m_0^- + s - i),[\gamma]} + C_3 \|B(y, D, \gamma)u\|^2_{m_0^- - 1, s+1/2,[\gamma]} \\ &\quad\quad - C_4(\varepsilon + \delta + |\gamma|^{-1}) \Big(|\gamma| \, \|u\|^2_{m-1,s,[\gamma]} + \sum_{j=0}^{m} \langle D_1^j u \rangle^2_{(m-1+s-j),[\gamma]} \Big) \end{aligned}$$

holds in the zone (1).

Theorem 1 is derived from the results of §4, and Theorem 2 follows in fact from the results of §5; these two theorems will be proved rigorously below. The most difficult point is the estimate involving the operators F_{ji}, $j > 0$. We have the following theorem.

THEOREM 3. *Let conditions* (A) *and* (B) *hold. Then for each* λ, $0 < \lambda < 1$, *there is a symbol* $Q^{(H)} \in O[\xi_1]^{m-1}$ (*of course,* $Q^{(H)}$ *depends on* λ) *such that for* $|s| \leqslant S$ *and* $\gamma < \gamma_0(S, \delta)$ *the inequality*

$$\operatorname{Im}[P(y,D)Q^{(H)}(y,D)u]_{(s),[\gamma]} \geqslant C_0 \sum_{j=1}^{l}\sum_{i=0}^{m_j^-}\langle F_{ji}(D,\gamma)u\rangle^2_{(m_j^-+s-1),[\gamma]} - C_5(\lambda,\varepsilon,\delta,\gamma)\left(\|u\|^2_{m-1,s,[\gamma]} + \sum_{j=0}^{m}\langle D_1^j u\rangle^2_{(m-1+s-j),[\gamma]}\right) \tag{14}$$

holds in the zone (1), *where the constant* C_5 *has the form*

$$C_5(\lambda,\varepsilon,\delta,\gamma) = C_{51}\lambda + C_{52}(\lambda)\varepsilon + C_{53}(\lambda,\varepsilon)\delta + C_{54}(\lambda,\varepsilon,\delta)|\gamma|^{-1}$$

and the constant C_0 *can be chosen so that it does not depend on* λ.

Theorems 1, 2, and 3 are formulated in the form suitable for their proof. As to inequality (2), its proof requires some crude consequences of inequalties (10), (13), and (14). Assuming that the above theorems are proved we derive the following assertions. To simplify the notation, we set

$$|||u|||^2_{s,\gamma} = |\gamma|\,\|u\|^2_{(s),[\gamma]} + \sum_{j=1}^{m-1}\langle D_1^j u\rangle^2_{(s),[\gamma]}. \tag{15}$$

THEOREM 1′. *Under the conditions of Theorem* 1 *the inequality*

$$|\gamma|^{-1}\|Pu\|^2_{0,s,[\gamma]} \geqslant C_1'|\gamma|\,\|u\|^2_{m-1,s,[\gamma]} - C_2'\sum_{j=0}^{m-1}\langle D_1^j u\rangle^2_{(s),\gamma} \tag{10′}$$

holds.

PROOF. By the Schwarz inequality, an expression of the type

$$(\theta|\gamma|)^{-1}\|Pu\|^2_{0,s,[\gamma]} + \theta|\gamma|\,\|P^{(1)}u\|^2_{0,s,[\gamma]} \leqslant (\theta|\gamma|)^{-1}\|Pu\|_{0,s,[\gamma]} + \theta C_6|\gamma|\,\|u\|^2_{m-1,s,[\gamma]}$$

is an upper bound for the left-hand side of (10). Selecting θ from the condition $\theta C_6 < C_1/2$ we arrive at (10′).

THEOREM 2′. *Under the conditions of Theorem* 2 *with some* $\theta > 0$ *the inequality*

$$(\theta\gamma)^{-1}\|Pu\|^2_{0,s,[\gamma]} \geqslant \sum_{i=0}^{m_0-1}\langle F_{0i}(D,\gamma)u\rangle^2_{(m_0-i+s),[\gamma]} - C_4'(\varepsilon+\delta+|\gamma|^{-1})|||u|||^2_{s,[\gamma]} \tag{13′}$$

holds.

PROOF. By (12), the operator with symbol $Q^{(P)}$ can be represented as

$$Q^{(P)}(y,D,\gamma) = C(y,D,\gamma)B(y,D,\gamma) + R_{m-2}, \tag{16}$$

where R_{m-2} is a differential operator of order $m-2$ with respect to the variable x_1, i. e.,

$$R_{m-2} = \sum_{j=0}^{m-2} r_{m-2-j}D_1^j,$$

where r_{m-2-j} are operators of order $m-2-j$ with respect to the variables y'. Therefore, the left-hand side of (13) can be rewritten as

$$\text{(17)} \qquad \operatorname{Im}[Pu, Q^{(P)}u]_{(s),\gamma} = \operatorname{Im}[Pu, CBu]_{(s),\gamma} + \operatorname{Im}[Pu, R_{m-2}u]_{(s),\gamma}.$$

By modifying the arguments in Proposition 2 of 4.3 we can show that

$$\text{(17}'\text{)} \qquad \operatorname{Im}[Pu, R_{m-2}u]_{(s),\gamma} \leqslant C_7|\gamma|^{-1}|||u|||^2_{(s),\gamma}.$$

Furthermore, for an arbitrary $\theta > 0$ we have

$$\text{(17}''\text{)} \qquad \begin{aligned} \operatorname{Im}[Pu, CBu]_{(s),\gamma} &\leqslant \theta^{-1}\|Pu\|^2_{0,s-1/2,[\gamma]} + \theta\|CBu\|_{0,s+1/2,|\gamma|} \\ &\leqslant (\theta|\gamma|)^{-1}\|Pu\|^2_{0,s,[\gamma]} + \theta C_8\|Bu\|_{m_0^- -1,s+1/2,|\gamma|}. \end{aligned}$$

The substitution of (17′), (17″), and (17) into (13) with θ found from the condition $\theta C_8 \leqslant C_3$ results in (13′).

THEOREM 3′. *Under the conditions of Theorem 3 with an arbitrary $\kappa > 0$ the inequality*

$$\text{(14}'\text{)} \quad (\kappa|\gamma|)^{-1}\|Pu\|^2_{0,s,[\gamma]} \geqslant C_0\sum_{j=1}^{l}\sum_{i=0}^{m_j^-}\langle F_{ji}u\rangle^2_{(m_j^-+s-i),[\gamma]} - C_5'(\lambda,\varepsilon,\delta,\kappa,\gamma)|||u|||^2_{(s),[\gamma]}$$

holds, where

$$\text{(18)} \qquad C_5'(\lambda,\varepsilon,\delta,\kappa,\gamma) = C_{51}'\lambda + C_{52}'(\lambda)\varepsilon + C_{53}'(\lambda,\varepsilon,\delta)|\gamma|^{-1} + C_{54}'(\lambda)\kappa.$$

PROOF. As in the foregoing theorems, by the Schwarz inequality, the left-hand side of (14) has an upper bound

$$(\kappa|\gamma|)^{-1}\|Pu\|_{0,s,[\gamma]} + \kappa\|Q^{(H)}\|_{0,s,[\gamma]}.$$

Since the symbol $Q^{(H)}$ belongs to $S[\xi_1]^{m-1}$ and, generally, depends on λ, we arrive at (14′).

Inequalities (10′), (13′), and (14′) and Proposition 6.2 allow us to complete the proof of (2). We multiply (10′) by a positive number ρ and then add together (10′), (13′), and (14′). Adding the sum of the norms of the boundary operators to the resulting inequality (and changing the notation of the constants) we finally obtain

$$\text{(19)} \qquad \begin{aligned} &\text{const}\,|\gamma|^{-1}\|Pu\|_{0,s,[\gamma]} + \sum\langle B_ju\rangle^2_{m-1+s-\beta_j,[\gamma]} \\ &\geqslant(\rho C - k_1\lambda - k_2(\lambda)\varepsilon - k_3(\lambda,\varepsilon,\delta)|\gamma|^{-1})|\gamma|\,\|u\|^2_{m-1,s,[\gamma]} \\ &\quad + \Big[\sum_{j=0}^{l}\sum_{i=1}^{m_j^-}\langle F_{ji}(D)u\rangle^2_{(m_j^-+s-i),[\gamma]} + \sum_{j=1}^{\mu}\langle B_j(X_0,D)u\rangle^2_{(m-1+s-\beta_j),[\gamma]}\Big] \\ &\quad + \sum\Big(\langle B_j(X,D)u\rangle^2_{(m-1+s-\beta_j),[\gamma]} - \sum_{j=1}^{\mu}\langle B_j(X_0,D)u\rangle^2_{(m-1+s-\beta_j),[\gamma]}\Big) \\ &\quad - (\rho C_1 + k_1'\lambda + k_2'(\lambda)\varepsilon + k_3'(\lambda,\varepsilon,\delta)|\gamma|^{-1})\sum_{j=1}^{m-1}\langle D_1^ju\rangle^2_{(m-1+s-j),[\gamma]}. \end{aligned}$$

According to Proposition 6.2, the expression in square brackets has a lower bound

$$C_2 \sum_{j=0}^{m-1} \langle D_1^j u\rangle^2_{m-1+s-j,[\gamma]}.$$

We now select the constant ρ proceeding from the condition $\rho C_1 < C_2/2$ and then consecutively choose λ, ε, and $|\gamma|^{-1}$ so that the constant in the first term on the righ-hand side becomes positive. Furthermore, since the support of $\widehat{u}'(x_1, \eta', \gamma)$ belongs to the neighborhood (1), the third term on the right-hand side has a lower bound

$$-C_3\varepsilon \sum_{j=0}^{m-1} \langle D_1^j u\rangle^2_{m-1+s-j,[\gamma]}.$$

Reducing, if necessary, the constants λ, ε, and $|\gamma|^{-1}$ once again, we find a lower bound

$$\text{const} \sum_{j=0}^{m-1} \langle D_1^j u\rangle^2_{(m-1+s-j),[\gamma]}$$

for the sum of the second, third, and fourth terms in (19), which proves (2).

6.4. Proof of Theorem 1 from 6.3. 1) According to condition 2) in 6.1, we have $P(y, D) = P(D) + P'(y, D)$, where the moduli of the coefficients of the symbol P' do not exceed a small constant δ. Therefore,

$$\begin{aligned} -\operatorname{Im}[P(y, D)u, P^{(1)}(y, D)u]_{(s),[\gamma]} &= -\operatorname{Im}[P(D)u, P^{(1)}(D)u]_{(s),[\gamma]} \\ &- \operatorname{Im}[P'(y, D)u, P^{(1)}(D)u]_{(s),[\gamma]} - \operatorname{Im}[P'(y, D)u, P'^{(1)}(D)u]_{(s),[\gamma]}. \end{aligned}$$

By Proposition 4.3, the sum of the second and third terms on the righ-hand side has a lower bound

$$-(c_1\delta + c_2|\gamma|^{-1})|||u|||^2_{(s),[\gamma]},$$

and therefore it suffices to prove the theorem in question for the case of constant coefficients.

2) Passing to partial Fourier's transform with respect to the variables y', we repeat the arguments in Lemma 5.3 and thus conclude that to prove (10) in the case of constant coefficients it suffices to prove the corresponding inequality for ordinary differential operators on the half-line:

$$\begin{aligned} &-\operatorname{Im} \int_0^\infty P(D, \xi', \tau) V(z) \overline{P^{(1)}(D, \xi', \tau) V(z)}\, dz \\ &\geqslant c_1|\gamma| \sum_{j=0}^{m-1} \Lambda^{2(m-1-j)}(\eta', \gamma) \int_0^\infty |D^j V(z)|^2\, dz \\ &\quad - c_2 \sum_{j=0}^{m-1} \Lambda^{2(m-1-j)}(\eta', \gamma)|(D_1^j V)(0)|^2, \end{aligned} \tag{20}$$

where $D = -i\partial/\partial z$.

3) According to the identity (4.29), the left-hand side of (20) is equal to

$$|\gamma| \sum_{j,k=0}^{m-1} \operatorname{Re} G_{jk}(\eta',\gamma) \int_0^\infty D^{m-1-j} V(z) \overline{D^{m-1-k} V(z)}\, dz \\ - \frac{1}{2} \sum_{j,k} \operatorname{Re} H_{jk}(\eta',\gamma) D^{m-1-j} V(0) \overline{D^{m-1-k} V(0)}.$$

Here the second term on the right-hand side has a lower bound

$$-\operatorname{const} \sum_{j=1}^{m-1} \Lambda^{2m-2-2j} |D^j V(0)|^2.$$

Thus, to prove (20) it suffices to establish the inequality

$$\begin{aligned} &\operatorname{Re} \sum g_{jk}(\eta',\gamma) \int_0^\infty D^{m-1-j} V(z) \overline{D^{m-1-k} V(z)}\, dz \\ &\geqslant c_1 \sum_{j=0}^{m-1} \Lambda^{2m-2-2j} \int_0^\infty |D^j V(z)|^2\, dz - c_2 \Lambda^{-1} \sum_{j=0}^{m-1} \Lambda^{2m-2-2j} |D^j V(0)|^2. \end{aligned} \tag{21}$$

Furthermore, by homogeneity, it suffices to prove (21) for $\Lambda(\eta',\gamma) = (|\eta|^2 + \gamma^2)^{1/2} = 1$. Indeed, assume that the inequality

$$\begin{aligned} &\operatorname{Re} \sum g_{jk}(\eta',\gamma) \int_0^\infty D^{m-1-j} V(z) \overline{D^{m-1-k} V(z)}\, dz \\ &\geqslant c_1 \sum \int_0^\infty |D^j V(z)|^2\, dz - c_2 \sum |D^j V(0)|^2, \quad \Lambda(\eta',\gamma) = 1, \end{aligned} \tag{22}$$

was already proved for this case. By replacing (η',γ) in (22) with $(\eta'/\Lambda, \gamma/\Lambda)$ and $V(z)$ with $\Lambda^{m-3/2}(\eta',\gamma)\, V(z\Lambda^{-1}(\eta',\gamma))$ and making change of variable $z \to \Lambda z$ under the integral sign we derive (21).

4) We first establish (22) for the functions $V = V_0$ that vanish at $z = 0$ together with their derivatives up to order $m-1$:

$$V_0(0) = DV_0(0) = \cdots = D^{m-1} V_0(0) = 0. \tag{23}$$

In this case the function $V_0(z)$ can be extended by zero to $z < 0$, and we retain the notation $V_0(z)$ for the extension. Let us substitute $V_0(z)$ into the left-hand side of (22). The resulting expression does not change if the integral over the half-line is replaced by the integral over the entire line $-\infty < z < \infty$. Passing to the Fourier transform we write the left-hand side of (22) in the form

$$\int_{-\infty}^{\infty} \operatorname{Re} \sum g_{jk}(\eta',\gamma) \lambda^{2m-2-j-k} |\widehat{V}(\lambda)|^2\, d\lambda. \tag{24}$$

In view of (4.28) and by the strong hyperbolicity of the polynomial P for $|\eta'|^2+\gamma^2=1$, $\gamma<0$, we have

$$\sum g_{jk}(\eta',\gamma)\lambda^{2m-2-j-k} = -\operatorname{Im} P(\lambda,\eta',\gamma)\overline{P^{(1)}(\lambda,\eta',\gamma)}/|\gamma| \geqslant \text{const}\sum_{j=0}^{m-1}\lambda^{2j}.$$

By virtue of this inequality, an expression of the form

$$(24') \qquad \sum_{j=0}^{m-1}\int_{-\infty}^{\infty}\lambda^{2j}|\widehat{V}_0(\lambda)|^2\,d\lambda = \sum_{j=0}^{m-1}\int_0^{\infty}|D^j V_0(z)|^2\,dz$$

is a lower bound of (24), i. e., inequality (22) is proved for the functions V_0 satisfyiing (23).

5) We now complete the derivation of the desired inequality. Let a smooth function $V(z)$ belong to $L_2(\mathbb{R}_+)$ together with the derivatives up to order $m-1$. We represent it in the form

$$(25) \qquad V(z) = V_0(z) + W(z), \quad W(z) = \chi(z)\sum_{k=0}^{m-1} z^k D^k V(0)/k!,$$

where $\chi(z)$ is a function of compact support on the line that is equal to 1 in a negh-borhood of the point 0. Then it becomes clear that V_0 satisfies (23), and, by what was proved, we have

$$(26) \quad \int_0^{\infty}\operatorname{Re}\sum g_{jk}D^{m-1-j}(V-W)\overline{D^{m-1-k}(V-W)}\,dz \geqslant c\sum_{j=0}^{m-1}|D^j(V-W)|^2\,dz,$$

whence

$$(27) \qquad \begin{aligned} &\operatorname{Re}\int_0^{\infty}\sum g_{jk}D^{m-1-j}V(z)\overline{D^{m-1-k}V(z)}\,dz \\ &\quad \geqslant c\int_0^{\infty}\sum_{j=0}^{m-1}|D^j V(z)|^2\,dz + 2\operatorname{Re}\int_0^{\infty}\sum g_{jk}D^{m-1-j}V\overline{D^{m-1-k}W} \\ &\quad - 2c\operatorname{Re}\sum g_{jk}D^j V\overline{D^j W}\,dz - \int_0^{\infty}\operatorname{Re}\sum g_{jk}D^{m-1-j}W\overline{D^{m-1-k}W}\,dz \\ &\quad + c\sum_{j=0}^{m-1}|D^j W|^2\,dz. \end{aligned}$$

The structure of the function $W(z)$ readily implies that the sum of the fourth and fifth terms on the righ-hand side has a lower bound $-\text{const}\sum|D^j V(0)|^2$. To find a lower bound for the sum of the second and third terms on the right-hand side we employ the inequalities

$$2\,\mathrm{Re}\int_0^\infty D^j V\overline{D^j W}\,dz \geqslant -\kappa\int_0^\infty |D^j V|^2\,dx - \kappa^{-1}\int_0^\infty |D^j W|^2\,dz.$$

Selecting a sufficiently small κ we arrive at (22).

6.5. Proof of Theorem 2 from 6.3. First of all we note that according to (11), (11′), and (12), the following factorizations hold for the symbols P and $Q^{(P)}$:

$$P(y,\xi,\sigma+i\gamma) = P_0^-(y,\eta,\gamma)B(y,\eta,\gamma),$$
$$Q^{(P)}(y,\xi,\sigma+i\gamma) = C(y,\eta,\gamma)B(y,\eta,\gamma).$$

The factorizations of the symbols P and $Q^{(P)}$ induce the factorizations of the operators:

$$P(y,D_1,D_2,\dots,D_{n+1}+i\gamma) = P_0^-(y,\gamma,D)B(y,\gamma,D) + R_{m-1,\gamma},$$
$$Q^{(P)}(y,D_1,D_2,\dots,D_{n+1}+i\gamma) = C(y,\gamma,D)B(y,\gamma,D) + R_{m-2,\gamma}.$$

By Proposition 2 in 4.2, we have

$$\begin{aligned} J &\overset{\text{def}}{=} \mathrm{Im}[P(y,D)u, Q^{(P)}(y,D)u]_{(s),[\gamma]} \\ &= \mathrm{Im}[P_0^-(y,D)B(y,D)u, C(y,D)B(y,D)u]_{(s),[\gamma]} \\ &\quad + O(|\gamma|^{-1})|||u|||^2_{(s),[\gamma]} \\ &\overset{\text{def}}{=} J_1 + O(|\gamma|^{-1})|||u|||^2_{(s),[\gamma]}. \end{aligned} \tag{28}$$

Further estimates are in many respects similar to those in the parabolic case (Theorem 5.3), because in the neighborhood (1) the imaginary parts of the roots $\lambda_{0j}(y,\eta',\gamma)$ of the symbol $P_0^-(y,\lambda,\eta',\gamma)$ satisfy inequalities of the type

$$-\,\mathrm{Im}\,\lambda_{0j}(y,\eta',\gamma) \geqslant C_0\Lambda(\eta',\gamma). \tag{29}$$

However, we cannot apply directly the arguments of Theorem 5.3 because in the case of variable coefficients the support of the partial Fourier transform of the function $B(y,D)u$ does not belong to the neighborhood where (29) holds. To overcome this difficulty, we use the standard microlocal analysis technique. Consider a "truncating" function $\varphi(\eta',\gamma)$, i. e., a function that belongs to C^∞ with respect to all the variables, equals to 1 identically in the support of $\widehat{u}(x_1,\xi',\sigma+i\gamma)$, and vanishes outside a neighborhood of the support. Set

$$B(y,D)u = {}_{[\gamma]}\varphi(D)B(y,D)u + {}_{[\gamma]}(1-\varphi(D)B(y,D))u \overset{\text{def}}{=} w + Ru. \tag{30}$$

In the case of a constant symbol $B(y,\eta,\gamma)$ the opertor R is identically zero, and in the general case it is an infinitely smoothing operator with respect to the variable y' for the functions in question. Furhermore, since the coefficient in the highest derivative with respect to x_1 in the operator $B(y,D)$ is constant, the expression Ru contains no derivatives of order $m-m_0^-$ with respect to x_1. Therefore for every $\theta\in\mathbb{R}$ there is a constant $K(\theta)$ such that

$$\|Ru\|_{(k,s),[\gamma]} \leqslant K(\theta)\|u\|_{(k+m-m_0^--1,s-\theta),[\gamma]}.$$

Consequently, by modifying the arguments of Proposition 2 in 4.2 we can prove that the modulus of the difference between the right-hand side of (28) and the form

$$J_1 = [P_0^-(y, D)w, C(y, D)w]_{(s),[\gamma]}$$

does not exceed $C(\varepsilon)|\gamma|^{-1}|||u|||^2_{(s),[\gamma]}$. Comparing this with (18) we see that

$$J \geqslant J_1 - C_1(\varepsilon)|\gamma|^{-1}|||u|||^2_{(s),[\gamma]}. \tag{31}$$

The form J_1 admits application of the arguments in Theorem 5.3. Set

$$P_{0j}^-(y, D_1, \eta', \gamma) = \prod_{k\neq j}(D_1 - \lambda_{0k}^-(y, \eta', \gamma)).$$

Then we have

$$C(y, D_1, \eta', \gamma) = \sum P_{0j}^-(y, D_1, \eta', \gamma).$$

We denote by $P_0^-(D)$, $P_{0j}(D)$, and $C(D)$ the symbols $P_0^-(y, D)$, etc. considered for $y = (0, y'^0)$. By Proposition 1 in 4.2, we have

$$\begin{aligned} J_1 \geqslant{}& \operatorname{Im}[P_0^-(D)w, C(D)w]_{(s),[\gamma]} \\ &- C_2\delta\left(\|w\|_{m_0^- -1,s+1/2,[\gamma]} + \sum_{j=0}^{m-1}\langle D_1^j u\rangle^2_{m-1+s-j,[\gamma]}\right). \end{aligned} \tag{30'}$$

Integrating by parts and applying (29) we get

$$\begin{aligned} \operatorname{Im}[P_0^-(D)w, C(D)w]_{(s),[\gamma]} &= \sum_{j=1}^{m_0^-}\operatorname{Im}[(D_1 - \lambda_{0j}^-(D'))P_{0j}(D)w, P_{0j}^-(D)w]_{(s),[\gamma]} \\ &\geqslant C_3\|w\|_{m_0-1,s+1/2,[\gamma]} + \sum_{j=1}^{m_0^-}\langle P_{0j}^-(D)w\rangle^2_{(s),[\gamma]}. \end{aligned}$$

We now substitute this lower bound into (31) assuming that $C_2\delta < C_3/2$ and thus arrive at the inequality

$$J_1 \geqslant C_3\|w\|_{m_0-1,s+1/2,[\gamma]} + \sum_{j=1}^{m_0^-}\langle P_{0j}^-(D)w\rangle^2_{(s),[\gamma]} - C_5\delta\sum_{j=0}^{m-1}\langle D_1^j u\rangle_{(m-1+s-j),[\gamma]}. \tag{32}$$

By the theorem on traces, we have

$$\|w\|_{m_0-1,s+1/2,[\gamma]} \geqslant \text{const}\sum_{i=0}^{m_0^- -2}\langle D_1^i w\rangle_{(m_0-1+s-i),[\gamma]}. \tag{33}$$

On the other hand, since the oprators P_{0j}^{-} are resolved with respect to the derivative $D_1^{m_0^- -1}$, we have

$$\sum_{j=1}^{m_0}\langle P_{0j}^{-}(D)w\rangle_{(s),[\gamma]}^2 \geqslant \kappa\langle D_1^{m_0^- -1}w\rangle_{(s),[\gamma]} \tag{34}$$
$$-\kappa\,\mathrm{const}\sum_{j=1}^{m_0^- -2}\langle D_1^i w\rangle_{(m_0-1+s-i),[\gamma]}.$$

In view of (32), (33), and (34), choosing a sufficiently small κ we obtain

$$J_1 \geqslant C_7\|w\|_{m_0-1,s+1/2,[\gamma]}^2 + C_8\sum_{i=1}^{m_0^-}\langle D_1^{i-1}w\rangle_{(s),[\gamma]}^2 - C_9\delta\sum_{j=0}^{m-1}\langle D_1^j u\rangle_{(m-1+s-j),[\gamma]}^2. \tag{35}$$

According to definition (30) and inequality (30′) with $\theta = 1/2$, we have

$$\begin{aligned}\|w\|_{m_0-1,s+1/2,[\gamma]}^2 &\geqslant \frac{1}{2}\|B(y,D)u\|_{m_0-1,s+1/2,[\gamma]}^2 - \|Ru\|_{m_0-1,s+1/2,[\gamma]}^2\\ &\geqslant \frac{1}{2}\|B(y,D)u\|_{m_0-1,s+1/2,[\gamma]}^2 - K^2\|u\|_{m_0-1,s,[\gamma]}^2\\ &\geqslant \frac{1}{2}\|B(y,D)u\|_{m_0-1,s+1/2,[\gamma]}^2 - K^2|\gamma|^{-1}|||u|||_{(s),[\gamma]}^2.\end{aligned} \tag{36}$$

We now proceed to the second term on the right-hand side of (35). As before, let $B(D_1)$ and $B(y,D)$ denote the operators corresponding to the symbols $B(X_0,\xi_1)$ and $B(X,\xi_1)$. Then we have

$$\begin{aligned}D_1^i w &= D_1^i\varphi(D)B(y,D)u + D_1^i Ru\\ &= D_1^i B(D_1)u + D_1^i(B(D) - B(D_1))u\\ &\quad + D_1^i\varphi(D)(B(y,D) - B(D))u + D_1^i Ru.\end{aligned}$$

Recall that the first term on the right-hand side is equal to $F_{0j}(D_1)u$. Thus,

(37)

$$\sum_{i=0}^{m_0^-}\langle D_1^i w\rangle_{(s),[\gamma]} \geqslant \sum_{i=0}^{m_0^-}\langle F_{0i}u\rangle_{(m_0+1+s-i),[\gamma]} - \mathrm{const}(\varepsilon+\delta)\sum_{i=0}^{m-1}\langle D_1^i u\rangle_{(m-1+s-i),[\gamma]}$$
$$- \mathrm{const}\,|\gamma|^{-1}|||w|||_{(s),[\gamma]}^2.$$

Furthermore, the substitution of (36) and (37) results in

$$\begin{aligned}J_1 \geqslant& C_{10}\|B(y,D)u\|_{(m_0-1,s+1/2),[\gamma]}^2 + C_{11}\sum_{i=1}^{m_0^-}\langle F_{0i}u\rangle_{(m_0-1+s-i),[\gamma]}\\ &- C_{12}(\varepsilon+\delta+|\gamma|^{-1})|||u|||_{(s),[\gamma]}^2.\end{aligned}$$

Taking (28) into account, we complete the proof of the theorem.

6.6. Proof of Theorem 3 from 6.3 (preliminary remarks). As in the case of Theorems 1 and 2, in proving Theorem 3 it is allowable to "freeze" the coefficients and to pass to the Fourier transform with respect to the variables y'. In other words, the assertion of Theorem 3 can be reduced to the following theorem.

THEOREM. *Let a symbol $P(y,\eta)$ satisfy conditions* (A) *and* (B) *of* §1 *and let it be representable in the form* (*cf.* 6.1)

$$P(y,\eta) = P(\eta) + P'(y,\eta), \quad P(\eta) = P(0, y'^0, \eta), \qquad [P'] \leqslant \delta. \tag{38}$$

Then for all $\lambda \in (0,1)$ there is a symbol $Q^{(H)}(y,\eta,\gamma) \in S[\xi_1]^{m-1}$ of the type (38):

$$\begin{gathered} Q^{(H)}(y,\eta,\gamma) = Q^{(H)}(\eta) + Q'^{(H)}(y,\eta,\gamma), \\ Q^{(H)}(\eta) = Q^{(H)}(0, y'^0, \eta, 0), \quad [Q^{(H)\prime}] \leqslant K\delta, \end{gathered} \tag{38'}$$

such that the inequality

$$\begin{aligned} I(V,V) &\overset{\text{def}}{=} \int_0^\infty \operatorname{Im}\left[P(D_1,\xi',\eta+i\gamma)V(x_1)\overline{Q^{(H)}(D_1,\eta',\gamma)V(x_1)}\right] dx_1 \\ &\geqslant C_0 \sum_{j=1}^{l}\sum_{i=1}^{m_j^-} |F_{ji}(D_1)V(0)|^2 \\ &\quad - (C\lambda + C'(\lambda)\varepsilon)\Big(|\gamma| \sum_{j=0}^{m-1} \int_0^\infty |D_1^j V(x_1)|^2\,dx_1 + \sum_{j=0}^{m-1} |D_1^j V(0)|^2\Big) \end{aligned} \tag{39}$$

holds for $(\eta',\gamma) \in U$, where U is the hyperbolic neighborhood (1).

The central point in the proof of the theorem is the consideration of the special case when $l = 1$ and the corresponding real root λ_1^0 is of multiplicity no less than two. We will prove the following

PROPOSITION. *Under the conditions of the theorem let the symbol $P(y,\eta)$ admit representation* (3) *in the neighborhood of the point $X_0 = (0, y'^0, \xi'^0, \tau'^0, 0)$. Let it be possible to associate a symbol $Q_j^{(H)}(y,\xi_1,\eta',\gamma) \in S[\xi_1]^{m_j-1}$ of the type* (38′) *with every symbol $P_j(y,\xi_1,\eta',\gamma)$ and an arbitrary $\lambda \in (0,1)$ so that for $(\eta',\gamma) \in U$ the inequality*
(40)

$$\begin{aligned} I_j(W,W) &\overset{\text{def}}{=} \operatorname{Im} \int_0^\infty P_j(D_1,\eta',\gamma)W(x_1)\overline{Q_j^{(H)}(D_1,\eta',\gamma)W(x_1)}\,dx_1 \\ &\geqslant C_{0j} \sum_{i=1}^{m_j^-} |(D_1 - \lambda_j^0)^{m_j-i} W(0)|^2 \\ &\quad - (C_j\lambda + C_j(\lambda)\varepsilon)\Big(|\gamma| \sum_{k=0}^{m_j-1} \int |D^k W(x_1)|^2\,dx_1 + \sum_{k=0}^{m_j-1} |D_1^k W(0)|^2\Big) \end{aligned}$$

holds. Then the assertion of the theorem holds.

To derive the theorem from the proposition we set

$$Q^{(H)}(y,\eta,\gamma)=\sum_{j=1}^{l}Q_j^{(H)}(y,\eta,\gamma)\widehat{P}_j(y,\eta,\gamma).$$

Replacing W in (40) by $\widehat{P}_j(D_1,\eta',\gamma)V$ and performing summation of the resulting inequalities with respect to j from 1 to l we find a lower bound for the left-hand side of (39):

$$\begin{aligned}&C_0\sum_{j=1}^{l}\sum_{i=1}^{m_j^-}\left|(D_1-\lambda_j^0)^{m_j-i}\widehat{P}_j(D_1,\eta',\gamma)V(0)\right|\\ &\quad-(C\lambda+C(\lambda)\varepsilon)\left(|\gamma|\sum_{k=0}^{m-1}\int_0^\infty|D^kV(x_1)|^2\,dx_1+\sum_{k=0}^{m-1}|D_1^kV(0)|^2\right).\end{aligned}\tag{40'}$$

The difference between the coefficients of the polynomials $\widehat{P}_j(X,\xi_1)$ and $\widehat{P}_j(X_0,\xi_1)$ in the neighborhood (1) is of the order of ε. Therefore, changing the coefficient $c(\lambda)$ in the second term on the right-hand side of (40′), we can replace the expression $((D_1-\lambda_j^0)\widehat{P}_j(D_1,\eta',\gamma)V)(0)$ by $(F_{ij}(D_1)V)(0)$ and thus derive (39).

We make the following important

REMARK. When proving the proposition we can assume that the root λ_j^0 is of multiplicity no less than two, i. e., $m_j\geqslant 2$. Indeed, if λ_j^0 is a simple root, then the degree of the polynomial $P_j(X,\lambda)$ is equal to 1, and it coincides either with $P_j^+(X,\lambda)$ or with $P_j^-(X,\lambda)$. In the first case we have $m_j^-=0$, and this polynomial does not contribute to the right-hand side of (39), i. e., we obtain $Q_j^{(H)}\equiv 0$. In the other case, when P_j coincides with P_j^- (i. e., $P_j^+\equiv 1$), we can set $Q_j^{(H)}\equiv 1$.

Indeed, if $P_j=\lambda-\lambda_j(\eta',\gamma)$, $\operatorname{Im}\lambda_j<0$, then

$$\begin{aligned}&\operatorname{Im}\int_0^\infty P_j(D_1,\eta',\gamma)V(x_1)\overline{V(x_1)}\,dx_1\\ &=\frac{1}{2}|V(0)|^2-\operatorname{Im}\lambda_j(\eta',\gamma)\int_0^\infty|V(x_1)|^2\,dx_1\\ &\geqslant\frac{1}{2}|V(0)|^2=\frac{1}{2}|F_{j0}(D_1)V(0)|^2.\end{aligned}$$

To prove the proposition for operators $P_j(x,D)$ whose symbols have real roots of high multiplicity at $X=X_0$ we need more detailed information about the structure of expansion (4) in the neighborhood of this point.

6.7. Additional information about expansion (4). Denote by $\mu_j(X)$ the mean of the roots of the polynomial $P_j(X,\lambda)$. By (5),

$$\mu_j(X)\in O^1,\quad \mu_j(X_0)=\lambda_j^0,\qquad j=1,\dots,l.\tag{41}$$

The symbol $P_j(X, \lambda)$ can be rewritten in the form[2]

$$P_j(X, \lambda) = (\lambda - \mu_j(x))^{m_j} + a_{2j}(X)(\lambda - \mu_j(x))^{m_j - 2} + \cdots + a_{m_j,j}(X).$$

According to (41), the coefficients $a_{sj}(X)$ are smooth functions, and we have

$$a_{2j}(X_0) = \cdots = a_{m_j,j}(X_0) = 0. \tag{43}$$

LEMMA 1. *If $P(y, \eta)$ is a strictly hyperbolic symbol having a real root of multiplicity no less than two at the point X_0, then*

$$\left.\frac{\partial a_{m_j,j}}{\partial \gamma}\right|_{X=X_0} = id_j, \tag{44}$$

where d_j is a real nonzero number.

PROOF. Since the symbol $P(y, \eta, \sigma)$ has simple roots with respect to σ, the relation $P(X_0, \lambda_j^0) = 0$ implies that

$$\left.\frac{\partial P(y, \xi, \sigma)}{\partial \sigma}\right|_{X=X_0, \xi_1=\lambda_j^0} = c_j,$$

where c_j is a real nonzero constant. The differentiation of (4) with respect to γ yields

$$ic_j = \left[\prod_{k\neq j} P_k(X_0, \lambda_k^0)\right] \left.\frac{\partial P_j(X, \lambda)}{\partial \gamma}\right|_{X=X_0, \xi_1=\lambda_j^0}.$$

Since $\lambda_k^0 \neq \lambda_j^0$ for $k \neq j$ and the roots of P are pairwise complex conjugate numbers, the expression in square brackets is real and nonzero. Noting that

$$\left.\frac{\partial P_j(X, \lambda)}{\partial \gamma}\right|_{X=X_0, \lambda=\lambda_j^0} = \left.\frac{\partial a_{m_j,j}(X_0)}{\partial \gamma}\right|_{X=X_0}$$

by (43), we complete the proof of the lemma.

PROPOSITION. *Let conditions* (A) *and* (B) *hold and let the symbol P be representable in the form* (4), *i. e.,*

(a) *if m_j is even, then $m_j^+ = m_j^- = m_j/2$;*

(b) *if m_j is odd, then*

$$m_j^+ = (m_j + 1)/2, \qquad m_j^- = (m_j - 1)/2 \qquad \text{for} \quad d_j > 0,$$
$$m_j^- = (m_j - 1)/2, \qquad m_j^+ = (m_j + 1)/2 \qquad \text{for} \quad d_j < 0,$$

where d_j is the constant in (44).

PROOF. Since the numbers $m_j^{\pm}$ do not depend on the values of ξ' and γ in the neighborhood in question and we have $\gamma < 0$, it suffices to calculate them along the ray $X_{0\gamma} = (0, y'^0, \xi'^0, \gamma)$. The following asymptotic formula holds for the roots of the symbol at small γ. Let $\lambda_{jk}(X)$, $k = 1, \ldots, m_j$, denote the roots (with respect to λ) of the symbol $P(X, \lambda)$. We have

[2]Recall that, in view of the remark in 6.6, it is assumed that $m_j \geqslant 2$.

LEMMA 2. *Under the conditions of the proposition,*

$$\lambda_{jk}(X_{0\gamma}) = \lambda_j^0 + |d_j\gamma|^{1/m_j} \exp\left(i\frac{2\pi k + \arg d_j + \pi/2}{m_j}\right) + O(|\gamma|^{2/m_j}). \tag{45}$$

PROOF. The symbol $P(X, \lambda)$ is resolved with respect to the highest power of λ, and therefore its roots depend continuously on the coefficients and, consequently, on the parameters γ, i. e., we have

$$\lambda_{jk}(X_{0\gamma}) = \lambda_j^0 + o(1), \quad \lambda_{jk}(X_{0\gamma}) - \mu(X_{0\gamma}) = o(1)$$

for $\gamma \to -0$. Substituting this into the equation

$$P_j(X_{0\gamma}, \lambda_{jk}(X_{0\gamma})) = 0$$

and taking into account that P_j is a polynomial of the type (42) and that, by (43), we have $a_{kj}(X_\gamma) = O(\gamma)$, $j = 2, \dots, m_j$, we obtain

$$(\lambda_{jk}(X_{0\gamma}) - \lambda_j^0 + o(1))^{m_j} + a_{m_j,j}(X_{0\gamma}) = o(\gamma). \tag{46}$$

By the Lagrange mean value theorem,

$$a_{m_j,j}(X_{0\gamma}) = a_{m_j,j}(X_0) + (X_{0\gamma} - X_0)\frac{\partial a_{m_j,j}(X_0)}{\partial \gamma} + O(\gamma^2) = i\gamma d_j + O(\gamma^2).$$

Substituting this expression into (46) we find

$$\lambda_{jk}(X_{0\gamma}) = \lambda_j^0 + (-i\gamma d_j)^{1/m_j} + O(\gamma^{2/m_j}), \quad \gamma \to -0,$$

where $k = 1, \dots, m_j$ correspond to the different values of the root of multiplicity m_j. Since $-\gamma = |\gamma|$, we arrive at (45). According to Lemma 1, d_j is a real number, and therefore $\arg d_j$ is equal to 0 or π, and hence formulas (45) imply the assertion of the proposition.

6.8. Proof of Proposition 6.6. The proof is rather lengthy, and we split it into several stages.

6.8.1. We first of all simplify the notation. It can be assumed, without loss of generality, that the polynomial has exactly one real root of multiplicity m at the point X_0 so that instead of P_j, λ_j^0, F_{ji}, $\mu_j(X)$ $a_{lj}(X)$, m_j, and $m_j^\pm$ it is allowed to write P, λ^0, $\mu(X)$ $a_l(X)$, m, and $m^\pm$, respectively. The transformation $V(x_1) \longmapsto V(x_1)\exp(i\lambda^0 x_1)$ reduces the desired assertion to the case $\lambda^0 = 0$.

Thus, we consider a given symbol

$$P(X, \xi_1) = (\xi_1 - \mu(X))^m + a_2(X)(\xi_1 - \mu(X))^{m-2} + \dots + a_m(X) \tag{47}$$

such that

$$a_2(X_0) = \dots = a_m(X_0) = \mu(X_0) = 0, \qquad \frac{\partial a_m(X_0)}{\partial \gamma} = id \quad (d = \overline{d}). \tag{48}$$

We look for a symbol $Q(y,\eta',\gamma,\xi) \in O[\xi_1]^{m-1}$ of the type (38) such that

$$I(V,V) = \operatorname{Im} \int_0^\infty P(\eta',\gamma,D_1)V(x_1)\overline{Q(\eta',\gamma,D_1)V(x_1)}\,dx_1$$

$$\geqslant c_0 \sum_{j=m^+}^{m-1} |D_1^j V(0)|^2 \tag{49}$$

$$- (c_1\lambda + c_2(\lambda)\varepsilon)\left(|\gamma| \sum_{j=0}^{m-1} \int_0^\infty |D_1^j V(x_1)|^2\,dx_1 + \sum_{j=0}^{m-1} |D_1^j V(0)|^2\right)$$

for the given $\lambda \in (0,1)$.

6.8.2. As was noted in 4.5, the integration by parts of the left-hand side of (49) results in

$$I(V,V) = -|\gamma| \operatorname{Re} \int_0^\infty G(\eta',\gamma,D_1,\overline{D_1})V(x_1)\overline{V(x_1)}\,dx \tag{50}$$

$$+ \frac{1}{2} H(\eta',\gamma,D_1,\overline{D_1})V(0)\overline{V(0)}.$$

Hence, it suffices to select a symbol Q such that the inequalities

(51)

$$H(\eta',\gamma,D_1,\overline{D_1})V(0)\overline{V}(0) \geqslant c_0 \sum_{j=m^+}^{m-1} |D_1^j V(0)|^2 - (c_1\lambda + c_2(\lambda)\varepsilon) \sum_{j=0}^{m_+-1} |D_1^j V(0)|^2,$$

(52)

$$\operatorname{Re} \int_0^\infty G(\eta',\gamma,D_1,\overline{D_1})V(x_1)\overline{V(x_1)}\,dx \leqslant (c_3\lambda + c_4(\lambda)\varepsilon) \sum_{j=1}^{m-1} \int_0^\infty |D_1^j V(x_1)|^2\,dx$$

hold. Let $G(D_1,\overline{D_1})$ and $H(D_1\overline{D_1})$ denote the forms $G(\eta',\gamma,D_1,\overline{D_1})$ and $H(\eta',\gamma, D_1,\overline{D_1})$ for $\gamma = 0$ and $\eta' = \eta'^0$. Then in the region in question we have

$$G(\eta',\gamma,D_1,\overline{D_1}) = G(D_1,\overline{D_1}) + G'(\eta',\gamma,D_1,\overline{D_1}),$$
$$H(\eta',\gamma,D_1,\overline{D_1}) = H(D_1,\overline{D_1}) + H'(\eta',\gamma,D_1,\overline{D_1}),$$

where G' and H' are bilinear forms of the derivatives D_1^{m-1-j} and $\overline{D_1}^{m-1-k}$, $j,k \leqslant m-1$, whose coefficients are of the order of ε in the region (1). Thus, if the inequalities

$$H(D_1,\overline{D_1})V(0)\overline{V(0)} > c_0 \sum_{j=m^+}^{m-1} |D_1^j V(0)|^2 - c_5\lambda^2 \sum_{j=0}^{m^+-1} |D_1^j V(0)|^2, \tag{53}$$

$$\int_0^\infty G(D_1,\overline{D_1})V(x_1)\overline{V(x_1)}\,dx_1 \leqslant c_6\lambda^2 \sum_{j=0}^{m-1} \int_0^\infty |D_1^j V(x_1)|^2\,dx_1 \tag{54}$$

are satisfied, then (53) and (54) will hold, and thus Proposition 6.6 will be proved.

6.8.3. To derive (53) and (54) we need some explicit formulas for H and G. Set

$$P(\xi_1,\xi',\sigma+i\gamma)=P_R(\eta',\gamma,\xi_1)+i\gamma P_J(\eta',\gamma,\xi_1).$$

Then P_R and P_J are homogeneous polynomials of degrees m and $m-1$, respectively, in all the variables, and, by virtue of (47) and (48), we have

$$P_R(\xi_1)=P_R(X_0,\xi_1)=P(X_0,\xi_1)=\xi_1^m, \tag{55}$$

$$P_J(\xi_1)=P_J(X_0,\xi_1)=\sum_{j=2}^{m}\frac{\partial a_j(X_0)}{\partial\gamma}\xi_1^{m-j}. \tag{56}$$

Accordingly, we write the symbols $Q\in O[\xi_1]^m$ in the form

$$Q(\eta',\gamma,\xi_1)=Q_R(\eta',\gamma,\xi_1)+i\gamma Q_J(\eta',\gamma,\xi_1), \tag{57}$$

where $Q_R\in S[\xi_1]^{m-1}$ and $Q_J\in S[\xi_1]^{m-2}$.

LEMMA. *The symbols* $H(\Xi_1,\overline{\Xi}_1)$ *and* $G(\Xi_1,\overline{\Xi}_1)$ *of the forms* $H(D_1,\overline{D}_1)$ *and* $G(D_1,\overline{D}_1)$ *defined in* 6.8.2 *satisfy the relations*

$$H(\Xi_1,\overline{\Xi}_1)=\frac{\Xi_1^m Q_R(\overline{\Xi}_1)-\overline{\Xi}_1^m Q_R(\Xi)}{2(\Xi_1-\overline{\Xi}_1)}, \tag{58}$$

$$\begin{aligned}2G(\Xi_1,\overline{\Xi}_1)=&\Xi_1^{m-1}\overline{\Xi}_1 Q_I(\overline{\Xi}_1)+\Xi_1 Q_I(\Xi_1)\overline{\Xi}_1^{m-1}\\&-P_J(\Xi_1)Q_R(\overline{\Xi}_1)-Q_R(\Xi_1)P_J(\overline{\Xi}_1).\end{aligned} \tag{59}$$

Assuming that the lemma is proved we proceed to the choice of Q.

6.8.4. We look for Q in the form

$$Q(X,\xi_1)=\sum_{k=1}^{m^-}\delta_k\lambda^{-2(m^- -k)}(\xi_1-\mu(X))^{m+1-2k}-c\lambda^{-\alpha}Q_0, \tag{60}$$

where

$$Q_0(X,\xi_1)=(\sigma-\sigma^0+i\gamma)(\xi_1-\mu(X))^{m-2}, \tag{61}$$

and the numbers δ_j, c, and α will be chosen in the course of the derivation of (53) and (54). According to (60) and (61), we have

$$Q_R(\xi_1)=\sum_{k=1}^{m^-}\delta_k\lambda^{-2(m^- -k)}\xi_1^{m+2-2k}, \tag{62}$$

(62′)

$$Q_J(\xi_1)=-\kappa\sum_{k=1}^{m^-}(m+1-2k)\delta_k\lambda^{-2(m^- -k)}\xi_1^{m+2-2k}-c\lambda^{-\alpha}\xi_n^{m-2}.$$

The substitution of (62) into (58) gives

$$H(\Xi_1,\overline{\Xi}_1)=\frac{1}{2}\sum_{k=1}^{m^-}\delta_k\lambda^{-2(m_- -k)}\sum_{s=0}^{2k-2}\Xi_1^{m-1-s}\overline{\Xi}_1^{m+1-2k+s}. \tag{63}$$

Denote by $\widehat{H}(\lambda, z, \overline{z})$ the Hermitian form obtained from (63) by replacing Ξ_1^j by z_j. We will show that for some suitably chosen constants δ_k the inequality

$$\widehat{H}(\lambda, z, \overline{z}) > k_0 \sum_{j=m^+}^{m-1} \lambda^{2(m^+-1)}|z_j|^2 - k_1 \sum_{j=0}^{m^+-1} \lambda^{2(m^+-j)}|z_j|^2 \tag{64}$$

holds. Replacing z_j by $D_1^j V(0)$ in (64) we will obtain (53). If z_j in (64) is replaced by $\lambda^j z_j$, then the structure (63) of the left-hand side implies that (64) is equivalent to the inequality

$$\widehat{H}(1, z, \overline{z}) > k_0 \sum_{j=m^+}^{m-1} |z_j|^2 - k_1 \sum_{j=0}^{m_1^+} |z_j|^2. \tag{64'}$$

According to (63), the left-hand side of (64′) has the form

$$\begin{aligned} 2\widehat{H}(1, z, \overline{z}) = {} & \delta_1 |z_{m-1}|^2 + \delta_2 (z_{m-1}\overline{z}_{m-3} + |z_{m-2}|^2 + z_{m-3}\overline{z_{m-1}}) + \cdots \\ & + \delta_{m^-}(z_{m-1}\overline{z}_{m-2m^-+1} + \cdots + |z_{m-m^-}|^2 + \cdots + z_{m-2m+1}\overline{z_{m-1}}). \end{aligned}$$

Choosing consecutively $1 = \delta_1 < \delta_2 < \cdots$ we get (64′).

6.8.5. We now choose c and α in (60) so that inequality (54) is satisfied. Denote by Q_I' the sum on the right-hand side of (62′); by virtue of (59), we obtain

$$\begin{aligned} 2G(D_1\overline{D}_1) = {} & -2c\lambda^{-\alpha} D_1^{m-1}\overline{D}_1^{m-1} + D_1^{m-1}\overline{D}_1 Q_I'(D_1) \\ & + D_1 Q_I'(D_1)\overline{D}_1^{m-1} - P_J(D_1)Q_R(\overline{D}_1) - Q_R(D_1)P_J(\overline{D}_1). \end{aligned} \tag{59''}$$

We now write

$$\begin{aligned} & \operatorname{Re} \int_0^\infty G(D_1, \overline{D_1}) V(x_1)\overline{V(x_1)}\, dx_1 \\ & = -2c\lambda^{-\alpha} \int_0^\infty |D_1^{m-1} V(x_1)|^2 + \operatorname{Re} \int_0^\infty D_1^{m-1} V(x_1) Q_I'(\overline{D}_1)\overline{D_1 V(x_1)}\, dx_1 \\ & \quad - \operatorname{Re} \int_0^\infty P_J(D_1) V(x_1)\overline{Q_R(D_1) V(x_1)}\, dx_1. \end{aligned} \tag{65}$$

Furthermore, we will establish the following inequalities:

(66)
$$\operatorname{Re} \int_0^\infty D_1^{m-1} V \overline{Q_I'(D_1) D_1 V}\, dx_1 \leqslant c_1 \lambda^{-\alpha_1} \int_0^\infty |D_1^{m-1} V|^2\, dx_1 + c_1' \lambda^2 \int_0^\infty |V|^2\, dx_1,$$

(67)
$$-\operatorname{Re} \int_0^\infty P_J(D_1) V \overline{Q_R(D_1) V}\, dx_1 \leqslant c_2 \lambda^{-\alpha_2} \int_0^\infty |D_1^{m-1} V|^2\, dx_1 + c_2' \lambda^2 \int_0^\infty |V|^2\, dx_1.$$

Choosing $c > c_1 > c_2$ and $\alpha \geqslant \max(\alpha_1, \alpha_2)$ we see that the right-hand side of (65) has an upper bound expressed in terms of the left-hand side of (54).

6.8.6. To prove inequalities (66) and (67) we need the interpolation inequalities below.

LEMMA. *Let $V(x_1) \in H^{m-1}(\mathbb{R}_+)$ and let $i, j \leqslant m-1$, $i+j < 2m-2$. Then for every $\varepsilon > 0$ we have*

$$\left|\int_0^\infty D_1^i V(x_1)\overline{D_1^j V(x_1)}\,dx_1\right| \leqslant \operatorname{const}\left(\varepsilon^{-\frac{2m-2}{i+j}}\int_0^\infty |D_1^{m-1}V(x_1)|^2\,dx_1 + \varepsilon^{\frac{2m-2}{2m-2-i-j}}\int_0^\infty |V(x_1)|^2\,dx_1\right). \tag{68}$$

PROOF. Setting $a = \xi_1^{i+j}\varepsilon^{-1}$, $b = \varepsilon$, $p = \frac{2m-2}{i+j}$, and $q = \frac{2m-2}{2m-2-i-j}$ in the elementary inequality

$$|ab| \leqslant \frac{1}{p}|a|^p + \frac{1}{q}|b|^q, \quad \frac{1}{p} + \frac{1}{q} = 1,$$

we obtain

$$|\xi_1|^{i+j} \leqslant \frac{i+j}{2m-2}\varepsilon^{-\frac{2m-2}{i+j}}|\xi_1|^{2m-2} + \frac{2m-2-i-j}{2m-2}\varepsilon^{\frac{2m-2}{2m-2-i-j}}.$$

If $W(x_1) \in H^{m-1}(\mathbb{R})$, then the multiplication of the above inequality by $|\widehat{W}(\xi_1)|^2$, subsequent integration with respect to ξ_1, and application of the Parseval relation result in

$$\left|\int_{-\infty}^\infty D_1^i W\overline{D_1^j W}\,dx_1\right| \leqslant \varepsilon^{-\frac{2m-2}{i+j}}\int_{-\infty}^\infty |D_1^{m-1}W|^2\,dx_1 + \varepsilon^{\frac{2m-2}{2m-2-i-j}}\int_{-\infty}^\infty |W|^2\,dx_1.$$

It now remains to note that the function $V(x_1) \in H^{m-1}(\mathbb{R}_+)$ can be extended to the entire real line so that the extended function satisfies the inequalities (see Lemma 5.4)

$$\int_{-\infty}^\infty |D_1^k W(x_1)|^2\,dx_1 \leqslant c_k \int_0^\infty |D_1^k V(x_1)|^2\,dx_1.$$

6.8.7. We proceed to inequality (66). The right-hand side of (66) is an expression of the form

$$\sum_{k=1}^{m^-} \delta_k' \lambda^{-2(m^- - k)} \operatorname{Re}\int_0^\infty D_1^{m-1}V(x_1)\overline{D_1^{m-2k}}\,dx_1. \tag{69}$$

Setting $i = m-1$ and $j = m-2k$ in (68) we obtain

$$\operatorname{Re}\int_0^\infty D_1^{m-1}V\overline{D_1^{m-2k}V}\,dx \leqslant \operatorname{const}\left(\varepsilon^{-\frac{2m-2}{2m-2k-1}}\int_0^\infty |D_1^{m-1}V|^2\,dx_1 + \varepsilon^{\frac{2m-2}{2k-1}}\int_0^\infty |V(x_1)D|^2\,dx_1\right). \tag{70}$$

Multiplying both sides of (70) by λ^{-2m^-+2k} we choose ε satisfying the relation

$$\lambda^{-2m^-+2k}\varepsilon^{\frac{2m-2}{2k-1}} = \lambda^2.$$

Then we obtain

$$\varepsilon = \lambda^{(2m^- -2k+2)(2k-1)/(2m-2)},$$

and inequality (70) takes the form

$$\tag{70'}\begin{aligned} &\lambda^{-2(m^- -k)} \operatorname{Re} \int_0^\infty D_1^{m-1} V \overline{D_1^{m-2k} V}\, dx \\ &\quad \leqslant \text{const}\left(\lambda^{\beta_k} \int_0^\infty |D_1^{m-1} V|^2\, dx_1 + \lambda^2 \int_0^\infty |V|^2\, dx_1\right), \end{aligned}$$

where

$$\beta_k = 2m_- - 2k + \frac{(2m-2k+2)(2k-1)}{2m-2k-1}, \qquad k = 1, \dots, m^-.$$

Taking $\alpha_1 = \max \beta_k$ we arrive at (66).

6.8.8. Inequality (67) is more intricate, and its proof is based on the application of Proposition 6.7. Let us replace the operator Q_R by its expression (62) and rewrite the left-hand side of (67) as

$$\tag{71}\begin{aligned} &-\sum_{k=1}^{m^- -1} \delta_k \lambda^{-2(m^- -k)} \operatorname{Re} \int_0^\infty P_J(D_1) V \overline{D_1^{m+1-2k} V}\, dx_1 \\ &\qquad -\delta_{m^-} \operatorname{Re} \int_0^\infty P_J(D_1) V \overline{D_1^{m+1-2m_-} V}\, dx_1. \end{aligned}$$

The first sum in (71) is estimated in the same way as the left-hand side of (66). Furthermore, if $m+1 > 2m^-$, then the second integral in (71) can also be estimated in the same manner. Consider the case $m+1 = 2m^-$. We write $P_J(D_1) = P_J'(d_1) + d$. Then the second term in (71) assumes the form

$$\tag{72} -\delta_{m^-} \operatorname{Re} \int_0^\infty P_J'(D_1) V \overline{V}\, dx_1 - \delta_{m^-} d \int_0^\infty |V(x_1)|^2\, dx.$$

The first term in (72) can be estimated in the same way as the expressions in 6.8.7. By Proposition 6.7, if $m+1 = 2m^-$, then $d < 0$, and the first term in (72) is positive. Discarding this term we only strengthen the inequality, and hence (67) is proved.

6.8.9. So, to complete the proof of the desired inequality, it remains to prove Lemma 6.8.3.

PROOF OF LEMMA 6.8.3. To the left-had side of (49) the following form corresponds:

$$\begin{aligned}
&\frac{1}{2i}\left[P(\eta',\gamma,D_1)\overline{Q}(\eta',\gamma,\overline{D}_1) - Q(\eta',\gamma,D_1)\overline{P}(\eta',\gamma',\overline{D}_1)\right]\\
&\quad = \frac{1}{2i}\Big[P_R(\eta',\gamma,D_1)Q_R(\eta',\gamma,\overline{D}_1) - Q_R(\eta',\gamma,D_1)P_R(\eta',\gamma',\overline{D}_1)\\
&\qquad\qquad + \gamma^2 P_J(\eta',\gamma,D_1)Q_J(\eta',\gamma',\overline{D}_1) - \gamma^2 Q_J(\eta',\gamma,D_1)P_J(\eta',\gamma',\overline{D}_1)\Big]\\
&\qquad + \frac{1}{2i}\Big[i\gamma P_J(\eta',\gamma,D_1)Q_J(\eta',\gamma,\overline{D}_1) - i\gamma P_R(\eta',\gamma,D_1)Q_J(\eta',\gamma',\overline{D}_1)\\
&\qquad\qquad + i\gamma Q_R(\eta',\gamma,D_1)P_J(\eta',\gamma,\overline{D}_1) - i\gamma Q_J(\eta',\gamma,D_1)P_R(\eta',\gamma',\overline{D}_1)\Big]\\
&\quad = J_1(\gamma,\xi,D_1,\overline{D}_1) + \gamma J_2(\gamma,\xi,D_1,\overline{D}_1).
\end{aligned}$$

Note that

$$J_1(\gamma,\eta',\xi_1,\xi_1) = 0 \quad \forall \xi_1 \in \mathbb{R}.$$

It follows (see **[H2,** §8.2]) that after the integration we obtain

$$\int_0^\infty J_1(\gamma,\xi,D_1,\overline{D}_1)V(x_1)\overline{V(x_1)}\,dx_1 = H_0(\gamma,\eta',D_1,\overline{D_1})V(0)\overline{V(0)},$$

where

$$\begin{aligned}
H_0(\gamma,\xi,\Xi_1,\overline{\Xi}_1) = &\frac{1}{2(\Xi_1-\overline{\Xi}_1)}\Big[P_R(\gamma,\eta',\Xi_1)Q_R(\gamma,\eta',\overline{\Xi}_1)\\
&- Q_R(\gamma,\eta',\Xi_1)P_R(\gamma,\eta',\overline{\Xi}_1)\\
&+ \gamma^2 P_J(\eta',\gamma,\Xi_1)Q_J(\eta',\gamma,\overline{\Xi}_1) - \gamma^2 Q_J(\eta,\gamma,\Xi_1)P_J(\eta',\gamma,\overline{\Xi}_1)\Big].
\end{aligned}$$

The form J_2 includes the terms $P_R V Q_J \overline{V}$ and $Q_J V P_R \overline{V}$ containing the mth derivatives of V and $\overline{V}$. Integrating by parts we obtain integrated terms with γ as coefficients. In other words,

$$H(\gamma,\xi,\Xi_1,\overline{\Xi}_1) = H_0(\gamma,\xi,\Xi_1,\overline{\Xi}_1) + \gamma^2 H_1(\gamma,\xi,\Xi_1,\overline{\Xi}_1).$$

Setting $\gamma = 0$ and $\eta' = \eta'^0$ and taking the formula for $H_0(\gamma,\eta',\Xi_1,\overline{\Xi}_1)$ into account we arrive at (58). As has already been noted, the form $G(\gamma,\eta',D_1,\overline{D}_1)$ is obtained from $J_2(\gamma,\eta',D_1,\overline{D}_1)$ by integrating by parts the terms containing D_1^m or $\overline{D}_1^m$. Setting $\gamma = 0$ and $\eta' = \eta'^0$ in J_2 after the integration we obtain (59).

§7. Energy estimate for the case of more general boundary conditions

In §3 we reduced the proof of the existence theorem for the mixed problem to the derivation of inequalities (2.9) (the "direct estimate") and (3.20) (the "dual estimate"). Sections 5 and 6 were devoted to detailed derivation of (2.9). The present section splits into two parts. In the first part we show that a slight modification of the arguments in §§5 and 6 makes it possible to prove an estimate for the mixed problem with boundary conditions of the type (3.18). In particular, we present a natural generalization of uniform Lopatinskii's condition. In the second part we prove that uniform Lopatinskii's condition for the original problem induces a similar condition for the problem with conditions (3.18).

7.1. Statement of the problem. We consider the equation

$$P(y,D)u(y) = f(y), \quad y \in \mathbb{R}^{n+1}_+, \tag{1}$$

in the half-space $\{x_1 \geqslant 0\}$ with boundary conditions

$$B_j(y',D)u\big|_{x_1=0} + \sum_{k=1}^{\kappa} C_{jk}(y',D')\rho_k(y') = g_j(y'), \qquad j = 1,\dots,\mu, \tag{2}$$

and zero initial conditions written in the form

$$u(y) = f(y) = \rho_k(y') = g_j(y') = 0 \quad \text{for} \quad t<0. \tag{3}$$

As before, $B_j(y,D)$ are differential operators of orders $\beta_j < m$, $j = 1,\dots,\mu$, and $C_{jk}(y',D')$ are differential operators with respect to the variables y', and it is assumed that there are some (not necessarily positive) integers $\alpha_1,\dots,\alpha_\kappa$ such that[3]

$$\deg C_{jk} \leqslant \beta_j + \alpha_k, \qquad j = 1,\dots,\mu, \quad k = 1,\dots,\kappa. \tag{4}$$

As before, we will assume that the coefficients of the operators under study belong to C^∞ and are stabilized for large $|y|$ (accordingly, for large $|y'|$).

It will also be assumed that conditions (A) and (B) of §1 hold. We will present analogs of conditions (C) and (D) of §1 for problem (1), (2). As in §1, these conditions will relate only to the principal homogeneous parts P_0 of the operators under consideration. And, as before, we will factorize the symbol of an operator: $P_0 = P_{0+}P_{0-}$.

Condition (C) in §1 should be replaced by

Condition $(\widehat{C})$. $\mu = \kappa + m^+$, $m^+ = \deg P_{0+}$.

We now proceed to the statement of Lopatinskii's condition. As in §1, X will denote the set of parameters $(y', \xi', \sigma + i\gamma)$, and for the principal symbol of an operator P the notation $P_0(X,\xi_1)$ will be used. Furthermore, as in §1, we will denote by $\mathfrak{M}(X)$ the space of solutions $V(z)$ to the equation

$$P_0(X,D_z)V(z) = 0, \qquad z \geqslant 0, \tag{5}$$

and $\mathfrak{M}_+(X)$ will denote the subspace of solutions bounded on the half-line $z \geqslant 0$.

We will say that problem (1), (2) satisfies Lopatinskii's condition if equation (5) with boundary conditions

$$B_{0j}(X,D_z)V(0) + \sum_{k=1}^{\kappa} C_{0jk}(X)R_k = c_j, \qquad j = 1,\dots,\mu, \tag{6}$$

possesses a unique solution $V(z) \in \mathfrak{M}_+(X)$. According to Proposition 2 in 1.3, every solution $V(z) \in \mathfrak{M}_+(X)$ is a solution to the equation

$$P_{0+}(X,D_z)V(z) = 0, \tag{5_+}$$

and in this case the boundary condition (6) becomes

$$B'_{0j}(X,D_z)V(0) + \sum_{k=1}^{\kappa} C_{0jk}(X)R_k = c_j, \tag{6_+}$$

[3] If $\beta_j + \alpha_k < 0$, then $C_{jk} \equiv 0$.

where

$$B'_{0j}(X,\lambda)=\sum_{k=1}^{m^+} b_{jk}(X)\lambda^{m^+-k}$$

is the remainder on dividing $B_{0j}(X,\lambda)$ by $P_{0+}(X,\lambda)$.

We define the matrix

$$F(X)=\|T_{jk}(X)\|_{j,k=1,\dots,\mu=m^++\kappa}, \tag{7}$$

where

$$T_{jk}(X)=\begin{cases} b_{jk}(X), & j,k=1,\dots,m^+,\\ C_{j,k-m^+}(X), & k=m^++1,\dots,\mu\end{cases} \tag{7'}$$

for $j=1,\dots,\mu$. Denoting by W the column composed of $V(0),\dots,D_z^{m^+-1}V(0)$, $R_1,\dots,R_\kappa$ and by C the column of the right-hand sides of (6) we can rewrite (6_+) in the form of an algebraic linear system of equations

$$F(X)W=C. \tag{8}$$

Since the solution to (5_+) can be reconstructed uniquely from the germ $V(0),\dots,$ $D_z^{m^+-1}V(0)$, the unique solvability of problem (5), (6) in $\mathfrak{M}_+(X)$ is equivalent to the unique solvability of system (8), i. e., to the nondegeneracy of the matrix (7):

$$|\det F(X)|>0. \tag{9}$$

In turn, condition (9) is equivalent to the inequality

$$\begin{aligned}&\sum_{j=1}^{m^+}|(D_z^i V)(0)|+\sum_{k=1}^{\kappa}|R_k|\\ &\quad\leqslant K(X)\sum_{j=1}^{\mu}\Big|B_j(X,D_z)V(0)+\sum_{k=1}^{\kappa}C_{jk}(X)R_k\Big| \quad \forall V(z)\in\mathfrak{M}^+(X).\end{aligned} \tag{10}$$

We say that problem (1), (2) satisfies uniform Lopatinskii's condition if for any values of the parameter X in (1.8) the left-hand side of (9) is uniformly separated from zero or, which is the same, the constants $K(X)$ in (10) have an upper bound K not depending on X.

Thus, conditions (A) and (B) of §1 and condition $(\widehat{C})$ will be supplemented with

Condition $(\widehat{D}.)$ Problem (1), (2) satisfies uniform Lopatinskii's condition.

7.2. A priori estimate for problem (1), (2). We have the following

THEOREM. *Let problem* (1), (2) *satisfy conditions* (A), (B), $(\widehat{C})$, *and* $(\widehat{D})$. *Then for all* $S>0$ *there is* $\gamma_0=\gamma_0(S)$ *such that for* $|s|\leqslant S$ *and* $\gamma\leqslant\gamma_0$ *the inequality*

$$\begin{aligned}&|\gamma|\,\|u\|^2_{m,s-1,[\gamma]}+\sum_{j=0}^{m-1}\langle D_1^j u\rangle^2_{(m-1+s-j),[\gamma]}+\sum_{k=1}^{\kappa}\langle\rho_k\rangle_{(m-1+s+\alpha_k),[\gamma]}\\ &\quad\leqslant C\Big(|\gamma|^{-1}\|P(y,D)u\|^2_{0,s,[\gamma]}\\ &\qquad+\sum_{j=1}^{\mu}\Big\langle B_j(y,D)u+\sum_{k=1}^{\kappa}C_{jk}(y',D')\rho_k\Big\rangle^2_{(m-1+s-\beta_j),[\gamma]}\Big)\end{aligned} \tag{11}$$

holds.

PROOF OF THEOREM 7.2. It follows the scheme of the proof of Theorem 5.1. We briefly remind the reader of the main stages of the proof and indicate the additional details relating to the change in the form of the boundary conditions.

1) We first of all note that it suffices to prove the inequality obtained from (11) by replacing the term $|\gamma|\,\|u\|_{m,s-1,[\gamma]}$ on the right-hand side by $|\gamma|\,\|u\|_{m-1,s,[\gamma]}$ (cf. 5.1).

2) Proposition 5.1 admits a trivial modification in application to inequality (11). Therefore we can repeat literally the remarks in the end of 5.1. Thus, it can be assumed that the operators P, B_j, and C_{jk} contain no lower terms, the operators B_j and C_{jk} have constant coefficients, and the symbol $P(y,\eta)$ has the form (5.4).

3) Inequality (11) can be microlocalized (cf. 5.2) so that the proof reduces to showing that the corresponding estimates hold in the parabolic and hyperbolic zones (see (5.7) and (5.8)).

7.3. Proof of inequality (11) in the parabolic zone. 1) In the zone (5.7), (5.8) inequality (11) goes into (cf. (5.1) and (5.1p))

$$\begin{aligned}&\|u\|_{m-1,s+1/2,[\gamma]} + \langle D_1^{m-1}u\rangle_{(s),[\gamma]} + \sum_{k=1}^{\kappa}\langle\rho_k\rangle_{m-1+s+\alpha_k,[\gamma]}\\ &\quad\leqslant C\Bigg(\|P(y,D)u\|_{0,s-1/2}\\ &\qquad+\sum_{j=1}^{\mu}\Big\langle B_j(y',D)u+\sum_{k=1}^{\kappa}C_{jk}(y',D')\rho_k\Big\rangle_{(m-1+s-\beta_j),[\gamma]}\Bigg).\end{aligned}\tag{11p}$$

REMARK. We have dropped the norms

$$\langle D_1^j u\rangle^2_{m-1+s-j,[\gamma]},\qquad j=0,\dots,m-2,$$

in inequality (11p) because they are estimated via the first term on the left-hand side of (11p) (see Proposition 1 in 5.3).

2) We will derive (11p) from an analog of inequality (5.10). Namely, if the symbol $Q(y,\eta)$ has the form (5.12), then the inequality

$$\begin{aligned}&\|u\|^2_{m-1,s+1/2,[\gamma]} + \langle D_1^{m-1}u\rangle^2_{(s),[\gamma]} + \sum_{k=1}^{\kappa}\langle\rho_k\rangle^2_{m-1+s+\alpha_k,[\gamma]}\\ &\quad\leqslant\Bigg(\operatorname{Im}[Pu,Qu]_{(s),[\gamma]}+\sum_{j=1}^{\mu}\Big\langle B_ju+\sum_{k=1}^{\kappa}C_{jk}\rho_k\Big\rangle^2_{m-1+s-\beta_j,[\gamma]}\Bigg)\end{aligned}\tag{12}$$

holds.

3) Literal repetition of the arguments in Theorem 5.3 reduces the proof of (12) to the case of symbols P and Q not depending on y. Passing to the Fourier-Laplace

transform with respect to the variables y' we arrive at an inequality on the half-line:
(13)

$$C_0\Big(\sum_{j=0}^{m-1}\Lambda^{2(m-1-j)+1}\int_0^\infty |D^jV(z)|^2\,dz+|D^{m-1}V(0)|^2+\sum_{k=1}^{\kappa}\Lambda^{2(m-1+\alpha_k)}|R_k|^2\Big)$$
$$\leqslant \operatorname{Im}\int_0^\infty P(D,\eta',\gamma)V(z)\overline{Q(D,\eta',\gamma)V(z)}\,dz$$
$$+\sum_{j=1}^{m}\Lambda^{2(m-1-\beta_j)}\Big|B_j(D,\eta',\gamma)V(0)+\sum_{k=1}^{\kappa}C_{jk}(\eta',\gamma)R_k\Big|^2$$
$$\forall V(z)\in H^{(m)}(\mathbb{R}_+),\quad (R_1,\dots,R_\kappa)\in C^\kappa.$$

4) The proof of (13) is similar to that of (5.13). The only distinction is that Proposition 2 of 5.3 should be replaced by the following

PROPOSITION. *If the operators P, B_j, and C_{jk}, $j=1,\dots,\mu$, $k=1,\dots,\kappa$, satisfy Lopatinskii's condition, then the inequality*
(14)

$$C\sum_{j=0}^{m-1}\Lambda^{2(m-1-j)}\int_0^\infty |D^jV(z)|^2\,dz+\sum\Lambda^{2(m-1+\alpha-1/2)}|P_k|^2$$
$$\leqslant\int_0^\infty |P_l(D,\eta',\gamma)V(z)|^2\,dz+\sum_{j=1}^{\mu}\Big|(B_j(D,\eta',\gamma)V)(0)+\sum_{k=1}^{\kappa}C_{jk}(D,\eta',\gamma)R_k\Big|^2$$

holds.

PROOF. In the case of elliptic problems without parameters (and even in the case of pseudodifferential operators satisfying the transmission condition) this assertion was proved in **[Es2]** . In the case of problems with parameters the proof is simplified, and it is outlined in the end of Appendix to the present section.

7.4. Proof of inequality (11) in the hyperbolic zone. The proof follows the scheme of §6. The major distinction is that instead of Proposition 6.2 one should apply the following

PROPOSITION. *Let uniform Lopatinskii's condition $(\widehat{D})$ hold. Then for an arbitrary set $R=(R_1,\dots,R_\kappa)\in\mathbb{C}^\kappa$ and the germ $\mathcal{V}^m=(V(0),\dots,D^{m-1}V(0))$ the inequality*

$$C_1\Big(\sum_{j=0}^{m-1}|D^jV(0)|+\sum|R_k|\Big)\leqslant\sum_{j=0}^{l}\sum_{i=1}^{m_j^-}|F_{ij}(D,\eta',\gamma)V(0)|$$
$$+\sum_{j=1}^{\mu}\Big|B_j(D,\eta',\gamma)V(0)+\sum C_{jk}(\eta',\gamma)R_k\Big| \tag{15}$$

holds, where (η',γ) belong to the zone (6.1), $\gamma^2+|\eta'|^2=1$, and the constant C_1 does not depend on η' and γ and on the size of the hyperbolic zone.

PROOF. 1) First of all we note that (15) is a two-sided inequality, which implies its stability relative to small perturbations of the right-hand side. Hence, if the size ε of the zone in question is sufficiently small, then it suffices to prove (13) for a single point, say, for $\gamma = 0$, $\eta' = \eta'^0$. Denoting $F_{ij}(X, \lambda)$ and $B_j(X, \lambda)$ as $F_{ij}(\lambda)$ and $B_j(\lambda)$, respectively, we reduce the proof of (12) to showing that the inequality

$$(15') \qquad \begin{aligned} C_1'\Big(\sum_{j=0}^{m-1} |D^j V(0)| + \sum_{k=1}^{\kappa} |R_k|\Big) \\ \leqslant \sum_{j=0}^{l}\sum_{i=1}^{m_j^-} |F_{ij}(D)V(0)| + \sum_{j=1}^{\mu}\Big|(B_j(D)V)(0) + \sum_{k=1}^{\kappa} C_{jk}R_k\Big| \end{aligned}$$

holds.

2) We introduce the notation

$$E_j(X, \lambda) = P_+(X, \lambda)\lambda^j, \quad E_j(\lambda) = E_j(X_0, \lambda), \qquad j = 0, 1, \dots .$$

To derive (15′) it suffices to prove the inequality

$$(16) \qquad \begin{aligned} C_2\Big(\sum_{j=0}^{m-1} |D^j V(0)| + \sum_{k=1}^{\kappa} |R_k|\Big) \\ \leqslant \sum_{j=1}^{m^- -1} |E_j(D)V(0)| + \sum_{j=1}^{\mu}\Big|(B_j(D)V)(0) + \sum_{k=1}^{\kappa} C_{jk}R_k\Big|. \end{aligned}$$

Indeed, as was already mentioned (see the proof of Proposition 6.2), the polynomials $F_{ij}(\lambda)$ are divisible by $P_+(\lambda)$, and the quotients $F_{ij}(\lambda)/P_+(\lambda)$ are linearly independent. Since the degrees of these quotients do not exceed $m_- - 1$ and their number is equal to m_-, any monomial λ^j, $j = 0, \dots, m_- - 1$, can be represented as a linear combination of F_{ij}/P_+. Therefore the polynomials $F_{ij}(\lambda)$ are linear combinations of $F_{ij}(\lambda)/P_+(\lambda)$, and the right-hand side of (16) has an upper bound of the type of the right-hand side of (15′).

3) We proceed to the proof of (16). Denote by $\mathcal{C}^m$ the space of germs $(V(0), \dots, D^{m-1}V(0))$ and by $\mathcal{C}_+^m$ the subspace of $\mathcal{C}^m$ determined by the equations

$$(17) \qquad (E_j(D)V)(0) = 0, \qquad j = 0, \dots, m_- - 1.$$

Note that if $B_j'(\lambda)$ denotes the remainder on dividing $B_j(\lambda)$ by $P_+(\lambda)$, then we have

$$(18) \qquad (B_j(D)V)(0) = (B_j'(D)V)(0), \quad \forall (V(0), \dots, D^{m-1}V(0)) \in \mathcal{C}_+^m.$$

This inequality and the uniform nonsingularity of the matrix (7) yield inequality (16) in the subspace $\mathcal{C}_+^m$.

REMARK. Equations (17) separate out the germs at zero of the solutions belonging to the subspace $\mathfrak{M}^+$, and (16) goes into inequality (10).

4) Denote by $\mathcal{C}_-^m$ the complementary subspace of $\mathcal{C}_+^m$, which is of dimension m_+. The definition of this finite-dimensional space implies that there is $C_3 > 0$ such that

$$\sum_{j=0}^{m^- -1} |(E_j(D)V)(0)| > C_3 \sum_{j=0}^{m-1} |D^j V(0)| \quad \forall (V(0), \dots, D^{m-1}V(0)) \in \mathcal{C}_-^m.$$

Furthermore, for any $\lambda \in (0,1)$ we have

$$\sum_{j=1}^{\mu}\Big|B_j(D)V(0)+\sum_{k=1}^{\kappa}C_{jk}R_k\Big| \geqslant \sum_{j=1}^{\mu}\lambda\Big|\sum C_{jk}R_k\Big| - C_4\lambda\sum_{j=0}^{m-1}|D^jV(0)|.$$

According to condition $(\widehat{D})$, the matrix (7) is nonsingular, and therefore the rank of the matrix

$$\|C_{jk}(\eta',\gamma)\|_{k=1,\dots,\kappa,j=1,\dots,\mu}$$

is equal to κ, whence

$$\sum_{j=1}^{\mu}\Big|\sum_{k=1}^{\kappa}C_{jk}(\eta'\gamma)R_k\Big| > C_5\sum_{k=1}^{\kappa}|R_k|.$$

Comparing these inequalities and choosing a sufficiently small λ we prove (16) for $(V(0),\dots,D^{m-1}V(0)) \in \mathcal{C}_-^m$. The proposition is proved

Combining Theorems 1′, 2′, and 3′ of 6.3 with the above proposition we prove (11).

7.5. Proof of the estimate for the transposed problem. When proving the basic estimates (5.1) we assumed that $\gamma < 0$. However, the desired inequality (3.21) involves $\gamma > 0$. As can easily be seen, the change of sign of γ results in a change in conditions (C) and (D) and, accordingly, $(\widehat{C})$ and $(\widehat{D})$, but it does not produce any significant effect on the proof of (11). To distinguish between the cases of positive and negative γ we will include the sign of γ in the notation of the conditions, i. e., we will write $(C^{\pm})$, $(D^{\pm})$, $(\widehat{C}^{\pm})$, and $(\widehat{D}^{\pm})$. Thus, Theorem 7.1 will remain true under the replacement of γ by $-\gamma$ and conditions $(\widehat{C}^-)$ and $(\widehat{D}^-)$ by $(\widehat{C}^+)$ and $(\widehat{D}^+)$, respectively. Therefore the desired inequality (3.21) is a consequence of an assertion of a purely algebraic nature. We have the following

THEOREM. *Let conditions* (A), (B), $(\widehat{C}^-)$, *and* $(\widehat{D}^-)$ *hold for the original problem* (2.5), (2.6). *Then the transposed problem satisfies conditions* (A), (B), $(\widehat{C}^+)$, *and* $(\widehat{D}^+)$.

The verification of the first three conditions encounters no difficulties. Since $P_0(y,\eta)$ is a real symbol, the principal homogeneous part of $P^*(y,\eta)$ coincides with $P_0(y,\eta)$ so that conditions (A) and (B) hold for the operator P^*.

Verification of condition $(\widehat{C}^+)$. Let $m_\pm(\gamma)$ be the number of the roots of the symbol $P_0(y',\lambda,\xi',\sigma+i\gamma)$ in the half-plane $\pm\operatorname{Im}\lambda > 0$. The number of the boundary conditions (3.12) is equal to m and the number of the additional functions in the boundary conditions is equal to $m^+(\gamma)$, where $\gamma < 0$. Thus, condition $(\widehat{C}^+)$ means that

$$m = m_+(-\gamma) + m_+(\gamma), \qquad \gamma < 0. \tag{19}$$

However, since P_0 is real, by passing to the complex conjugate quantities in the factorization

$$P_0(y',\lambda,\xi',\sigma+i\gamma) = P_{0+}(y',\eta',\gamma,\lambda)P_0^-(y',\eta',\gamma,\lambda)$$

we obtain the factorization for $-\gamma$:

$$\begin{aligned}P_0(y',\lambda,\xi',\sigma-i\gamma) &= \overline{P_{0-}(y,\eta',\gamma,\lambda)}\;\overline{P_{0+}(y,\eta',\gamma,\lambda)}\\ &= P_{0+}(y,\eta',-\gamma,\lambda)P_{0-}(y,\eta',-\gamma,\lambda).\end{aligned}$$

Hence, we have $m_+(-\gamma) = m_-(\gamma)$, which implies (19).

The remaining part of this section is devoted to the derivation of condition $(\widehat{D}^+)$ from (C^-) and (D^-). In the course of the proof we will employ some ideas of Sakamoto **[S2]**.

7.6. Uniform Lopatinskii's condition for problem (3.17), (3.18). We break the proof of Theorem 7.5 into several stages.

7.6.1. Let us once again state the assertion to be checked. To this end we have to write down the principal symbols of the boundary operators entering (3.18). When determining the principal symbols at a point $y' = y'^0$ we can consider operators having constant coefficients ("frozen" at that point) and involving no lower terms. In this case the symbols of the operators in (3.18) are found from the relation (see (3.16) and Lemma 3 in 3.3)

$$\begin{aligned}(\overline{P}(D)\theta_+ - \theta_+\overline{P}(D))u - \sum_{j=1}^{\mu} \overline{B}_j(D)(\rho_j(y') \times \delta(x_1))\\ = \sum_{l=1}^{m}\Big[(E_l(D)u)(0) + \sum_{j=1}^{\mu} H_{lj}(D')\rho_j\Big] D_1^{l-1}\delta(x_1).\end{aligned}$$

We substitute $u = \exp(\gamma t)h(y')W(x_1)$ and $\rho_j(y') = \exp(\gamma t)h(y')R_j$ into this relation and pass to the Fourier transform with respect to y'. Assuming that $\widehat{h}(\xi', \tau)$ does not vanish, we arrive at

$$\begin{aligned}&\big(P(D_1, \eta', \sigma - i\gamma)\theta_+(x_1) - \theta_+(x_1)P(D_1, \xi', \sigma - i\gamma)\big) W(x_1)\\ &\quad - \sum_{j=1}^{\mu} R_j \overline{B}_j(D_1, \xi', \sigma - i\gamma)\delta(x_1)\\ &= \sum_{l=1}^{m}\Big[(E_l(D_1, \xi', \sigma - i\gamma) W)(0) + \sum H_{lj}(\xi', \sigma + i\gamma)R_j\Big] D^{l-1}\delta(x_1).\end{aligned} \tag{20}$$

Thus, we have to check the fulfillment of uniform Lopatinskii's condition for the following problem on the half-line:

$$P(D, \eta', \sigma - i\gamma)W(z) = 0, \quad D = -id/dz, \tag{21}$$

$$E_l(D, \eta', \sigma - i\gamma)W(0) + \sum_{k=1}^{\mu} R_k H_{lk}(\xi', \sigma - i\gamma) = h_l, \quad l = 1, \dots, m, \tag{22}$$

where $\gamma < 0$, $\eta'^2 + \gamma^2 = 1$ and $\mu = m_+(\gamma) = m_-(-\gamma)$.

7.6.2. Consider a given nondegenerate matrix $C(\eta', \gamma) = \|C_{jk}(\eta', \gamma)\|_{j,k=1,\dots,m}$ whose elements are continuous and homogeneous functions of degree zero in the variables (η', γ) such that

$$C^{-1} \leqslant |\det C(\eta', \gamma)| < C \quad \forall(\eta', \gamma),\ \gamma \leqslant 0.$$

We set

$$'E_l(\lambda, \eta', -\gamma) = \sum_{r=1}^{m} C_{lr}(\eta', \gamma) E_r(\lambda, \xi', \sigma - i\gamma),$$

$$'H_{lj}(\lambda, \eta', -\gamma) = \sum_{r=1}^{m} C_{lm}(\eta', \gamma) H_{rj}(\lambda, \xi', \sigma - i\gamma).$$

It is clear that problem (21), (22) and the problem obtained from it by replacing E_l and H_{lj} by $'E_l$ and $'H_{lj}$ satisfy or do not satisfy uniform Lopatinskii's condition simultaneously. We will say that two problems of this kind are equivalent.

We will also need an equivalent representation of uniform Lopatinskii's condition for the original problem (2.5), (2.6).

LEMMA. *Let $P_+(X, \lambda)$, $B_1(X, \lambda), \dots, B_\mu(X, \lambda)$, $\mu = \deg P_+$, $X = (\eta', \gamma)$, be a given set of polynomials. Then the following conditions are equivalent.*

(i) *The polynomials $B_1(X, \lambda), \dots, B_\mu(X, \lambda)$ are uniformly linearly independent* mod $P_+(X, \lambda)$ *for $\eta'^2 + \gamma^2 = 1$.*

(ii) *There is a set of polynomials (in λ) $H_j(X, \lambda')$, $j = 1, \dots, m-\mu$, $\deg H_j < m-\mu$, with coefficients continuous in X such that the polynomials $\{B_j(X, \lambda)\}$, $j = 1, \dots, m$, where $B_j(X, \lambda) = H_{j-\mu}(X, \lambda) P_+(X, \lambda)$ for $j > \mu$, are uniformly linearly independent for $\eta'^2 + \gamma^2 = 1$.*

The assertion of the lemma is obviously true.

7.6.3. We now consider a transformation of relations (20) that will result in the replacement of problem (21), (22) by an equivalent problem in the sense of 7.6.2. By virtue of the lemma, the monomials $1, \dots, \lambda^{m-1}$ can be expressed in terms of $B_1(X, \lambda), \dots, B_m(X, \lambda)$ using a uniformly nonsingular matrix. Therefore, the commutator of $P(D, \xi', \sigma - i\gamma)$ and $\theta_+(x_1)$ can be written in the form

$$\begin{aligned} &(P(D, \xi', \sigma - i\gamma)\theta_+(z) - \theta_+(z) P(D, \xi', \sigma - i\gamma)) W(z) \\ &\qquad = \sum_{l=1}^{m} (F_l(D, \eta', -\gamma) W)(0) \overline{B}_l(D, \xi', \sigma - i\gamma) \delta(z). \end{aligned} \tag{23}$$

Thus, inequality (20) becomes

$$\begin{aligned} &(P(D, \xi', \sigma - i\gamma)\theta_+(z) - \theta_+(z) P(D, \xi', \sigma - i\gamma)) W(z) \\ &\qquad - \sum_{j=1}^{\mu} R_j \overline{B}_j(D, \xi', \sigma - i\gamma) \delta(z) \\ &\qquad = \sum_{l=1}^{\mu} [F_l(D, \xi', -\gamma) W(0) + R_l] \overline{B}_l(D, \xi', \sigma - i\gamma) \delta(z) \\ &\qquad + \sum_{l=\mu+1}^{m} F_l(D, \xi', -\gamma) W(0) \overline{B}_l(D, \xi', \sigma - i\gamma) \delta(z). \end{aligned} \tag{20'}$$

Comparing the right-hand sides of (20) and (20′) we find a set of equivalent boundary conditions in the sense of 7.6.2:

$$F_l(D, \xi', -\gamma)W(0) + R_l = h_l, \qquad l = 1, \dots, \mu. \tag{24}$$

$$F_l(D, \xi', -\gamma)W(0) = h_l, \qquad l = \mu + 1, \dots, m. \tag{24′}$$

Uniform Lopatinskii's condition for equation (21) with boundary conditions (24), (24′) is a consequence of the following assertion.

PROPOSITION. *Let the polynomials $B_j(\eta', \gamma, \lambda)$ in Lemma* 7.6.2 *be uniformly linearly independent and let the symbols $F_j(X, \lambda)$ be defined by means of* (23). *Then*

(i) *the polynomials $F_j(X, \lambda)$ are uniformly linearly independent for $\eta'^2 + \gamma^2 = 1$*;

(ii) *the polynomials $F_j(\lambda, \eta', -\gamma)$, $j = 1, \dots, \mu$, are divisible by $P_+(\lambda, \eta', -\gamma)$.*

Assuming that the proposition is proved we complete the proof of Theorem 7.5.

Comparing assertions (i) and (ii) of the proposition we see that the polynomials $F_j(\lambda, \eta', -\gamma)$, $j = \mu+1, \dots, m$, are linearly independent[4] mod $P_+(\lambda, \eta', -\gamma)$. Furthermore, every solution $W(z)$ to problem (21), (24), (24′) bounded for $z > 0$ is a solution to the problem

$$P_+(D, \eta', -\gamma)W(z) = 0, \qquad z > 0, \tag{21′}$$

$$F'_l(D, \eta', \gamma)W(0) = h_l, \qquad l = \mu + 1, \dots, m, \tag{24″}$$

where F'_l are the remainders on dividing F_l by $P_+(\lambda, \eta', -\gamma)$. It is the uniform linear independence of the polynomials F'_l, $l = \mu + 1, \dots, m$, that implies the fulfillment of uniform Lopatinskii's condition for problem (21′), (24″). Hence, problem (21′), (24″) is solvable for any right-hand sides, and an estimate uniform with respect to the parameters holds for it. Therefore the original problem (21), (24), (24′) also possesses this property, i.e., uniform Lopatinskii's condition holds for it.

7.6.4. Before proceeding to the proof of the proposition we establish a relationship between the operators P, F_l, and B_l. To this end we apply the functional (23) of finite support to a function $V(z)$. This results in

$$\int_0^\infty \left[W(z)P(-D, \xi', \sigma - i\gamma)V(z) - V(z)P(D, \xi', \sigma + i\gamma)W(z)\right] dz$$
$$= \sum_{l=1}^m F_l(D, \eta', -\gamma)W(0)\overline{B}_l(D, \xi', \sigma + i\gamma)V(0).$$

Let us substitute $W(z) = \exp(i\lambda z)$ and $V(z) = \exp(-i\rho z)$, where $\operatorname{Im}(\lambda - \rho) > 0$, into this relation. In this case the integrals on the left-hand side are absolutely convergent, and therefore the integration by parts leading to this relation is legitimate. As a result, we find

$$i\frac{P(\lambda, \xi', \sigma - \gamma) - P(\rho, \xi', \sigma - i\gamma)}{\lambda - \rho} = \sum_{l=1}^m F_l(\lambda, \eta', -\gamma)\overline{B}_l(\rho, \xi', \sigma - i\gamma). \tag{25}$$

Formula (25) has been derived for the case $\operatorname{Im}(\lambda - \rho) > 0$. Since the left-hand and right-hand sides are polynomials in variables ρ and λ, formula (25) holds for all λ, $\rho \in \mathbb{C}$.

[4]Recall that $\deg P_+(\lambda, \eta', -\gamma) = \deg P_-(\lambda, \eta', \gamma) = m_-(\gamma) = m - \mu$.

7.6.5. Proof of Proposition 7.6.3 (i). We define square matrices $\mathcal{B}$ and $\mathcal{F}$ by means of the relations

$$\begin{pmatrix} \overline{B}_1(\lambda) \\ \vdots \\ \overline{B}_1(\lambda) \end{pmatrix} = \mathcal{B} \begin{pmatrix} 1 \\ \vdots \\ \lambda^{m-1} \end{pmatrix}, \quad \begin{pmatrix} F_1(\lambda) \\ \vdots \\ F_m(\lambda) \end{pmatrix} = \mathcal{F} \begin{pmatrix} 1 \\ \vdots \\ \lambda^{m-1} \end{pmatrix}.$$

Denoting by $D(\lambda)$ the column vector $(\lambda^{m-1}, \dots, 1)$ we rewrite the right-hand side of (25) as a scalar product:

$$\langle \mathcal{F}D(\lambda), \mathcal{B}D(\rho)\rangle = \langle D(\lambda), {}^t\mathcal{F}\mathcal{B}D(\rho)\rangle. \tag{25'}$$

On the other hand, we can write

$$P(\lambda, \xi', \sigma + i\gamma) = \sum_{j=0}^{m} a_j(\xi', \sigma - i\gamma)\lambda^{m-j}, \qquad a_0 \equiv 1.$$

It follows that the left-hand side of (25) is equal to

$$\sum_{j=0}^{m-1}\sum_{i=0}^{m-j-1} a_j \lambda^{m-j-1-i}\rho^i. \tag{26}$$

We introduce the matrix

$$\mathcal{A} = \begin{pmatrix} 0 & 0 & \dots & 0 & a_0 \\ 0 & 0 & \dots & a_0 & a_1 \\ \vdots & \vdots & \ddots & \vdots & \vdots \\ 0 & a_0 & \dots & a_{m-3} & a_{m-2} \\ a_0 & a_1 & \dots & a_{m-2} & a_{m-1} \end{pmatrix}$$

and rewrite (26) in the form

$$\langle D(\lambda), \mathcal{A}D(\rho)\rangle. \tag{25''}$$

The comparison of (25′) and (25″) implies

$$i\mathcal{A} = {}^t\mathcal{F}\mathcal{B}, \quad \mathcal{F} = {}^t(\mathcal{A}\mathcal{B}^{-1}).$$

Since we have $\det \mathcal{A} = a_0^m = 1$, and $|\det \mathcal{B}|$ possesses uniform upper and lower bounds, $|\det \mathcal{F}|$ possesses the same property, which proves the desired assertion.

7.6.6. Proof of Proposition 7.6.3 (ii). We multiply both sides of (25) by $\rho^{k-1}/2\pi i P_-(\eta', -\gamma, \rho)$ and integrate the resulting relation over a contour Γ encircling the zeros of $P_-(\eta', -\gamma, \rho)$ and lying at a positive distance from the zeros of $P_+(\eta, -\gamma, \rho)$. The integration results in

$$\begin{aligned} &\frac{P(\lambda, \xi', \sigma + i\gamma)}{2\pi i}\int_\Gamma \frac{\rho^{k-1}\,d\rho}{(\lambda - \rho)P_-(\eta' - \gamma, \rho)} - \frac{1}{2\pi i}\int_\Gamma \frac{P_+(\eta', -\gamma, \rho)\rho^{k-1}}{\lambda - \rho}\,d\rho \\ &\qquad = \sum_{j=1}^{m} b_{kj}(\eta', -\gamma)F_j(\lambda, \eta', -\gamma), \end{aligned} \tag{27}$$

where

$$(28) \qquad b_{kj}(\eta', -\gamma) = -\frac{1}{2\pi i}\int\limits_{\Gamma} \frac{\overline{B}_j(\rho, \xi', \sigma - i\gamma)}{P_-(\eta', -\gamma, \rho)}\, d\rho.$$

If λ lies outside Γ, then the second integral on the left-hand side of (27) is equal to zero. Furthermore, since the function $\rho^{k-1}/P_-(\eta', -\gamma, \rho)$ is holomorphic in the exterior of Γ for $k \leqslant \mu = \deg P_+(\eta', \gamma, \rho) = \deg P_-(\eta', -\gamma, \rho)$, the Cauchy theorem implies that the integral is equal to $\lambda^{k-1}/P_-(\eta', -\gamma, \lambda)$, and the first term on the left-hand side is equal to $\lambda^{k-1}P_+(\eta', -\gamma, \lambda)$.

We now consider the right-hand side of (27). According to the definition (see Lemma 7.6.2), we have $B_j(\rho, \xi', \sigma + i\gamma) = H_{j-\mu}(\eta', -\gamma, \rho)P_-(\eta', \gamma, \rho)$ for $j > \mu$, whence $\overline{B}_j(\lambda, \xi', \sigma - i\gamma) = \overline{H_{j-\mu}(\eta', -\gamma, \lambda)}P_-(\eta', -\gamma, \rho)$, i. e., the integrand expression in (28) has no poles inside the contour Γ, which implies $b_{kj} = 0$ for $j > \mu$. Thus, (27) takes the form

$$(29) \qquad \sum_{j=1}^{m_+} b_{kj}(\eta', -\gamma)F_j(\lambda, \eta', \gamma) = \lambda^{k-1}P_+(\eta', -\gamma, \lambda).$$

By virtue of uniform Lopatinskii's condition in §1, the matrix $\|b_{kj}(\eta', -\gamma)\|$ is uniformly nonsingular so that upon solving system (29) we can represent $F_j(\lambda, \eta', -\gamma)$ as a product of some polynomials in λ by P_+. The proof of the theorem is completed.

Appendix to §7

1. We now extend the results of §5 to the case of more general problems in §7 and prove inequality (7.14).

Thus, we consider a problem on the half-line:

$$(1) \qquad Q(D, \Lambda)V(z) = F(z), \qquad z > 0,$$

$$(2) \qquad (B_j(D, \Lambda)V)(0) + \sum_{k=1}^{\kappa} C_{jk}(\Lambda)R_k = g_k, \qquad k = 1, \dots, \mu,$$

where $Q(\lambda, \Lambda)$ and $B_j(\lambda, \Lambda)$ are homogeneous polynomials in λ and Λ of degrees m and $\beta_j < m$. The polynomial Q satisfies the ellipticity condition with a parameter:

$$(3) \qquad Q(\lambda, \Lambda) \neq 0, \quad \lambda^2 + \Lambda^2 \neq 0, \quad \Lambda \geqslant 0.$$

We will assume that there are numbers $\alpha_1, \dots, \alpha_\mu$ such that

$$(4) \qquad C_{jk}(\Lambda) = c_{jk}\Lambda^{\beta_j + \alpha_k},$$

where $\alpha_1, \dots, \alpha_\mu$ can be negative and $c_{jk} = 0$ if $\beta_j + \alpha_k < 0$.

We state Lopatinskii's condition for problem (1), (2). By virtue of (3), the polynomial Q has no real roots, and therefore it admits the factorization

$$(5) \qquad Q(\lambda, \Lambda) = Q^+(\lambda, \Lambda)Q^-(\lambda, \Lambda), \quad Q^\pm(\lambda, \Lambda) \neq 0, \quad \pm\operatorname{Im}\lambda > 0.$$

It is assumed that

$$(6) \qquad \mu = \kappa + m_+, \quad m_\pm = \deg Q^\pm.$$

As in Appendix to §5, we denote by $B_j'(\lambda,\Lambda)$, $j=1,\dots,\mu$, the remainders on dividing $B_j(\lambda,\Lambda)$ by $Q^+(\lambda,\Lambda)$. Let us write

$$B_j'(\lambda,\Lambda)=\sum_{k=1}^{m_+} b_{jk}(\Lambda)^{m_+-k}, \quad b_{jk}(\Lambda)=b_k\Lambda^{b_j-m_++k}. \tag{7}$$

As in 7.1, we set $\mathcal{T}=\|T_{jk}\|$, $j,k=1,\dots,\mu$,

$$T_{jk}=\begin{cases} b_{jk}, & k=1,\dots,m_+, \quad j=1,\dots,\mu,\\ c_{j,k-m}, & k=m_++1,\dots, \quad m_++\kappa=\mu.\end{cases}$$

Lopatinskii's condition for problem (1), (2) is

$$\det\|T_{jk}\|\neq 0. \tag{8}$$

We now show that under the above conditions the inequality

(9)
$$C\Big(\sum_{j=0}^{m-1}\Lambda^{2(m-1-j)}\int_0^\infty |D^jV(z)|^2\,dz+\sum_{k=1}^{\kappa}\Lambda^{2(m-1-\alpha_k)-1}|R_k|^2\Big)$$
$$\leqslant\int_0^\infty |Q(D,\Lambda)V(z)|^2\,dz+\sum_{j=1}^{\mu}\Lambda^{2(m-1-\beta_j)-1}\Big|(B_j(D,\Lambda)V)(0)+\sum C_{jk}(\Lambda)R_k\Big|^2$$

holds.

2. As in Appendix to §5, first we consider the case when $V(z)$ is a solution to the homogeneous equation (1), i. e., $Q(D,\Lambda)V(z)=0$. Since $V(z)$ is bounded for $z\to+\infty$, we have

$$Q^+(D,\Lambda)V(z)=0, \quad (B_jV)(0)=(B_j'V)(0), \quad j=1,\dots,\mu. \tag{10}$$

Note that the fact that $\mathcal{T}$ is a nonsingular matrix implies that it has a nonsingular minor of maximum order:

$$\|b_{jk}\|, \quad j=1,\dots,\mu, \quad k=1,\dots,m_+.$$

To simplify the notation, assume that this minor corresponds to the first m_+ rows. Then the system of operators

$$\{Q,B_1,\dots,B_{m_+}\}$$

satisfies Lopatinskii's condition, and therefore the inequality (9) in Appendix to §5 holds for it:

$$C_0\sum_{j=0}^{m-1}\int_0^\infty \Lambda^{2(m-1-j)}|D^jV(z)|^2\leqslant\sum_{j=1}^{m_+}\Lambda^{2(m-1-\beta_j)-1}|(B_j(D,\Lambda)V)(0)|^2. \tag{11}$$

We extend the summation on the right-hand side to the values of j from 1 to μ and express $(B_j V)(0)$ from (2) to obtain

$$C_0 \sum_{j=0}^{m-1} \int_0^{\infty} \Lambda^{2(m-1-j)} |D^j V(z)|^2 \, dz - C_1 \sum_{k=1}^{\kappa} \Lambda^{2(m-1-\alpha_k)-1} |R_k|^2 \leqslant \sum_{j=1}^{\mu} \Lambda^{2(m-1-\beta_j)-1} |g_j|^2. \tag{12}$$

Furthermore, if relations (10) hold, then we have

$$\sum b_{jk} \Lambda^{b_j - m_+ + k} (D^{m_+ - k} V)(0) + \sum c_{jk} \Lambda^{\alpha_k + \beta_j} R_k = g_j, \quad j = 1, \ldots, \mu.$$

If Lopatinskii's condition is fulfilled, then we can solve the above system and represent $R_1, \ldots, R_\kappa$ and $(D^{m_+ - k} V)(0)$, $k = 1, \ldots, m_+$, as linear combinations of $g_1, \ldots, g_\mu$, which implies the inequality

$$\sum_{k=1}^{\kappa} \Lambda^{2(m-1+\alpha_k)-1} |R_k|^2 + \sum_{k=0}^{m_+ - 1} \Lambda^{2(m-1-k)-1} |(D^k V)(0)|^2 \leqslant K \sum \Lambda^{2(m-1-\beta_j)-1} |g_j|^2.$$

Multiplying this inequality by a suitably chosen constant and adding it to (12) we derive inequality (9) for the solution to the homogeneous equation.

3. In the general case the solution to problem (1), (2) is sought in the form $V = V^* + W$, where V^* is a solution to equation (1) and W is the solution to a problem of the type (1), (2) with zero right-hand side. More precisely,

$$Q(D, \Lambda) W(z) = 0, \quad z > 0,$$
$$(B_j(D, \Lambda) W)(0) + \sum C_{jk}(\Lambda) R_k = g_j - h_j,$$

where

$$h_j = (B_j(D, \Lambda) V^*)(0).$$

Now, using the already proved inequality (9) for the function W and modifying the arguments in 4 of Appendix to §5 we establish inequality (9) to the full extent.

§8. Mixed problem in a cylindrical domain

We come back to the statement of the mixed problem in the introduction to the present chapter. Let $\Omega \in \mathbb{R}^n$ be a given domain with the smooth boundary $\partial\Omega$. We will consider problem (0.1), (0.2), (0.3) in the cylinder $\Omega \times \mathbb{R}_t$. The coefficients of the operators entering (0.1) and (0.2) will be assumed to belong to C^∞, and if Ω is an unbounded region, then let the coefficients be constant in a neighborhood of infinity. By virtue of these assumptions, the boundary $\partial\Omega$ can be covered by a finite number of neighborhoods in each of which there exists a diffeomorphism of $\overline{\Omega}$ onto a neighborhood of the half-space. Thus, in a neighborhood of an arbitrary point on the boundary there is a local coordinate system with axis x_1 directed along the normal to $\partial\Omega$; the local coordinates on $\partial\Omega$ will be denoted x', and let $y' = (x', t)$.

Denote by P_0 and B_{0j} the principal symbols of the operators A and B_j calculated in the chosen coordinate system. We assume that the symbols $(P_0, B_{01}, \ldots, B_{0\mu})$ satisfy conditions (A), (B), (C), and (D) of §1. It is clear that these conditions are invariant under (nonsingular) transformations of the above coordinate systems.

We can define the spaces $H^{m,s}_{[\gamma]}(\Omega\times\mathbb{R})$ and $H^{(s)}_{[\gamma]}(\partial\Omega\times\mathbb{R})$ in the standard way using local charts. Since all estimates in §§2–7 were derived with the aid of localization, they can be extended in a trivial manner to the case of a manifold with boundary. Therefore we have the following

THEOREM. *The conditions below are equivalent for the cylinder* $\Omega\times\mathbb{R}$ *and the operators* $P, B_1, \dots, B_\mu$ *on it.*

(I) *Conditions* (A), (B), (C), and (D) *of* §1 *hold.*

(II) *For all* $S > 0$ *there is* $\gamma_0 = \gamma_0(S)$ *such that for* $|s| \leqslant S$ *and* $\gamma \leqslant \gamma_0$ *problem* (0.1), (0.2), (0.3) *with right-hand sides* $f \in H^{0,s+1}_{[\gamma]}(\Omega\times\mathbb{R})_+$ *and* $g_j \in H^{m+s-\beta_j}_{[\gamma]}(\partial\Omega\times\mathbb{R})_+$, $j = 1, \dots, \mu$, *possesses a unique solution* $u \in H^{m,s}_{[\gamma]}(\Omega\times\mathbb{R})_+$ *which satisfies the inequality*

$$|\gamma|\,\|u, \Omega\times\mathbb{R}\|^2_{m,s,[\gamma]} + \sum_{j=0}^{m-1} \langle D_1^j u, \partial\Omega\times\mathbb{R}\rangle^2_{(m+s-j),[\gamma]}$$
$$\leqslant C\left(|\gamma|^{-1}\|Pu, \Omega\times\mathbb{R}\|^2_{0,s+1,[\gamma]} + \sum_{j=1}^{\mu}\langle B_j u, \partial\Omega\times\mathbb{R}\rangle^2_{(m+s-\beta_j),[\gamma]}\right).$$

Let us discuss condition (B) of the theorem. According to this condition, for each point $\omega \in \partial\Omega\times\mathbb{R}$ the direction of the normal N_ω is a noncharacteristic direction for the operator P, and the number of roots of the equation

$$P_0(\omega; \xi + \lambda N_\omega, \tau) = 0, \qquad \operatorname{Im}\tau < 0, \quad \zeta \in \partial\Omega,$$

in the upper half-plane does not depend on ω, ζ, and τ. As is seen from the example of a strictly hyperbolic operator of odd order with constant coefficients, there exist operators that cannot satisfy condition (B) in any bounded region. We now state a sufficient condition for a strictly hyperbolic operator to satisfy condition (B) in an arbitrary bounded domain.

LEMMA. *Let* $P(\xi, \tau)$ *be a strictly hyperbolic homogeneous polynomial of degree* m *and let*

$$P(\xi, 0) \neq 0, \quad |\xi| \neq 0. \tag{1}$$

Then

(i) *the polynomial* P *has an even order, i. e.,* $\deg P = m = 2l$;

(ii) *for* $\operatorname{Im}\tau < 0$ *the equation (with respect to* λ)

$$P(\lambda, \xi', \tau) = 0 \tag{2}$$

has l *roots in each of the (upper and lower) half-planes.*

PROOF. (i) is an obvious consequence of (1). Let equation (2) have m_+ roots λ in the upper half-plane for $\operatorname{Im}\tau < 0$ and let $m_+ \neq l$. Then for $\operatorname{Im}\tau > 0$ this equation has $m - m_+ \neq m_+$ roots in the upper half-plane. It follows that for any real values of τ and, in particular, for $\tau = 0$ there must exist real roots, which contradicts (1). The lemma is proved.

Now let the principal symbol $P_0(y;\xi,\tau)$ of the operator P satisfy condition (A) in the cylinder $\Omega \times \mathbb{R}$ and let

$$|P_0(y',\xi,0)| > \delta > 0, \quad |\xi| \neq 0. \tag{3}$$

Then, by virtue of the lemma, the operator P has an even order and condition (B) holds at all points on the boundary. The first boundary value problem for this operator

$$\langle D, N\rangle u\big|_{\partial\Omega\times\mathbb{R}} = g_j, \quad j = 1,\dots,m/2,$$

where N is the normal vector to the boundary, satisfies conditions (A)–(D). By the above theorem, the first boundary value problem for a strictly hyperbolic operator of even order satisfying condition (3) is uniquely solvable.

CHAPTER IV

Mixed Problem for q-Parabolic and q-Hyperbolic Equations

Introduction

In the present chapter we complete the study of mixed problem for the three classes of equations that were considered in Chapter II in connection with the Cauchy problem. The theory of mixed problem for q-parabolic equations is much simpler than the similar theory for hyperbolic equations, and in this case the proof of the basic a priori estimate differs from the proof of Theorem III.5.3 only in some insignificant purely technical details (relating to the weight q). In studying the mixed problem for q-hyperbolic equations we use the exponential weight with respect to the spatial variables, and this requires the consideration of dual variables in the complex domain, which leads to a situation that in many respects resembles the one in Chapter II. In the derivation of the basic estimate we also have to carry out microlocalization and consider separately the "parabolic" and "hyperbolic" zones. However, contrary to Chapter III, in the hyperbolic zone we deal with perturbation of pairwise distinct real roots and therefore avoid the significant technical difficulties of III.6.3 relating to the existence of multiple real roots. Note that the theory of mixed problem for q-hyperbolic equations was constructed in **[GV4–GV8]**.

The presentation of the material in this chapter is organized as follows. In §1 we formulate the main solvability theorem for the mixed problem in the case of q-parabolic equations. Its proof involves no new ideas (as compared to hyperbolic equations), and therefore we only briefly outline the plan of the proof. Combining this theorem with the results of Chapter III we arrive at an analog of the theorem in II.5.5 for the case of mixed problem.

In §2 the mixed problem is posed for q-hyperbolic equations, some function spaces required for its investigation are introduced, and the main theorem is stated. Since the major part of the proof of the solvability theorem for the mixed problem in the case of q-hyperbolic equations is similar to the proof of the corresponding theorems for hyperbolic equations in Chapter III, we confine ourselves to some brief remarks. A more important distinction between q-hyperbolic and hyperbolic equations manifests itself in the derivation of the energy estimate, although in this case as well there are many "parallel" considerations. Section 3 is devoted to the derivation of an estimate for the mixed problem in the case of q-hyperbolic equations. The concluding §4 can in a sense be regarded as a summary for the entire book. It contains the statement of the "general" solvability theorem for the mixed problem whose "special" cases are the corresponding theorems for hyperbolic, q-parabolic, and q-hyperbolic equations.

§1. Mixed problem for q-parabolic equations

1.1. Statement of the problem. Basic assumptions. We will consider the mixed problem

$$P(y, D)u(y) = f(y), \qquad y \in \mathbb{R}^{n+1}_+, \tag{1}$$

$$B_j(y', D)u(y)\big|_{x_1=0} = g_j(y'), \qquad j = 1, \dots, \mu, \tag{2}$$

$$u(y) = f(y) = g_1(y') = \cdots = 0, \qquad t < 0. \tag{3}$$

As in Chapter III, it will be assumed that the coefficients of the operators under consideration belong to C^∞ and are constant for large $|y|$ (accordingly, for large $|y'|$).

We set a number $q \geqslant 1$ and denote by $P_0(y, \eta)$ and $B_{0j}(y, \eta)$ the q-principal parts of the symbols $P(y, \eta)$ and $B_j(y, \eta)$.

Condition (A). $q = 2b$ is an even number and $P_0(y, \xi, \tau)$ is a uniformly q-parabolic symbol (for the definition see II.5.1).

According to condition (A) (see the proposition in II.5.1), we have

$$P_0(y, \xi, \tau) \neq 0, \quad |\xi| + |\tau| \neq 0, \quad \operatorname{Im} \tau \leqslant 0.$$

It follows that $P_0(y, \xi, 0)$ does not vanish for $|\xi| \neq 0$, and therefore all $N \in \mathbb{R}^n$ are noncharacteristic directions for the symbol $P(y, \eta)$. In particular, condition (A) implies

Condition (B). $N = (1, 0, \dots, 0)$ is a noncharacteristic direction at the points of the plane $\{x_1 = 0\}$.

By condition (B), the symbol can be written in the form

$$P_0(y, \xi, \tau) = \xi_1^r + \sum_{j=1}^{r} P_{0j}(y, \xi', \tau)\xi_1^{r-j}, \tag{4}$$

where $P_{0j}(y, \xi', \tau)$ are q-homogeneous symbols (provided that the weight q is assigned to τ).

Consider the algebraic equation

$$P_0(y', \lambda, \xi', \tau) \overset{\text{def}}{=} P_0(X, \lambda) = 0, \tag{5}$$

and let $m_\pm(X)$ be the number of its roots in the open half-plane $\pm \operatorname{Im} \lambda > 0$.

LEMMA. *If condition* (A) *holds, then*

$$m_+(y', \xi', \tau) = m_-(y', \xi', \tau) = r/2, \tag{6}$$

$$\forall y \in \mathbb{R}^{n+1}, \quad |\xi'| + |\tau| > 0, \quad \operatorname{Im} \tau \leqslant 0. \tag{6'}$$

PROOF. Since P_0 is a quasi-homogeneous symbol, the sets of numbers (λ, ξ', τ) and $(-\lambda, -\xi', \tau)$ simultaneously satisfy equation (5) so that

$$r_+(y, \xi', \tau) = r_-(y, -\xi', \tau).$$

On the other hand, if (6′) holds, then, by virtue of condition (A), equation (5) has no real roots, and therefore the numbers $r_\pm(y', \xi', \tau)$ are constant; we will denote them as $r_\pm$. Since we have $r_+ + r_- = r$, inequality (6) holds, and, in particular, r is an even number.

REMARK. We have $r = qm$, where m is the degree of P with respect to the variable τ.

In view of the lemma, we will assume that the following condition holds.

Condition (C). $\mu = r/2$, where μ is the number of boundary conditions (2).

Since the symbol $P_0(y, \lambda, \xi', \tau)$ has no real roots, it admits the factorization:

$$P_0(X, \lambda) = P_{0+}(X, \lambda)P_{0-}(X, \lambda),$$

where $P_{0\pm}(X, \lambda)$ are polynomials in λ of degree $r/2$ with roots in the half-plane $\pm\lambda > 0$. Denote by $B'_{0j}(X, \lambda)$ the remainders on dividing $B_{0j}(X, \lambda)$ by $P_{0+}(X, \lambda)$, and let

$$B'_{0j}(X, \lambda) = \sum_{k=1}^{\mu} b_{jk}(X)\lambda^{\mu-k}.$$

We set

$$\operatorname{Lop}(P_0, B_{01}, \dots, B_{0\mu})(X) = \det \mathcal{B}(x), \quad \mathcal{B}(x) = \|b_{jk}.(X)\|.$$

Condition (D). There is $K > 0$ such that

$$|\operatorname{Lop}(P_0, B_{01}, \dots, B_{0\mu})(y', \xi', \tau)| \geqslant K, \qquad y' \in \mathbb{R}^n, \quad |\xi'|^2 + |\tau|^{2q} = 1, \quad \operatorname{Im}\tau \leqslant 0.$$

1.2. Basic spaces. We will need some spaces whose only distinction from those in III.2.1 is that the weight $q > 1$ is assigned to the dual variable to t. Thus, $H_{[\gamma]}^{(r,s|q)}(\mathbb{R}^{n+1})$ will denote the space of functions (distributions) with finite norm

$$(7) \qquad \left(\iint\limits_{\operatorname{Im}\tau=\gamma} (1 + |\xi|^2 + |\tau|^{2/q})^r (1 + |\xi'|^2 + |\tau|^{2/q})^s |\widehat{u}(\xi, \tau)|^2 \, d\xi \, d\tau \right)^{1/2},$$

and $H_{[\gamma]}^{(r,s|q)}(\mathbb{R}_+^{n+1})$ will be the space of restrictions to the half-space $\{x_1 \geqslant 0\}$. For a positive integer r the expression

$$(7') \qquad \begin{aligned} &\|u, \mathbb{R}_+^{n+1}\|_{(r,s|q),[\gamma]} \\ &= \left(\sum_{j=0}^{r} \int_0^{\infty} \iint\limits_{\operatorname{Im}\tau=\gamma} (1 + |\xi'|^2 + |\tau|^{2/q})^{r+s-j} |D_1^j \widehat{u}'(x_1, \xi', \tau)|^2 \, d\xi' d\tau \, dx_1 \right)^{1/2} \end{aligned}$$

can be taken as the norm in $H_{[\gamma]}^{(r,s|q)}(\mathbb{R}_+^{n+1})$. Below we will omit the symbol $\mathbb{R}_+^{n+1}$ in the notation of the norm.

We will also need the space $H_{[\gamma]}^{(s|q)}(\mathbb{R}^n)$ of functions (distributions) with support in the plane $\{x_1 = 0\}$, and the corresponding norm is

$$(8) \qquad \langle g \rangle_{(s|q),[\gamma]} = \left(\iint\limits_{\operatorname{Im}\tau=\gamma} (1 + |\xi'|^2 + |\tau|^{2/q})^s |\widehat{g}(\xi', \tau)|^2 \, d\xi \, d\tau \right)^{1/2}.$$

REMARK. The notation of the norm (8) for $q > 1$ does not agree with the notation in II.2.7, according to which this norm should have been denoted as $\langle g \rangle_{(sq|q),[\gamma]}$. However, to combine the a priori estimates in Chapter III and in the present chapter, we will use the suggested new notation.

1.3. THEOREM. *For a set of operators $\{P, B_1, \dots, B_\mu\}$ the conditions below are equivalent.*

(I) *Conditions* (A), (B), (C), *and* (D) *of* 1.1 *hold.*

(II) *For every S there is $\gamma_0 = \gamma_0(S)$ such that for $|s| \leqslant S$ and any set of right-hand sides $f \in H_{[\gamma]}^{(0,s|q)}(\mathbb{R}_+^{n+1})_+$, $g_j \in H_{[\gamma]}^{(r+s-\beta_j-1/2|q)}(\mathbb{R}^n)_+$, $j = 0, \dots, \mu$, there exists a unique solution $u \in H_{[\gamma]}^{(r,s|q)}(\mathbb{R}_+^{n+1})$, and the inequality*

(9)
$$\|u, \mathbb{R}_+^{n+1}\|_{(r,s|q),[\gamma]} \leqslant C\Big(\|P(y,D)u\|_{(0,s|q),[\gamma]} + \sum_{j=1}^{\mu} \langle B_j(y',D)u\rangle_{(r+s-\beta_j-1/2|q),[\gamma]}\Big)$$
$$\forall u \in H_{[\gamma]}^{(m,\infty)}(\mathbb{R}_+^{n+1})$$

holds.

REMARK. We present a weaker estimate that follows from (9) and coincides with (III.2.9) for $q = 1$:

(9′)
$$|\gamma|^{1/2q}\|u\|_{(r,s|q),[\gamma]} + \sum_{j=0}^{r-1}\langle D_1^j u\rangle_{(r+s-j|q),[\gamma]} \leqslant C\Big(|\gamma|^{-1/2q}\|Pu\|_{(0,s|q),[\gamma]} + \sum \langle B_j u\rangle_{(r+s-\beta_j|q),[\gamma]}\Big).$$

To derive (9′) from (9) one should replace s in (9) by $s + 1/2$, add the term

$$\sum_{j=0}^{r-1}\langle D_1^j u\rangle_{(r+s-j|q),[\gamma]}$$

to the left-hand side (by the embedding theorem, this term can be estimated via const $\|u\|_{(r,s+1/2|q),[\gamma]}$), and then apply the obvious inequalities

$$|\gamma|^{1/2q}\|u\|_{(r,s|q),[\gamma]} \leqslant \|u\|_{(r,s+1/2|q),[\gamma]},$$
$$\|Pu\|_{(0,s+1/2|q),[\gamma]} \leqslant |\gamma|^{-1/2q}\|Pu\|_{(0,s+1|q),[\gamma]}.$$

The proof of the above theorem follows the plan of the proof of Theorem III.2.2, which is the main subject of Chapter III. Here we confine ourselves to some brief remarks about the necessity of conditions (A), (B), (C), and (D) and their sufficiency for the validity of (9). The necessity of conditions (A), (B), (C), and (D) is derived following the scheme in III.2.3. In 1.5 we will prove their necessity for a somewhat weaker solvability assertion, which includes inequality (9′).

1.4. Scheme of the derivation of inequality (9). 1) Since (9) involves the large parameter $-\gamma$, it can be assumed, without loss of generality, that the operators P and $B_1, \dots, B_\mu$ coincide with their q-principal parts. Furthermore, on performing localization, we reduce the proof to the case when $u(y)$ has a small support. Then it can be assumed that the coefficients of the operators $P, B_1, \dots, B_\mu$ are nearly constant.

2) Since (9) is a two-sided estimate, the case of operators with almost constant coefficients can be easily reduced to operators with constant coefficients. Thus, we will

prove the following inequality (for convenience, we replace s with $s - 1/2$):

$$\|u\|_{(r,s-1/2|q),[\gamma]} + \sum_{j=0}^{r-1} \langle D_1^j u\rangle_{(r+s-j-1|q),[\gamma]} \leqslant C\left(\|P(D)u\|_{(0,s-1/2|q),[\gamma]} + \sum_{j=1}^{\mu} \langle B_j(D)u\rangle_{(r+s-1-\beta_j|q),[\gamma]} \right).$$

Using condition (B) in 1.1 and the relations

$$\|u\|^2_{(r,s-1/2|q),[\gamma]} = \|D_1^r u\|^2_{(0,s-1/2|q),[\gamma]} + \|u\|^2_{(r-1,s+1/2|q),[\gamma]}$$

we can derive the above inequality from a formally weaker estimate

$$\|u\|_{(r-1,s+1/2|q),[\gamma]} + \sum_{j=0}^{r-1} \langle D_1^j u\rangle_{(r+s-j-1|q),[\gamma]} \leqslant C\left(\|P(D)u\|_{(0,s-1/2|q),[\gamma]} + \sum_{j=1}^{\mu} \langle B_j(D)u\rangle_{(r+s-1-\beta_j|q),[\gamma]} \right). \tag{10}$$

In proving (10), we follow the scheme of III.5.3 and consider the form

$$J = \operatorname{Im}[P(D)u, Q(D)u]_{(0,s|q),[\gamma]},$$

where the notation (similar to that in III.5.3)

$$\begin{aligned} P(\xi_1, \xi', \tau) &= P_+(\eta', \gamma, \xi_1)P_-(\eta', \gamma, \xi_1), \\ Q(\xi_1, \eta', \tau) &= P_+(\eta', \gamma, \xi_1)\partial P_-(\eta', \gamma, \xi_1)/\partial \xi_1, \\ [u, v]_{(0,s|q),[\gamma]} &= [\Lambda_q^s(D')\exp(\gamma t)u, \Lambda_q^s(D')\exp(\gamma t)v], \\ \Lambda_q^s(\eta', \gamma) &= (1 + |\xi'|^2 + |\sigma + i\gamma|^{2/q})^{s/2} \end{aligned}$$

is used. Inequality (10) is a consequence of the inequality (cf. the derivation of (1p) from (1) in III.5.3)

$$\|u\|^2_{(r-1,s+1/2|q),[\gamma]} + \sum_{j=0}^{r-1} \langle D_1^j u(0,\cdot)\rangle_{(r-1,s-j|q),[\gamma]} \leqslant C\left(J + \sum_{j=1}^{\mu} \langle B_j(D)u\rangle_{(r+s-\beta_j|q),[\gamma]} \right). \tag{11}$$

Passing to the Fourier transform with respect to y' we can repeat literally the derivation of (III.5.13) and thus obtain (11).

1.5. Mixed problem for q-stable correct differential operators. Here, as in II.5.5, we state the general result by combining in one assertion the theory in the foregoing chapter and a modification of Theorem 1.3. It will be assumed (without special stipulation) that the coefficients of the differential operators in question belong to C^∞ and are constant outside a ball.

THEOREM. *Let a number $q > 1$, a differential operator $P(y; D)$, and a set of boundary operators $B_1(y, D), \dots, B_\mu(y, D)$ be given. Then the following conditions are equivalent:*

(I) *for any set of operators $\{Q(y, D), C_1(y, D), \dots, C_\mu(y, D)\}$, $\deg_q Q < \deg_q P$, $\deg_q C_j \leqslant \deg_q B_j$, $j = 1, \dots, \mu$, and any $S > 0$ there is $\gamma_0 = \gamma_0(S, Q, C_1, \dots, C_\mu)$ such that for $|s| \leqslant S$ and $\gamma \leqslant \gamma_0$ the mixed problem*

$$(P(y, D) + Q(y, D))u = f, \tag{12}$$

$$(B_j(y, D) + C_j(y, D))u\big|_{x_1=0} = g_j, \quad j = 1, \dots, \mu, \tag{13}$$

with $f \in H^{(0,s+1|q)}_{[\gamma]+}$ and $g_j \in H^{(r+s-\beta_j|q)}_{[\gamma]+}$, $j = 1, \dots, \mu$, possesses a unique solution $u \in H^{(r,s|q)}_{[\gamma]+}$, which satisfies the inequality (cf. (9′))

$$\begin{aligned} &|\gamma|^{1/2q}\|u\|_{(r,s|q),[\gamma]} + \sum_{j=0}^{r-1} \langle D_1^j u(0, \cdot)\rangle_{(r+s-j|q),[\gamma]} \\ &\qquad \leqslant C\left(|\gamma|^{-1/2q}\|f\|_{(0,s+1|q),[\gamma]} + \sum_{j=1}^{\mu} \langle g_j\rangle_{(r+s-\beta_j|q),[\gamma]}\right); \end{aligned} \tag{14}$$

(II) *either $q = 1$ and conditions (A), (B), (C), and (D) of* §III.1 *hold or q is even, $q \geqslant 2$, and conditions (A), (B), (C), and (D) of* 1.1 *hold.*

The implication (II)⇒(I) for $q = 1$ coincides with the main assertion of Theorem III.5.2. For q even condition (II) implies the stronger assertion (II) of Theorem 1.3. Thus, the proof of the theorem reduces to verifying the implication (I)⇒(II). By modifying the arguments of the proof of Proposition 1 in III.2.3 we derive the following

PROPOSITION. *Let condition (I) of the theorem hold and let $y^0 = (x_1^0, y'^0)$ be a point in $\mathbb{R}^{n+1}_+$. We set $P(D) = P_0(y^0, D)$ and $B_j(D) = B_{0j}(y'^0, D)$. Then the following assertions hold:*

(i) *if $x_1^0 > 0$, then the equation $P(D)u = f$ with an arbitrary right-hand side $f \in H^{(0,s+1|q)}_{[\gamma]}(\mathbb{R}^{n+1})_+$ has a unique solution $u \in H^{(r,s|q)}_{[\gamma]}(\mathbb{R}^{n+1})_+$ for $\gamma < 0$, and it satisfies the inequality*

$$|\gamma|^{1/q}\,{}'\|u\|_{(r,s|q),[\gamma]} \leqslant C'\|P(D)u\|_{(0,s+1|q),[\gamma]},$$

where (cf. III.2.3*) the norms ${}'\|\cdot\|_{(r,s|q),[\gamma]}$ are obtained from (7) by replacing the factors $(1 + |\xi|^2 + |\tau|^{2/q})^\tau$ and $(1 + |\xi'|^2 + |\tau|^{2/q})^s$ with $(|\xi|^2 + |\tau|^{2/q})^\tau$ and $(|\xi'|^2 + |\tau|^{2/q})^s$, respectively;*

(ii) *if $x_1^0 = 0$, then the mixed problem*

$$\begin{aligned} &P(D)u = f, \quad x_1 > 0, \\ &B_j(D)u\big|_{x_1=0} = g_j, \quad j = 1, \dots, \mu, \end{aligned}$$

has a unique solution $u \in H^{(r,s|q)}_{[\gamma]}(\mathbb{R}^{n+1}_+)_+$ for any $f \in H^{(0,s+1|q)}_{[\gamma]+}(\mathbb{R}^{n+1}_+)_+$ and $g_j \in H^{(r+s-\beta_j|q)}_{[\gamma]+}$, $j = 1, \dots, \mu$, which satisfies an inequality of the type (9) (replacement of $P(y, D)$ and $B_j(y, D)$ by $P(D)$ and $B_j(D)$ and the norms $\|\cdot\|_{(r,s|q),[\gamma]}$ and $\langle\cdot\rangle_{(s|q),[\gamma]}$ by ${}'\|\cdot\|_{(r,s|q),[\gamma]}$ and ${}'\langle\cdot\rangle_{(s|q),[\gamma]}$, respectively).

Following the scheme of the proof of Proposition 2 in III.2.3 we derive from (ii) an estimate for the polynomial P:

$$|\gamma|^{1/q}(|\tau|^{1/q}+|\xi|)^{\tau} \leqslant C|P(\xi,\tau)|(|\tau|^{1/q}+|\xi'|). \tag{15}$$

In turn, (15) implies a weaker estimate

$$|\gamma|^{1/q}(|\tau|^{1/q}+|\xi|)^{r-1} \leqslant C|P(\xi,\tau)|,$$

whence it follows that the polynomial $P(\xi,\tau)$ is q-stable correct, and therefore either $q=1$ and P is a strictly hyperbolic polynomial or q is even and P is a q-parabolic polynomial (Theorem I.2.5). If we replace $(\xi_1,\xi',\sigma+i\gamma)$ in (15) by $(\rho\xi_1,\xi',\sigma+i\gamma\rho^{q\varepsilon})$, $0<\varepsilon<1$, divide both sides by $\rho^{(m+\varepsilon)q}$, and pass to the limit as $\rho\to+\infty$, we derive condition (B). The statement (ii) is obtained by repeating literally the arguments in III.2.3.

§2. Mixed problem for q-hyperbolic equations

2.1. Statement of the problem. Conditions on the symbol of the operator. As in §1, we will consider problem (1.1), (1.2), (1.3). We retain the assumption of stabilization for the coefficients of the operators entering (1.1) and (1.2) and denote by $P_0(y,\eta)$ and $B_{0j}(y,\eta)$ their q-principal parts.

Condition (A). $q=2b+1>1$ is an odd number and $P_0(y,\eta)$ is a uniformly q-hyperbolic symbol (see the definition in II.6.2).

According to the definition in II.6.2, there exists a nonempty cone of q-hyperbolicity, which is denoted as $V(P)$. In what follows the condition below will be very important.

Condition ($B^{\pm}$). The direction $e_1=(1,0,\dots,0)$ belongs to the cone $\pm V(P)$.

LEMMA. *Let condition* ($B^{\pm}$) *hold. Then* e_1 *is not a characteristic direction for the operator* $P(y,D)$, *i. e.,*

$$P_0(y;1,0,\dots,0,0)\neq 0. \tag{1}$$

PROOF. According to assumption II.6.2 (iii), the inequality

$$v(|\xi_1+iv|+|\xi'|)^{q-1}(|\xi_1+iv|+|\xi'|+|\tau|^{1/q})^{r-1} \leqslant C|P_0(\xi_1+iv,\xi',\tau)|,$$
$$\pm v>0, \quad \operatorname{Im}\tau<-R|v|^q,$$

holds. Substituting $\xi'=0$ into this inequality we see that $P_0(y,\xi_1,0,\tau)$ is also a uniformly q-hyperbolic symbol. Since this symbol is q-homogeneous, it can be written as a product of polynomial factors:

$$P_0(y;\xi_1,0,\tau)=b(y)\prod_{j=1}^{m}(\tau-a_j(y)\xi_1^q).$$

The condition of uniform q-hyperbolicity means that there are positive constants a and b such that

$$b>|b(y)|>b^{-1}, \qquad \pm a_j(y)\geqslant a, \quad j=1,\dots,m.$$

It now remains to note that

$$P_0(y;1,0,\dots,0)=b(y)(-a_1(y))\cdots(-a_m(y)).$$

In view of the lemma, we can assume that the symbol P_0 is resolved with respect to the highest power of ξ_1, i. e.,

$$(2) \qquad P_0(y;\xi_1,\eta') = \xi_1^r + \sum_{j=1}^{n} P_{0j}(y;\eta')\xi_1^{r-j}, \quad r = mq.$$

As in §III.1 and §1, to formulate further conditions on problem (1.1), (1.2) we consider the algebraic equation

$$(3) \qquad P_0(X,\lambda) \stackrel{\text{def}}{=} P_0(y',\lambda,\xi',\tau) = 0, \quad X = (y',\xi',\operatorname{Re}\tau,\operatorname{Im}\tau).$$

2.2. Remarks on the roots of equation (3). 1) Denote by $\lambda_j^{\pm}(X)$, $j = 1,\dots,r_{\pm}(X)$, the roots of equation (3) in the half-plane $\pm\operatorname{Im}\lambda > 0$. If $\operatorname{Im}\tau \neq 0$, then, according to the definition of q-hyperbolic symbols, equation (3) has no real roots, and therefore we have $r_+(X) + r_-(X) = r$, and these numbers do not depend on the parameters when the variable τ runs over the open half-plane $\pm\operatorname{Im}\tau < 0$ and the other variables assume arbitrary values. Thus, when calculating the numbers $r_\pm$ we can set $\xi' = 0$. Equation (3) decomposes into m equations

$$\xi_1^q - (a_j(y))^{-1}\tau = 0, \quad \pm a_j(y) > a > 0, \quad j = 1,\dots,m.$$

For $\operatorname{Im}\tau < 0$, depending on the sign of $a_j(y)$, each of these equations has $b+1$ roots in one of the half-planes $\pm\operatorname{Im}\lambda > 0$ and b roots in the other. Therefore we have

$$(4^+) \qquad r_+ = mb, \qquad r_- = m(b+1), \quad e_1 \in V(P), \qquad \operatorname{Im}\tau < 0,$$
$$(4^-) \qquad r_+ = (m+1)b, \quad r_- = mb, \qquad e_1 \in -V(P), \quad \operatorname{Im}\tau < 0.$$

2) We now consider equation (3) for real τ. We denote by $\lambda_{0j}(y',\xi)$ the roots of the symbol P_0 (with respect to the variable τ) and replace (3) with the set of equations

$$(5) \qquad \sigma - \lambda_{0j}(y',\lambda,\xi') = 0, \qquad j = 1,\dots,m.$$

By condition ($B^\pm$),

$$\pm\frac{\partial\lambda_{0j}}{\partial\xi_1}(y_1',\xi_1,\xi') > 0.$$

Therefore for fixed (y',ξ') the function $\lambda_{0j}(y',\xi_1,\xi')$ is monotone increasing from $-\infty$ to ∞ (accordingly, monotone decreasing from ∞ to $-\infty$) as ξ_1 runs over the real axis. Hence, equation (5) has exactly one real root for each $j = 1,\dots,m$. Since for $|\xi| \neq 0$ the roots $\lambda_{0j}(y',\xi)$ are distinct, equation (3) has exactly m distinct real roots for real values of τ. Since the coefficients of this equation are real, the other roots of (3) are pairwise complex conjugate for $\operatorname{Im}\tau = 0$, i. e., mb roots lie in the upper half-plane $\operatorname{Im}\lambda > 0$, and the lower half plane contains the same number of roots.

3) We now fix (y',ξ',σ) and consider the roots of (3) as functions of the parameter $\gamma = \operatorname{Im}\tau$. When we pass from $\gamma = 0$ to a small value $\gamma < 0$, the complex roots are subject to small perturbations and remain at a positive distance from the real axis. Thus, for $\gamma < 0$ the upper and lower half-planes $\pm\operatorname{Im}\lambda > 0$ contain mb roots each.

The roots that are real for $\gamma = 0$ also move to the complex domain. If $e_1 \in V(P)$, then, by (4^+), these roots move to the lower half-plane. And if $e_1 \in -V(P)$, then, according to (4^-), they move to the upper half-plane. Since the real roots are pairwise distinct, they retain this property for a sufficiently small γ. Therefore the roots under consideration are smooth functions of γ, and their imaginary parts have zeros of order

$l \geqslant 1$ for $\gamma = 0$. Repeating the arguments in III.6.7 (see Lemma 2 in this subsection) we readily show that $l = 1$.

4) Consider the "hemisphere"

$$\Sigma^+ = \{(\eta', \gamma), \quad |\xi'|^2 + (\sigma^2 + \gamma^2)^{2/q} \overset{\text{def}}{=} \Lambda_q^2(\eta', \gamma) = 1,\ \gamma \leqslant 0\}. \tag{6}$$

By what was said above, we can split it into parabolic and hyperbolic zones:

$$\Sigma^+_{\text{parab}} = \{(\eta', \gamma) \in \Sigma^+, \quad \gamma < -\kappa < 0\}, \tag{6'}$$

$$\Sigma^+_{\text{hyperb}} = \{(\eta', \gamma) \in \Sigma^+, \quad -\kappa < \gamma < 0\}. \tag{6''}$$

The imaginary parts of all roots are nonzero in the zone (6′), i. e., there is $\varepsilon = \varepsilon(\kappa)$ such that

$$\pm \operatorname{Im} \lambda_j^{\mp}(y', \eta', \gamma) \geqslant \varepsilon \qquad j = 1, \dots, r_\pm, \quad (\eta', \gamma) \in \Sigma^+_{\text{parab}}. \tag{7}$$

As for the hyperbolic zone, here the cases $e_1 \in V(P)$ and $-e_1 \in V(P)$ should be considered separately. In the first case the imaginary parts of the roots λ_j^+ are separated from zero:

$$\operatorname{Im} \lambda_j^+(y', \eta', \gamma) \geqslant \varepsilon, \qquad j = 1, \dots, mb, \quad (\eta', \gamma) \in \Sigma^+_{\text{hyperb}}, \quad e_1 \in V(P). \tag{8}$$

The roots lying in the lower half-plane split into two groups the first of which contains mb roots whose imaginary parts are separated from zero. Indexing them in a suitable manner we have

$$\operatorname{Im} \lambda_j^-(y', \eta', \gamma) \leqslant -\varepsilon, \qquad j = 1, \dots, mb, \quad (\eta', \gamma) \in \Sigma^+_{\text{hyperb}}, \quad e_1 \in V(P). \tag{8'}$$

The remaining roots λ_j^-, $j = mb+1, \dots, m(b+1)$, are pairwise distinct, and there are $c_1, c_2 > 0$ such that

$$\begin{gathered} c_1\gamma \leqslant \operatorname{Im} \lambda_j^-(y', \eta', \gamma) \leqslant c_2\gamma, \\ j = mb+1, \dots, (m+1)b, \quad (\eta', \gamma) \in \Sigma^+_{\text{hyperb}}, \quad e_1 \in V(P). \end{gathered} \tag{8''}$$

In the case $e_1 \in -V(P)$ the roots λ_j^+ and λ_j^- interchange. The imaginary parts of λ_j^- are uniformly separated from zero, and the roots λ_j^+ split into two groups. Hence, the result in the present subsection can be stated as the following

PROPOSITION. *Let conditions* (A) *and* (B$^\pm$) *hold. Then equation* (3) *possesses the properties below.*

(i) *For each $\kappa > 0$ there is ε such that*

$$\pm \operatorname{Im} \lambda_j^{\mp}(y', \eta', \gamma) \geqslant \varepsilon \Lambda_q(\eta', \gamma), \qquad \gamma < -\kappa \Lambda_q(\eta', \gamma), \quad j = 1, \dots, r_\pm.$$

(ii$^\pm$) *If $e_1 \in \pm V(P)$, then for each κ there is ε such that*

$$\begin{gathered} \pm \operatorname{Im} \lambda_j^{\pm}(y', \eta', \gamma) \geqslant \varepsilon \Lambda_q(\eta', \gamma), \qquad j = 1, \dots, mb. \\ -\kappa \Lambda_q(\eta', \gamma) \leqslant \gamma \leqslant 0. \end{gathered} \tag{9}$$

(iii$^\pm$) *If κ is sufficiently small, then the roots $\lambda_j^{\mp}(y', \eta', \gamma)$, $j = 1, \dots, m(b+1)$, split into two groups. The first group contains mb roots for which inequality* (9) *is retained:*

$$\mp \operatorname{Im} \lambda_j^{\mp}(y', \eta', \gamma) \geqslant \varepsilon \Lambda_q(\eta', \gamma), \qquad -\kappa \Lambda_q(\eta', \gamma) \leqslant \gamma \leqslant 0. \tag{10}$$

The second group contains m pairwise distinct roots, and there are positive c_1 and c_2 such that

$$(10')\qquad \begin{gathered} c_1|\gamma|\Lambda_q^{1-q}(\eta',\gamma) \geqslant \mp \operatorname{Im}\lambda_j^{\mp}(y',\eta',\gamma) \geqslant c_2|\gamma|\Lambda_q^{1-q}(\eta',\gamma), \\ -\kappa\Lambda_q(\eta',\gamma) \leqslant \gamma \leqslant 0, \quad j = mb+1,\dots,m(b+1). \end{gathered}$$

REMARK. The roots of equation (3) can also be considered in the case $\gamma = \operatorname{Im}\tau > 0$. Since $\lambda_j(y,\eta',-\gamma) = -\lambda_j(y,\eta',\gamma)$, we have $r_\pm(y',\eta',-\gamma) = r_\mp(y',\eta',\gamma)$, and therefore if the roots $\operatorname{Im}\tau > 0$ in (9) and (10) are replaced by $\lambda_j^{\mp}$, then the proposition remains true.

2.3. Uniform Lopatinskii's condition. Examples. As in the cases of hyperbolic and q-parabolic equations, we will assume that the following conditions hold.

Condition (C). $\mu = r_+$, where μ is the number of boundary conditions (2.2).

Condition (D). There is $K \geqslant 0$ such that (see (III.1.14) and (III.1.15))

$$|\operatorname{Lop}(P_0, B_{01},\dots,B_{0\mu})(y',\xi',\tau)| \geqslant K, \quad y' \in \mathbb{R}^n, \quad |\xi'|^2 + |\tau|^{2/q} = 1, \quad \operatorname{Im}\tau \leqslant 0.$$

Now we consider simplest examples of problems satisfying conditions (A), ... , (D). As the operator P we take the 3-hyperbolic operator with the symbol

$$(11)\qquad P(\xi,\tau) = -\tau + \xi_1^3 + \xi_1(\xi_2^2 + \cdots + \xi_n^2).$$

Since $\partial P/\partial\xi_1 = 3\xi_1^2 + \xi_2^2 + \cdots + \xi_n^2 > 0$ for $|\xi| \neq 0$, the direction $(1,0,\dots,0)$ definitely belongs to the cone $V(P)$, and, according to (4^+), we have $r_+ = 1$, so that only one condition must be set on the boundary. For example, consider the oblique derivative problem

$$(12)\qquad B_1(D) = D_1 + b_2D_2 + \cdots + b_nD_n, \qquad b = (b_2,\dots,b_n) \in \mathbb{C}^{n-1}.$$

Since $P_+ = \xi_1 - \lambda_1^+$, we have

$$\operatorname{Lop}\{P, B_1\} = \lambda_1^+ + b_2\xi_2 + \cdots + b_n\xi_n$$

and

$$\operatorname{Im}\operatorname{Lop}\{B, B_1\}(\eta',\gamma) = \operatorname{Im}\lambda_1^+ - \sum_{j=2}^{n}\xi_j \operatorname{Im} b_j > \Bigl(\varepsilon - \bigl(\textstyle\sum(\operatorname{Im} b_j)^2\bigr)^{1/2}\Bigr)\Lambda_3(\eta',\gamma).$$

Lopatinskii's condition for problem (11), (12) holds if and only if $\varepsilon^2 > \sum(\operatorname{Im} b_j)^2$. In particular, it holds for real $b_2,\dots,b_n$.

As a second example we consider the 3-hyperbolic operator

$$(11')\qquad P(\xi,\tau) = \tau + \xi_1^3 + \xi_1(\xi_2^2 + \cdots + \xi_n^2).$$

In this case we have $(1,0,\dots,0) \in -V(P)$, so that $r_+ = 2$. We consider the first boundary value problem

$$(12')\qquad B_1(D) \equiv 1, \quad B_2(D) = D_1$$

for operator (11). Here we have $\operatorname{Lop}\{P, B_1, B_2\} \equiv 1$, i. e., uniform Lopatinskii's condition holds. We now consider some problems (cf. the Neumann problem for the wave equation) for which uniform Lopatinskii's condition does not hold. To this end we replace (12′) with the boundary conditions

$$(12'')\qquad B_1(D) = D_1, \quad B_2(D) = D_1^2.$$

Since $P_+ = (\xi_1 - \lambda_1^+)(\xi_1 - \lambda_2^+)$, we have $B_1'(\lambda) = \lambda$, $B_2'(\lambda) = (\lambda_1^+ + \lambda_2^+)\lambda + \lambda_1^+\lambda_2^+$ and

$$\operatorname{Lop}\{P, B_1, B_2\} = \det \begin{Vmatrix} 1 & 0 \\ \lambda_1^+ + \lambda_2^+ & \lambda_1^+\lambda_2^+ \end{Vmatrix} = \lambda_1^+\lambda_2^+.$$

Uniform Lopatinskii's condition cannot hold because, by Proposition 2.2 (iii$^-$), we have $\operatorname{Lop}\{P, B_1, B_2\} \to 0$ as $\gamma \to -0$.

2.4. Basic spaces. We will use spaces that generalize the spaces defined in §1 to the case of the weight $\exp(\gamma t + \nu x_1)$, where the sign of ν is determined by condition (B$^\pm$):

$$(\operatorname{sgn} \nu, 0, \dots, 0) \in V(P). \tag{13}$$

In other words, under condition (B$^+$) the case $\nu > 0$ is considered, and for condition (B$^-$) we assume that $\nu < 0$.

So, we denote by $H_{[\gamma,\nu]}^{(r,s|q)}$ the space of functions (distributions) with finite norm

$$\Bigl(\iint\limits_{\substack{\operatorname{Im}\tau=\gamma \\ \operatorname{Im}\zeta_1=\nu}} (1 + |\zeta_1|^2 + |\xi'|^2 + |\tau|^{2/q})^r (1 + |\xi'|^2 + |\tau|^{2/q})^s \times |\widehat{u}(\zeta_1, \xi', \tau)|^2 \, d\zeta_1 d\xi' d\tau\Bigr)^{1/2}, \tag{14}$$

and let $H_{[\gamma,\nu]}^{(r,s|q)}(\mathbb{R}_+^{n+1})$, $r > 0$, be the space of restrictions to the half-space $\{x_1 \geqslant 0\}$. For positive integers r we take

$$\begin{aligned} &\|u, \mathbb{R}_+^{n+1}\|_{(r,s|q),[\gamma,\nu]} \\ &= \Bigl(\sum_{j=0}^{r} \int_0^\infty \exp(2\nu x_1) \iint\limits_{\operatorname{Im}\tau=\gamma} (1 + |\xi'|^2 + |\tau|^{2/q})^{r+s-j} |D_1^j \widehat{u}(x_1, \xi', \tau)|^2 \, d\xi' d\tau \, dx_1\Bigr)^{1/2} \end{aligned} \tag{14'}$$

as the norm in $H_{[\gamma,\nu]}^{(r,s|q)}(\mathbb{R}_+^{n+1})$. We will consider the space $H_{[\gamma]}^{(s|q)}(\mathbb{R}^n)$ of functions on the plane $\{x_1 = 0\}$ with the norm (1.8). As before, Φ_+, $\Phi = H_{[\nu,\gamma]}^{(r,s|q)}$, and $H_{[\gamma]}^{(s|q)}(\mathbb{R}^n)$ denote the spaces of elements vanishing for $t < 0$.

THEOREM. *For a set of operators $\{P, B_1, \dots, B_\mu\}$ the following conditions are equivalent.*

(I) *Conditions* (A), (B$^\pm$), (C), *and* (D) *hold.*

(II) *There is $R > 0$ such that for all $S > 0$ there is $\nu_0(S)$ such that for $\pm\nu > \nu_0(S)$ (the sign of ν is determined from* (13$'$)) *and $\gamma < -R|\nu|^q$ problem* (1.1), (1.2) *has a unique solution $u \in H_{[\nu,\gamma]}^{(r,s|q)}(\mathbb{R}_+^{n+1})_+$ for the right-hand sides $f \in H_{[\gamma,\nu]}^{(0,s+1|q)}(\mathbb{R}_+^{n+1})_+$ and $g_j \in H_{[\gamma]}^{(r+s-\beta_j)}(\mathbb{R}^n)_+$, $j = 1, \dots, \mu$. In this case we have*

$$D_1^j u(0, y') \in H_{[\gamma]+}^{(m+s-j|q)}(\mathbb{R}^n), \qquad j = 1, \dots, r-1,$$

and the inequality

$$\begin{aligned} &|\nu|\, \|u\|^2_{(r,s|q),[\gamma,\nu]} + \sum_{j=0}^{r_1} \langle D_1^j u(0,\cdot)\rangle^2_{(r+s-j|q),[\gamma]} \\ &\leqslant C\left(|\nu|^{-1}\|P(y,D)u\|^2_{(0,s+1|q),[\gamma,\nu]} + \sum_{j=1}^{\mu} \langle B_j(y',D)u\rangle^2_{(r+s-\beta_j|q),[\gamma]}\right) \end{aligned} \tag{15}$$

holds.

As in the case of hyperbolic and parabolic equations, the proof of solvability of the mixed problem for a q-hyperbolic operator (i. e., the implication (I)$\Rightarrow$(II)) is carried out in three stages: the derivation of (15), the derivation of an estimate for the transposed problem, and the estimation in the norms involving the exponential weights $(\exp(-\gamma' t)+\exp(-\gamma'' t))^{-1}$. In the next section we will dwell on the derivation of (15). As to the other estimates, their proof does not involve any new ideas as compared to Chapter III, and we leave this to the reader. For detailed derivation of these estimates see **[GV4, GV7]**.

2.6. Necessity of conditions (A), (B$^\pm$), (C), and (D). By modifying the arguments in the proof of Proposition 1 in III.2.3 we prove the following analog of Proposition 1.5.

PROPOSITION. *Let condition* (I) *of Theorem* 2.3 *hold and let* $y^0 = (x_1^0, y'^0) \in \mathbb{R}^{n+1}$. *Set*

$$P(D) = P_0(y^0, D), \quad B_j(D) = B_{0j}(y'^0; D), \qquad j = 1, \dots, \mu.$$

Then the following conditions are equivalent.

(i) *If* $x_1^0 > 0$, *then for any right-hand side* $f \in H^{(0,s+1|q)}_{[\gamma,\nu]}(\mathbb{R}^{n+1})_+$, $|\nu| > 0$, $\gamma < -R|\nu|^q$, *the equation* $P(D)u = f$ *possesses a solution* $u \in H^{(r,s|q)}_{[\gamma,\nu]}(\mathbb{R}^{n+1})_+$ *for which the inequality*

$$|\nu|\,'\|u\|_{(r,s|q),[\gamma,\nu]} \leqslant C\,'\|P(D)u\|_{(0,s+1|q),[\gamma,\nu]}$$

holds, where the norm $'\|\cdot\|_{(r,s|q),[\gamma,\nu]}$ *is obtained from* (14) *by replacing the factors* $1+|\zeta_1|^2+|\xi'|^2+|\tau|^{2/q}$ *and* $1+|\xi'|^2+|\tau|^{2/q}$ *by* $|\zeta_1|^2+|\xi'|^2+|\tau|^{2/q}$ *and* $|\xi'|^2+|\tau|^{2/q}$, *respectively.*

(ii) *If* $x_1^0 = 0$, *then the mixed problem*

$$\begin{aligned} P(D)u &= f, & x_1 &> 0, \\ B_j(D)u\big|_{x_1=0} &= g_j, & j &= 1, \dots, \mu, \end{aligned}$$

has a unique solution $u \in H^{(r,s|q)}_{[\gamma,\nu]}(\mathbb{R}^n_+)_+$ *for any* $f \in H^{(0,s+1|q)}_{[\gamma,\nu]}(\mathbb{R}^n_+)_+$ *and* $g_j \in H^{(r+s-\beta_j|q)}_{[\gamma]+}$, $j = 1, \dots, \mu$, *for which a type* (15) *inequality holds when* $P(y,D)$ *and* $B_j(y,D)$ *are replaced by* $P(D)$ *and* $B_j(D)$ *and the norms* $\|\cdot\|_{(r,s|q),[\gamma,\nu]}$ *and* $\langle\cdot\rangle_{(s|q),[\gamma]}$ *by* $'\|\cdot\|_{(r,s|q),[\gamma,\nu]}$ *and* $'\langle\cdot\rangle_{(s|q),[\gamma]}$, *respectively.*

COROLLARY. (i) *The inequality*

$$\begin{aligned} &|\nu|(|\zeta_1| + |\xi'| + |\tau|^{1/q})^r \leqslant C_1 |P(\zeta_1, \xi', \tau)|(|\xi'| + |\tau|^{1/q}), \\ &\zeta_1 = \xi_1 + i\nu, \quad |\nu| > 0, \quad \operatorname{Im}\tau < -R|\nu|^q, \end{aligned} \tag{16}$$

holds.

(ii) *Let* $\Lambda_q(\eta',\gamma) = 1$ *and* $\gamma \leqslant 0$. *Then there is* $R > 0$ *such that if*

$$0 < \pm\nu \leqslant R^{-1/q}|\gamma|^{1/q}, \tag{17}$$

then for any set of numbers $(c_1, \dots, c_\mu) \in \mathbb{C}^\mu$ *the problem*

$$P(D, \xi', \sigma + i\gamma)V(z) = 0, \qquad z \geqslant 0, \tag{18}$$

$$(B_j(D, \xi', \sigma + i\gamma)V)(0) = c_j, \quad j = 1, \dots, \mu, \tag{19}$$

on the half-line has a unique solution $V(z)$, $\exp(\nu z)D^j V(z) \in L_2(\mathbb{R}_+)$, $j = 1, \dots, r-1$, *and this solution satisfies the inequality*
(20)

$$|\nu| \sum_{j=0}^{r-1} \int_0^\infty \exp(2\nu z)|D^j V(z)|^2\,dz + \sum_{j=0}^{r-1} |D^j V(0)|^2 \leqslant C_2 \sum_{j=1}^{\mu} |(B_j(D, \xi', \sigma + i\gamma)V)(0)|^2.$$

Assertion (i) is derived from the analogous assertion of the proposition by means of the Fourier transform; the related argument was used in Chapter II many times. Assertion (ii) is derived from the analogous assertion with the aid of partial Fourier transform with respect to y' following the scheme of the proof of Lemma III.5.3.

Necessity of conditions (A) *and* (B$^\pm$). Since q is an odd number, (16) and Theorem I.4.3 imply that $P(\xi, \tau)$ is a q-hyperbolic polynomial, and the vector $(\operatorname{sgn}\nu, 0, \dots, 0)$ belongs to the cone of q-hyperbolicity $V(P)$.

Necessity of condition (C). Denote by $\mathfrak{M}_\nu^+(\eta', \gamma)$ the space of the solutions $V(z)$ to the homogeneous equation (18) for which $\exp(\nu z)V(z) \in L_2(\mathbb{R}_+)$. According to part (ii) of the corollary, the boundary problem on $\mathfrak{M}_\nu^+$ with μ conditions (19) is uniquely solvable if

$$(\operatorname{sgn}\nu, 0, \dots, 0) \in V(P), \quad \gamma < -R|\nu|^q, \tag{21}$$

which implies that under condition (21) we have

$$\dim \mathfrak{M}_\nu^+(\eta', \gamma) = \mu. \tag{22}$$

Furthermore, since the coefficients of equation (20) are constant, we have

$$\dim \mathfrak{M}_\nu^+(\eta', \gamma) = r_\nu^+(\eta', \gamma), \tag{22$'$}$$

where the right-hand side is the number of the roots of equation (3) for which $\operatorname{Im}\lambda_j > \nu$.

If ν and γ satisfy condition (21), then the properties of q-hyperbolic symbols imply that equation (3) has no roots on the line $\operatorname{Im}\lambda = \nu$. It follows that the number $r_\nu^+(\eta', \gamma)$ does not depend on η'. Therefore, replacing (ξ', σ) with $(\rho\xi', \rho^q\sigma)$ and taking into account that (3) is q-homogeneous we conclude that

$$r_\nu^+(\gamma) = r_{\nu/\rho}^+(\gamma/\rho^q). \tag{22$''$}$$

Comparing (22), (22$'$), and (22$''$) as $\rho \to +\infty$ we see that $\mu = r_0^+(0)$, where the right-hand term is exactly the number r_+, i. e., condition (C) holds.

Necessity of condition (D). Note that if a function $V(z)$ satisfies (18) and belongs to $L_2(\mathbb{R}_+)$ with weight $\exp(\nu z)$, then it satisfies in fact an equation of lower order

$$P_+^\nu(D, \eta', \gamma)V(z) = 0,$$

where $P_+^\nu(\lambda, \eta', \gamma)$ denotes the product of those factors $\lambda - \lambda_j(\eta', \gamma)$ for which $\operatorname{Im} \lambda_j > \nu$. As was already noted, if (21) holds, then the symbol $P_+^\nu(\lambda, \eta', \gamma)$ coincides with $P_+(\lambda, \eta', \gamma)$. Hence, (18) and (19) imply that $V(z)$ is a solution to the problem

$$(18') \qquad P_+(D, \eta', \gamma)V(z) = 0,$$

$$(19') \qquad (B_j'(D, \eta', \gamma)V)(0) = c_j, \quad j = 1, \dots, \mu,$$

where the polynomial $B_j'(x, \eta', \gamma)$ is the remainder on dividing $B_j(\lambda, \eta', \gamma)$ by $P_+(\lambda, \eta', \gamma)$.

Inequality (20) for problem (18), (19) implies

$$(23) \qquad \sum_{j=0}^{\mu-1} |D^j V(0)| \leqslant C_2 \sum_{j=1}^{\mu} |(B_j'(D, \eta', \gamma)V)(0)|.$$

Since the dimension of the space $\mathfrak{M}_\nu^+$ of functions satisfying (18) and belonging to $L_2(\mathbb{R}_+)$ with weight $\exp(\nu z)$ is equal to μ, the germ $(V(0), \dots, D^{\mu-1}V(0))$ in inequality (21) can be replaced by an arbitrary point belonging to $\mathbb{C}^\mu$. Therefore, (23) implies that the polynomials $B_1'(\lambda, \eta', \gamma)$ are uniformly linearly independent.

§3. A priori estimates in the mixed problem for q-hyperbolic equations

In this section we derive inequality (2.15). As in the case of hyperbolic equations, we will in fact prove a more precise estimate including the quadratic form of the original operator P and an operator Q of order $r - 1$ (the "separating" operator of the mixed problem). Inequality (2.15) is a weakened corollary of an estimate for the quadratic form. To derive the estimates in this section we will need some analogs of the results of §III.4 in application to the q-homogeneous case. Since there is no fundamental difference between the proofs in this section and those in §III.4, we will confine ourselves to formulating the related assertions.

3.1. Auxiliary assertions. Denote by $S^{j|q}$ the class of symbols $b(y, \eta', \gamma)$ belonging to C^∞ with respect to the variables $y \in \mathbb{R}^{n+1}$ and (η', γ), $|\eta'| + |\gamma| \neq 0$, and admitting the estimates

$$|D_y^\alpha D_{\eta'}^\beta D_\gamma^p b(y, \eta', \gamma)| < C_{\alpha\beta p} \Lambda_q(\eta', \gamma)^{j-|\beta|-p}.$$

With each such function we can associate a pseudodifferential operator $b(y, D, \gamma)$ (see formula (III.4.3)) acting with respect to the variables y' and depending on the parameter x_1. Assertions 1)–4) of II.4.1 remain true for these operators if the norms $\langle\cdot\rangle_{s,\gamma}$ are replaced with $\langle\cdot\rangle_{(s|q),[\gamma]}$.

Let $S[\xi_1]^{r,l|q}$ denote the class of symbols of the type

$$L(y, \eta, \gamma) = \sum_{j=0}^{r} b_j(y, \eta', \gamma)\xi_1^{r-j}, \quad b_j \in S^{l+j|q}.$$

For $l = 0$ we write $S[\xi_1]^{r|q}$. Denote by $S_0[\xi_1]^{r,l|q}$ the set of those symbols L for which $b_0(y, \eta', \gamma) \equiv \text{const}$.

The proposition below is an analog of Lemma III.4.2.

PROPOSITION 1. *Let a symbol L belong to $S[\xi_1]^{r,l|q}$. Then the inequality*

$$\|L(y, D, \gamma)u\|_{(k,s|q),[\gamma,\nu]} \leqslant (\kappa[L]_0 + c|\nu|^{-1})\|u\|_{(r+k,s+l|q),[\gamma,\nu]}$$

holds.

As in §III.4, we consider quadratic forms of functions on $\mathbb{R}^{n+1}_+$, but instead of the weight $\exp(\gamma t)$ we take $\exp(\gamma t + \nu x_1)$. Set

$$[u, v]_{[\gamma,\nu]} = [\exp(\gamma t + \nu x_1)u, \exp(\gamma t + \nu x_1)v],$$
$$[u, v]_{(0,s|q),[\gamma,\nu]} = [\Lambda_q(D', \gamma)u, \Lambda_q^s(D', \gamma)v]_{[\gamma,\nu]}.$$

In the sequel we will always assume that $q = 2b + 1$ and $\gamma < -R|\nu|^q$. An analog of Propositions 1 and 2 in III.4.2 is

PROPOSITION 2. (i) *If $L \in S_0[\xi_1]^{k|q}$ and $M \in S_0[\xi_1]^{k-1|q}$, then*

$$(1)\quad \begin{aligned} |J| &\overset{\text{def}}{=} |[Lu, Mv]_{(0,s|q),[\gamma,\nu]}| \\ &\leqslant C(L, M, \nu)\left(\|u\|^2_{(k-1,s+1/2|q),[\gamma,\nu]} + \sum_{j=0}^{k-1}\langle D_1^j u(0,\cdot)\rangle^2_{(k-1-j+s|q),[\gamma,\nu]}\right), \end{aligned}$$

where (cf. (III.4.18))

$$(2)\quad C(L, M, \nu) = \kappa[L]_0[M]_0 + c|\nu|^{-1/2}.$$

(ii) *If $L \in S[\xi_1]^{k-\theta|q}$ and $M \in S[\xi_1]^{k-2+\theta|q}$, $\theta = 0, 1$, then*

$$(1')\quad |J| \leqslant c\left(\|u\|^2_{(k-1,s|q),[\gamma,\nu]} + \langle D_1^{k-1}u\rangle_{(s|q),[\gamma]}\sum_{j=0}^{k-2}\langle D^j u\rangle_{(k-2-j+s|q),[\gamma]}\right).$$

Following III.4.3, we define $Q^{j|q}$ as the set of symbols $b(y, \eta', \gamma) \in S^j$ that are q-homogeneous functions of ξ' and (σ, γ) whose imaginary parts vanish at $\gamma = 0$. Accordingly, let $O[\xi_1]^{k|q}$ denote the subclass of those symbols $L \in S_0[\xi_1]^{k|q}$ whose coefficients belong not to $S^{k-j|q}$, but to $O^{k-1|q}$. In the case of these classes Proposition 2(i) can be refined (cf. Proposition II.4.3).

PROPOSITION 3. *Let $L \in O[\xi_1]^k$ and let $M \in O[\xi_1]^{k-1}$. Then*

$$(3)\quad \begin{aligned} |J| \leqslant C(L, M, \nu)\Bigg(&|\nu|\,\|u\|^2_{(k-1,s|q),[\gamma,\nu]} + |\gamma|\,\|u\|^2_{(k-1,s-b|q),[\gamma,\nu]} \\ &+ \sum_{j=0}^{k-1}\langle D_1^j v(0,\cdot)\rangle^2_{(s+k-1-j|q),[\gamma,\nu]}\Bigg), \end{aligned}$$

where the constant on the right-hand side has the form (2).

3.2. Statement of the main result. In this section we will prove the following

THEOREM. *Let conditions* (A), (B$^{\pm}$), (C), *and* (D) *of* §1 *hold. Then there is a symbol* $Q(y,\xi,\gamma) \in S[\xi_1]^{r-1|q}$ *and a number* $R > 0$ *such that for every* $S > 0$ *there is* $\nu_0 = \nu_0(S)$ *such that for* $\pm\nu \geqslant \nu_0(s)$ *and* $\gamma < -R|\nu|^q$ *the inequality*
(4)

$$C\left(|\nu|\,\|u\|^2_{(r-1,s|q),[\gamma,\nu]} + |\gamma|\,\|u\|^2_{(r-1,s-b|q),[\gamma,\nu]} + \sum_{j=0}^{r-1}\langle D_1^j u(0,\cdot)\rangle^2_{(r-1-j+s|q),[\gamma,\nu]}\right)$$
$$\leqslant \operatorname{Im}[P(y,D)u, Q(y,D,\gamma)u]_{(s|q),[\gamma,\nu]} + \sum_{j=1}^{\mu}\langle B_j(y',D)u\rangle^2_{(r-1+s-\beta_j|q),[\gamma,\nu]}$$

holds.

First of all, we derive some corollaries from inequality (4) and, incidentally, obtain (1.15). We will need the notation

$$M(\nu,\gamma,\eta') = (|\nu| + |\gamma|\Lambda_q^{1-q}(\gamma,\xi'))^{1/2}. \tag{5}$$

Let us associate with symbol (5) the pseudodifferential operator

$$_{[\gamma]}M = \exp(-\gamma t)M(\nu,\gamma,D')\exp(\gamma t)$$

acting on the variable y'. The sum of the first two terms on the left-hand side of (4) can now be denoted as

$$\|_{[\gamma]}Mu\|^2_{(r-1,s|q),[\gamma,\nu]}.$$

COROLLARY 1. *Under the conditions of the theorem the inequality*

$$C_1\left(\|_{[\gamma]}Mu\|^2_{(r-1,s|q),[\gamma,\nu]} + \sum_{j=0}^{r-1}\langle D_1^j u\rangle^2_{(r-1-j+s|q),[\gamma,\nu]}\right)$$
$$\leqslant \|_{[\gamma]}M^{-1}P(y,D)u\|_{(0,s|q),[\gamma,\nu]} + \sum_{j=1}^{\mu}\langle B_j u\rangle^2_{(r-1-\beta_j+s|q),[\gamma]} \tag{6}$$

holds.

To prove the corollary one should apply the Schwarz inequality to estimate the quadratic form on the right-hand side of (4) in terms of

$$(4\varepsilon)^{-1}\|M^{-1}Pu\|^2_{(0,s|q),[\gamma,\nu]} + \varepsilon\|MQu\|^2_{(0,s|q),[\gamma,\nu]}.$$

Using Proposition 1 in the foregoing subsection we can estimate the second term via the first term on the left-hand side of (6) with a constant proportional to ε. Choosing a sufficiently small ε we arrive at (6). As in III.5.1, we use the fact that the direction of the axis x_1 is not characteristic to derive from Corollary 1 the following

COROLLARY 2. *Under the conditions of the theorem,*

$$C_2\left(_{[\gamma]}Mu\|^2_{(r,s|q),[\gamma,\nu]} + \sum_{j=0}^{r-1}\langle D_1^j u\rangle_{(r+s-j|q),[\gamma]}\right)$$
$$\leqslant \|_{[\gamma]}M^{-1}Pu\|^2_{(0,s+1|q),[\gamma,\nu]} + \sum_{j=1}^{\mu}\langle B_j u\rangle^2_{(r+s-\beta_j|q),[\gamma]}. \tag{7}$$

Noting that $|\nu| \leqslant M(\nu,\gamma,\eta')^2$ and $M(\nu,\gamma,\eta')^2 < |\nu|^{-1}$ we obtain

COROLLARY 3. *Under the conditions of the theorem, inequality* (1.15) *holds.*

Proceeding from Proposition 2(ii) of 3.1 and repeating the arguments in III.5.1 we can drop the q-lower terms in the operators P and $B_1,\dots,B_\mu$ and reduce the proof of the inequality to the case of functions $u(y)$ with support of a small diameter in a neighborhood of a point on the plane $\{x_1 = 0\}$. In other words, in what follows we will assume that

1) the support of $u(y)$ lies in a neighborhood of a point $y^0 = (0, y'^0)$;

2) the operators P, $B_1,\dots,B_\mu$ coincide with their q-principal parts P_0, $B_{01},\dots,B_{0\mu}$;

3) the symbols of the operators under consideration have the form

$$P(y,\eta) = P(\eta) + P'(y,\eta), \quad B_j(y',\eta) = B_j(\eta) + B_j'(y',\eta'), \tag{8}$$

and the norms $[P']_0, [B_1']_0,\dots,[B_\mu']_0$ can be made arbitrarily small;

4) the boundary operators $B_1,\dots,B_\mu$ have constant coefficients.

3.3. Microlocalization of inequality (4). The estimate in the parabolic zone. Let Σ^+ be the "hemisphere" (2.6) and let U be an open subset of Σ^+. Denote by K_q the "q-cone" constructed over U:

$$K_q(U) = \{\xi' = \rho\overline{\xi}', \sigma = \rho^q\overline{\sigma}, \gamma = \rho^q\overline{\gamma}, \quad (\overline{\xi}',\overline{\sigma},\overline{\gamma}) \in U, \quad \rho > 0\}.$$

For brevity, we will say that an inequality holds in U for a function $u(x_1,x',t)$ if it holds for those $u(y)$ whose partial Fourier transforms $\widehat{u}(x_1,\xi',\sigma+i\gamma)$ have the support belongiong to $K_q(U)$ for all $x_1 \in \mathbb{R}_+$. Repeating literally the arguments in III.5.2 we prove an analog of the proposition in this subsection in application to q-hyperbolic operators. Thus, to prove inequality (4) it suffices to establish it in the regions (2.6′) and (2.6″).[1]

In the parabolic zone the imaginary parts of all roots are separated from zero, and $|\gamma| > \kappa^{-1}\Lambda_q(\eta',\gamma)$. It follows that

$$|\gamma|\,\|u\|_{(r-1,s-b|q),[\gamma,\nu]} \approx \|u\|_{(r-1,s+1/2|q),[\gamma,\nu]},$$

where the equivalence constant does not depend on γ and ν. Thus, we have to find a symbol $Q(y,\eta',\gamma) \in S[\xi_1]^{r-1|q}$ such that the inequality

$$\begin{aligned} C\Bigl(\|u\|^2_{(r-1|s+1/2|q),[\gamma,\nu]} &+ \sum_{j=0}^{r-1}\langle D_1^j u\rangle^2_{(r-1-j+s|q),[\gamma]}\Bigr) \\ &\leqslant \operatorname{Im}[Pu, Qu]_{(s|q),[\gamma,\nu]} + \sum_{j=1}^{\mu}\langle B_j u\rangle^2_{(r-1-\beta_j+s|q),[\gamma]} \end{aligned} \tag{9}$$

holds for $\gamma < -R|\nu|^q$. According to Proposition 2(i) in 3.1, the right-hand side of (9) can be estimated in terms of the left-hand side. Therefore, using (8), we can reduce (9) to a similar inequality for the operators $P(D)$, $B_1(D),\dots$. Furthermore, as in the proof of inequality (III.5.1p) (see Lemma III.5.4), we can pass to the inequality on

[1] More precisely, for the regions (2.6′) and (2.6″) to cover (2.6) it is necessary to replace κ in (2.6′) and (2.6″) by κ' and κ'', respectively, where $\kappa'' > \kappa'$.

the half-line for the ordinary differential operator $P(D, \xi', \tau)$. Besides, on performing the transformation $U(z) = \exp(-\nu z)W(z)$ we arrive at the inequality

(10)
$$C_1\Big(\Lambda_q \sum_{j=0}^{r-1} \Lambda_q^{2(r-1-j)} \int_0^\infty |(D+i\nu)^j W(z)|^2\, dz + \sum_{j=0}^{r-1} \Lambda_q^{2(r-1-j)} |(D+i\nu)^j V(z)|^2\Big)$$
$$\leqslant \operatorname{Im} \int_0^\infty P(D+i\nu, \xi', \tau)V(z)\overline{Q(D+i\nu, \xi', \tau)V(z)}\, dz$$
$$+ \sum_{j=1}^{\mu} \Lambda_q^{2(r-1-\beta_j)} |B_j(D+i\nu, \xi', \tau)V)(0)|^2.$$

Since $|\nu| \leqslant R^{-1/q}|\gamma|^{1/q} \leqslant R^{-1/q}\Lambda_q(\eta', \gamma)$, where R is sufficiently large, inequality (10) is equivalent to the analogous inequality with $\nu = 0$.

If the symbol (III.5.12)

$$Q(y', \xi', \eta', \gamma) = P_+(y', \xi_1, \eta', \gamma)\partial P_-(y, \xi_1, \eta', \gamma)/\partial \xi_1 \tag{11}$$

is taken as Q, then, repeating almost literally the arguments in III.5.4 and II.5.4, we derive (10).

3.4. Estimate in the hyperbolic zone (the case of the fulfillment of condition ($\mathbf{B}^+$)). So, we now prove inequality (4) in the zone (2.6″) under the assumption that $(1, 0, \dots, 0) \in V(P)$. As in the case of the parabolic zone, we take (11) as the symbol Q.

3.4.1. According to Proposition 2.2(iii), the roots λ_j^-, $j = 1, \dots, mb$, satisfy (2.10), and the roots with $j = mb+1, \dots, m(b+1)$ are close to the real axes, distinct, and satisfy (2.10′). According to the division of the roots into two groups, we factorize the symbol P_-:

$$\begin{aligned} P_-(y, \xi_1, \eta', \gamma) &= R_-(y, \xi_1, \eta', \gamma)H_-(y, \xi_1, \eta', \gamma) \\ &= \prod_{j=1}^{mb}(\xi_1 - \lambda_j^-(y, \eta', \gamma)) \prod_{j=mb+1}^{m(b+1)} (\xi_1 - \lambda_j^-(y, \eta', \gamma)). \end{aligned} \tag{12}$$

The factorization (12) allows us to represent (11) as the sum

$$Q = Q_0 + Q_1 = P_+R_-\frac{\partial H_-}{\partial \xi_1} + P_+H_-\frac{\partial R_-}{\partial \xi_1}. \tag{13}$$

We will also assume that

$$A = P_+R_-, \quad C = P_+H_-$$

so that

$$Q_0 = A\frac{\partial H_-}{\partial \xi_1}, \quad Q_1 = C\frac{\partial R_-}{\partial \xi_1}.$$

If we denote by $M(y, \xi_1, \eta', \gamma)$ one of the previously introduced symbols Q, Q_0, Q_1, R_-, H_-, A, and B, then, by (2.8), it can be represented in the form

$$M(y, \xi_1, \eta', \gamma) = M(\xi_1, \eta', \gamma) + M'(y, \xi_1, \eta', \gamma),$$

where in this assertion $[M']_0$ can be small.

We note the following two assertions important below. First, we have

$$Q_0(y,\xi_1,\eta,\gamma)\in O[\xi_1]^{r-1|q}. \tag{14}$$

Indeed, since the roots λ_j^- corresponding to the symbol H_- are real for $\gamma=0$, we have

$$H_-\in O[\xi_1]^{m/q},\quad \partial H_-/\partial\xi_1\in O[\xi_1]^{m-1|q}.$$

Therefore $A=P/H_-$ belongs to $O[\xi_1]^{2bm|q}$, which implies (14). Furthermore, we have the following

LEMMA. *Let condition* (B^+) *and uniform Lopatinskii's condition hold and let* (η',γ) *belong to the zone* $(1.6'')$, *where* κ *is sufficiently small. Then the polynomials*

$$\{A_j(\lambda)=\lambda^{j-1}A(\lambda,\eta',\gamma)\},\quad j=1,\dots,m, \tag{15}$$

$$\{B_j(\lambda)=B_j(\lambda,\xi',\sigma+i\gamma)\},\quad j=1,\dots,\mu=mb, \tag{15$'$}$$

$$\{C_j(\lambda)=\lambda^{j-1}C(\lambda,\eta',\gamma)\},\quad j=1,\dots,mb, \tag{15$''$}$$

are linearly independent, and, consequently, they form a basis in the space of polynomials of degree less than $r=(2b+1)m$.

PROOF (cf. the proof of Proposition III.6.2). Assume the contrary, i.e., let there exist polynomials $a(\lambda)$ and $c(\lambda)$, $\deg a(\lambda)\leqslant m-1$, $\deg c(\lambda)\leqslant mb-1$, and constants $b_1,\dots,b_\mu$ such that

$$a(\lambda)A(\lambda)+c(\lambda)C(\lambda)+\sum b_jB_j(\lambda)\equiv 0. \tag{16}$$

Then the remainder of the division of polynomial (16) by $P_+(\lambda)$ is identically zero, i. e.,

$$\sum b_jB'(\lambda)\equiv 0.$$

Uniform Lopatinskii's condition implies that $b_1=\dots=b_\mu=0$, and dividing (16) by $P_+(\lambda)$ we obtain

$$a(\lambda)R_-(\lambda)+c(\lambda)H_-(\lambda)\equiv 0.$$

If κ in $(2.6')$ is sufficiently small, then, according to the definition of the polynomials R_- and H_-, they have no common roots, and, consequently, $R_-(\lambda)$ is indivisible by $H_-(\lambda)$. Since the degree of $a(\lambda)$ is less than that of H_-, we arrive at a contradiction, which proves the lemma.

3.4.2. We will estimate the forms corresponding to (13) separately. Denote by ω a constant that can be made arbitrarily small by choosing suitable constants δ (the size of the support of $u(y)$), κ (in $(1.6'')$), R^{-1}, $|\nu|^{-1}$, and $|\gamma|^{-1}$. It is easy to consider the dependence of ω on all these constants (as was done in §III.6), but we will not dwell on it. We will prove the following inequalities (using the notation (12),(15), … ,$(15'')$):

$$\begin{aligned}&C_1\Big(\|Mu\|^2_{(r-1,s|q),[\gamma,\nu]}+\sum_{j=0}^{m-1}\langle A_j(D,\gamma)u\rangle^2_{(m-1+s-j|q),[\gamma]}\Big)\\ &\qquad-\omega\sum_{j=0}^{r-1}\langle D_1^ju\rangle_{(r-1+s-j|q),[\gamma,\nu]}\\ &\leqslant \operatorname{Im}[Pu,Q_0u]_{(s|q),[\gamma,\nu]}+\sum_{j=1}^{\mu}\langle B_j(D,\gamma)u\rangle^2_{(r-1+s-\beta_j|q),[\gamma]},\end{aligned} \tag{17}$$

$$C_2\left(\|C(y,D,\gamma)u\|^2_{(mb-1,s+1/2|q),[\gamma,\nu]}+\sum_{j=0}^{mb-1}\langle C_j(D,\gamma)u\rangle^2_{(mb-1+s-j|q),[\gamma]}\right)$$

$$-\omega\left(\|Mu\|^2_{(r-1,s|q),[\gamma,\nu]}+\sum_{j=0}^{r-1}\langle D_1^j u\rangle_{(r-1+s-j|q),[\gamma]}\right) \tag{18}$$

$$\leqslant \operatorname{Im}[Pu, Q_1u]_{(s|q),[\gamma,\nu]}.$$

Inequality (4) is a consequence of (17) and (18). Indeed, let us add (17) and (18). Then the right-hand side of the resulting inequality coincides with that of (4). As for the left-hand side, we discard the term corresponding to the first term on the right-hand side of (18), add the last term on the right-hand side of (17), and then drop the term $-\omega\|Mu\|^2_{(r-1,s|q),[\gamma,\nu]}$ using the fact that the coefficient ω is small. As a result, we conclude that the right-hand side of (4) can be estimated from below by

$$C_3\left(\|Mu\|^2_{(r-1,s|q),[\gamma,\nu]}+\sum_{j=1}^{m}\langle A_j(D,\gamma)u\rangle^2_{(m-1+s-j|q),[\gamma]}\right.$$

$$\left.+\sum_{j=1}^{\mu}\langle B_j(D,\gamma)u\rangle^2_{(r-1+s-\beta_j|q),[\gamma]}+\sum_{j=1}^{mb}\langle C_j(D,\gamma)u\rangle^2_{(m-b-1+s-j|q),[\gamma]}\right)$$

$$-\omega\sum_{j=0}^{r+1}\langle D_1^j u\rangle^2_{(r-1+s-j|q),[\gamma]}.$$

By Lemma 3.4.1, the sum of the first, third, and fourth terms has a lower bound

$$C_4\sum_{j=0}^{r-1}\langle D_1^j u\rangle^2_{(r-1+s-j|q),[\gamma]}.$$

Since ω is small, we arrive at (4).

3.4.3. Proof of inequality (17). By (14), in proving (17) we can use Proposition 3 of 3.1 and (in the case of functions with small supports) reduce (17) to a similar estimate for the constant symbols P and Q_0. Furthermore, as was already done many times, we can pass to the estimate on the half-line for a function $U(z)$ belonging to $L_2(\mathbb{R}_+)$ with weight $\exp(\nu z)$. Performing the transformation $U(z)=\exp(-\nu z)V(z)$

we derive the inequality

$$
\begin{aligned}
(19)\qquad & C_1\Big(M^2(\nu,\eta',\gamma)\sum_{j=0}^{r-1}\Lambda_q^{2(r-1-j)}\int_0^\infty |(D+i\nu)^j V(z)|^2\,dz \\
&\qquad +\sum_{j=1}^{m}|(A_j(D+i\nu,\eta',\gamma)V)(0)|^2\Lambda_q^{2(m-1-j)}\Big) \\
&\qquad -\omega\sum_{j=0}^{r-1}\Lambda_q^{2(r-1-j)}|D_1+i\nu^j V(0)|^2 \\
&\leqslant \operatorname{Im}\int_0^\infty P(D_1+i\nu,\xi',\sigma+i\gamma)V(z)\overline{Q_0(D_1+i\nu,\eta',\gamma)V(z)}\,dz \\
&\qquad +\sum_{j=1}^{\mu}\Lambda_q^{2(r-1-\beta_j)}|(B_j(D+i\nu,\xi,\sigma+i\gamma)V)(0)|^2.
\end{aligned}
$$

Since we have $|\nu| \leqslant R^{-1/q}|\gamma|^{1/q}$, and R is sufficiently large, we can set $\nu = 0$ in the left-hand side of (19) and also in the second term on the right-hand side. Denoting by J the form in inequality (19) we rewrite it as

$$
\begin{aligned}
(19')\qquad & C_1'\Big(M^2(\nu,\eta',\gamma)\sum_{j=0}^{r-1}\Lambda_q^{2(r-1-j)}\int_0^\infty |D^j V(z)|^2\,dz \\
&\qquad +\sum_{j=1}^{m}\Lambda_q^{2(r-1-j)}|(D^{j-1}A(D,\eta',\gamma)V)(0)|^2\Big) \\
&\qquad -\omega\sum_{j=1}^{r}\Lambda_q^{2(r-1-j)}|D^j V(0)|^2 \\
&\leqslant J+\sum_{j=1}^{\mu}\Lambda_q^{2(r-1-j)}|(B_j(D,\xi',\sigma+i\gamma)V)(0)|^2.
\end{aligned}
$$

We now proceed to estimation of J. Set $H_-^{(k)} = H_-/(\lambda-\lambda_{k+mb}^-)$, $k = 1,\dots,m$, and

$$W(z) = A(D+i\nu,\eta',\gamma)V(z).$$

The form J is now written as

$$
\begin{aligned}
J = \sum_{k=1}^{m}(\nu-\operatorname{Im}\lambda_{k+mb}^-)\int_0^\infty |H_-^{(k)}(D+i\nu,\eta',\gamma)W(z)|^2\,dz& \\
+\frac{1}{2}\sum|(H_-^{(k)}(D+i\nu,\eta',\gamma)W)(0)|^2.&
\end{aligned}
$$

Since the roots λ_j^-, $j = mb+1,\dots,m(b+1)$, are distinct, the polynomials $H_-^{(k)}(\lambda,\eta',\gamma)$ form a basis (uniform with respect to η',γ) in the space of polynomials of degree at most $m-1$. Since $-\operatorname{Im}\lambda_{k+mb}^-(\eta',\gamma) > \kappa|\gamma|\Lambda_q^{1-q}(\eta',\gamma)$, we get the inequality

(20)

$$J \geqslant C_3\Big(M^2(v,\eta',\gamma)\sum_{j=0}^{m-1}\int_0^\infty \Lambda_q^{2(m-1-j)}|D^j W(z)|^2\,dz + \sum_{j=0}^{m-1}\Lambda_q^{2(m-1-j)}|D^j W(0)|^2\Big)$$

$$-\,\omega\Big(\sum_{j=0}^{r-1}|D^j V(0)|^2\Lambda_q^{2(r-1-j)} + M^2(v,\eta',\gamma)\sum_{j=0}^{r-1}\Lambda_q^{2(r-1-j)}|D^j V(z)|^2\,dz\Big).$$

Furthermore, the expression $A(\lambda,\eta',\gamma) = P_+(\lambda,\eta',\gamma)R_-(\lambda,\eta',\gamma)$ can be regarded as an elliptic symbol of order $2mb$ with large parameter Λ_q. The results in Appendix to §III.5 imply the inequality

(21)

$$\sum_{j=0}^{r-1}\Lambda_q^{2(r-1-j)}\int_0^\infty |D^j V(z)|^2\,dz \leqslant \sum_{j=0}^{m-1}\Lambda_q^{2(m-1-j)}\int_0^\infty |(D^j AV)(z)|^2\,dz$$

$$+\,\Lambda_q^{-1}\sum_{j=1}^{\mu}\Lambda_q^{2(r-1-\beta_j)}|(B_j(D,\xi',\sigma+iv)V)(0)|^2.$$

Substituting (21) into (20) and then into (19′) we derive (17).

3.4.4. Proof of inequality (18). To the factorizations $P = R_-C$ and $Q_1 = R_-^{(1)}C$ of the symbols, where $R_-^{(1)} = \partial R_-/\partial\xi_1$, there correspond factorizations of the related operators:

$$P(y,D) = R_-(y,D,\gamma)C(y,D,\gamma) + P_{r-1},$$

$$Q_1(y,D) = R_-^{(1)}(y,D,\gamma)C(y,D,\gamma) + P_{r-2},$$

where P_{r-1} and P_{r-2} are operators of q-orders no higher than $r-1$ and $r-2$, respectively. In view of Proposition 2(ii) in 3.1, when deriving inequality (18), we can replace its right-hand side by

$$\operatorname{Im}[R_-v, R_-^{(1)}v]_{(s|q),[\gamma,\nu]}, \qquad v = C(y,D,\gamma)u.$$

Now we are in a situation that was in fact considered in the derivation of the estimate in the parabolic zone. Replacing Cu on the left-hand side of (18) by v we arrive at an inequality whose right-hand side can be estimated in terms of the left-hand side. Therefore we can "freeze" the coefficients of the operators $R_-(y,D,\gamma)$ and $R_-^{(1)}(y,D,\gamma)$, pass to the Fourier transform with respect to y', and then repeat literally the above arguments.

3.5. Estimate in the hyperbolic zone (the case when condition (B^-) is satified). Now we assume that $-(1,0,\dots,0)\in V(P)$. In this case the roots that are close to the real axis have positive imaginary parts, and therefore P_+ admits of the natural factorization:

$$P_+(y,\xi_1,\eta',\gamma) = R_+(y,\xi_1,\eta',\gamma)H_+(y,\xi_1,\eta',\gamma)$$

$$= \prod_{j=1}^{mb}(\xi_1 - \lambda_j^+(y,\eta',\gamma)) \prod_{j=mb+1}^{m(b+1)}(\xi_1 - \lambda_j^+(y,\eta',\gamma)).$$

Let us take

$$Q(y,\xi_1,\eta',\gamma) = -\rho R_+\frac{\partial H_+}{\partial \xi_1} + P_+\frac{\partial P_-}{\partial \xi_1} = Q_0 + Q_1$$

as the symbol Q, where the constant ρ will be chosen later in the course of the proof. The symbol Q_0 belongs to $O[\xi_1]^{r-1|q}$, and the estimation of the corresponding form can be carried out in the same way as in 3.4.3. The only distinction is that, in contrast to 3.4.3, the boundary terms have the minus signs (which accounts for the inclusion of the small parameter ρ in the symbol). Thus, we have the inequality

$$(22)\quad \begin{aligned}(c\rho-\omega)\Big(\|Mu\|^2_{(r-1,s|q),[\gamma,\nu]} - \sum_{j=0}^{r-1}\langle D_1^j u\rangle^2_{(r-1+s-j|q),[\gamma]}\Big)\\ \leqslant \operatorname{Im}[Pu, Q_0u]_{(s|q),[\gamma,\nu]} + \sum\langle B_j u\rangle^2_{(r-1+s-\beta_j|q),[\gamma]}.\end{aligned}$$

The estimate for the form corresponding to Q_1 is similar to (18), where C should be replaced by P_+. As a result, we obtain

(23)

$$\begin{aligned}C_1\Bigg(\|P_+u\|^2_{(mb-1,s+1/2|q),[\gamma,\nu]} + \sum_{j=0}^{mb-1}\langle P_+(D,\gamma)D_1^j u\rangle^2_{(mb-1+s-j|q),[\gamma]}\Bigg)\\ -\omega\Big(\sum\langle D_1^j u\rangle^2_{(r-1+s-j|q),[\gamma]} + \|Mu\|^2_{(r-1|q),[\gamma,\nu]}\Big) \leqslant \operatorname{Im}[Pu, Q_1u]_{(s|q),[\gamma,\nu]}.\end{aligned}$$

Let us add together (22) and (23) and add to the left-hand side the second term on the right-hand side of (22). By virtue of uniform Lopatinskii's condition, the sum of this term and the second term on the left-hand side of (23) has a lower bound

$$C_2\sum_{j=0}^{r-1}\langle D_1^j u\rangle^2_{(r-1+s-j|q),[\gamma]}.$$

We now select ρ from the condition $c\rho < C_2/2$. Finally, choosing the parameters δ, κ, R^{-1}, and $|\nu|^{-1}$ so that $\omega(\delta,\kappa,\dots) < \rho/2$ we obtain the desired inequality. Estimate (4) is proved.

§4. Main theorem

By analogy with Chapter I, where the results were put together in §I.4, and Chapter II, where the results on the solvability of the Cauchy problem were put together in §II.6, in the present section we will state a unified theorem combining Theorems III.2.2, 1.3, and 2.5 about the solvability of the mixed problem for the three classes of operators under study.

4.1. We take a vector $N \in \mathbb{R}^n$ and consider the mixed problem in the half-space

$$(1)\qquad \mathbb{R}^{n+1}_{N+} = \{y=(x,t)\in\mathbb{R}^{n+1},\quad \langle N,x\rangle > 0\}.$$

Choose coordinates $\overline{x}$ in $\mathbb{R}^n_x$ so that the $\overline{x}_1$ directions coincide with the inner normal to the subspace (1) and $\overline{x}' = (\overline{x}'_2,\dots,\overline{x}'_n)$ are the coordinates in the plane $\langle x,N\rangle = 0$. We also set $\overline{y}' = (\overline{x}',t)$.

Thus, we consider the problem

$$P(y;D)u(y) = f(y), \quad y \in \mathbb{R}^{n+1}_{N+}, \tag{2}$$

$$B_j(\overline{y}', D)u(\overline{y}') = g_j(\overline{y}'), \quad j = 1, \dots, \mu, \tag{3}$$

$$u(x,t) = f(x,t) = 0, \quad g_j(\overline{x}', t) = \dots = g_\mu(\overline{x}', t) = 0, \quad t < 0.$$

It will be assumed without further stipulation that the coefficients of the operators under study belong to C^∞ and are constant outside some compact set. Also, let the order of the derivative with respect to $\overline{x}_1$ in the operator P exceed the corresponding orders of the operators $B_1, \dots, B_\mu$.

4.2. We now state the conditions under which problem (2), (3) will be considered. We take some $q \geqslant 1$ and denote by $P_0(y,\eta)$ and $B_{0j}(\overline{y}', \eta)$ the symbols of the q-principal parts of the operators in question. The conditions below refer to q, N, and the symbols P_0 and B_{0j}.

Condition (A). q is an integer, and one of the following conditions holds.

(A_1). $q = 1$, and $P_0(y,\eta)$ is a uniformly hyperbolic symbol.

(A_{2b}). $q = 2b$ is even, and $P_0(y,\eta)$ is a uniformly strictly q-parabolic symbol.

(A_{2b+1}). $q = 2b+1$ is odd, and $P_0(y,\eta)$ is a uniformly q-hyperbolic symbol.

Condition (B). The symbol $P_0(y,\eta)$ and the direction $N \in \mathbb{R}^n$ satisfy one of the following conditions.

(B_1). $q = 1$, and N is an arbitrary direction that is not characteristic for P_0.

(B_{2b}). $q = 2b$, and N is an arbitrary direction in $\mathbb{R}^n$.

($B^\pm_{2b+1}$). $q = 2b+1$ and $\pm N \in V(P_0)$, where $V(P_0)$ is the cone of q-hyperbolicity of the symbol P_0.

We denote by $\overline{\xi} = (\overline{\xi}_1, \overline{\xi}')$ the dual variables to $\overline{x}$, write the symbol P_0 in these variables, and consider the algebraic equation

$$P_0(X,\lambda) = 0, \quad X = (\overline{y}', \overline{\xi}', \sigma, \gamma), \qquad \gamma < 0. \tag{4}$$

According to (A), this equation has no real roots, and by (B), the roots are continuous functions of the parameter X. Therefore the numbers $r_\pm$ of the roots in the half-planes $\pm\,\mathrm{Im} > 0$ do not depend on X.

Condition (C). $r_+ = \mu$.

As above, since the symbol $P_0(X,\lambda)$ has no real roots in the range of the parameter X under consideration, we can factorize it as follows:

$$P_0(X,\lambda) = P_{0+}(X,\lambda)P_{0-}(X,\lambda)$$

and define Lopatinskii's determinant

$$\mathrm{Lop}(P_0, B_{01}, \dots, B_{0\mu})(\overline{y}', \overline{\xi}', \sigma, \gamma).$$

Condition (D). There is $K > 0$ such that

$$|\mathrm{Lop}(P_0, B_{01}, \dots, B_{0\mu})(\overline{y}', \overline{\xi}', \tau)| \geqslant K, \quad \overline{y}' \in \mathbb{R}^n, \quad |\overline{\xi}'|^2 + |\tau|^{2q} = 1, \quad \mathrm{Im}\,\tau < 0.$$

4.3. We will deal with spaces whose special cases for $N = (1, 0, \dots, 0)$ were considered in §2, and they will depend on the weight $\exp(\nu\langle N, x\rangle + \gamma t)$, where N satisfies condition (B); in the case of odd q (i. e., under condition (B_{2q+1})) the sign of ν will be chosen from the condition

$$\mathrm{sgn}\,\nu N \in V(P), \tag{5}$$

and in the case of even q or $q=1$ it will be arbitrary.

So, now we denote by $H^{(r,s|q)}_{[\gamma,\nu N]}(\mathbb{R}^{n+1})$ the space of functions with finite norm

$$(6)\qquad \left(\iint\limits_{\substack{\operatorname{Im}\gamma\\ \operatorname{Im}\zeta=\nu N}} (1+|\zeta+i\nu N|^2+|\tau|^{2/q})^r(1+|\xi'|^2+|\tau|^{2/q})^s|\widehat{u}(\zeta,\tau)|^2\,d\zeta\,d\tau\right)^{1/2},$$

and let $H^{(r,s|q)}_{[\gamma,\nu N]}(\mathbb{R}^{n+1}_{N+})$, where $r>0$ is an integer, be the space of restrictions to the subspace (1). We take

$$\|u\|_{(r,s|q)[\gamma,\nu N]}=\Big(\sum_{j=0}^{r}\exp(2\nu\overline{x}_1)\iint\limits_{\operatorname{Im}\tau=\gamma}(1+|\overline{\xi}'|^2+|\tau|^{2/q})^{r+s-j}\times|D^j_{\overline{x}_1}\widehat{u}(\overline{x}_1,\overline{\xi}',\tau)|^2\,d\overline{\xi}'d\tau\,d\overline{x}_1\Big)^{1/2}$$

as the norm in this space, where $\overline{x}$ and $\overline{\xi}$ are the above-mentioned coordinates. In the plane $\{\overline{x}_1=0\}$ we consider the spaces $H^{(s|q)}_{[\gamma]}(\mathbb{R}^n)$ with a norm of the type (1.8).

4.4. THEOREM. *For given $q\geqslant 1$ and $N\in\mathbb{R}^n$ and for a set of operators $P, B_1,\dots,B_\mu$ the conditions below are equivalent.*

(I) *For any operators $\{Q,C_1,\dots,C_\mu\}$, $\deg_q Q<\deg_q P$, $\deg_q C_j<\deg_q B_j$, $j=1,\dots,\mu$, there is $R>0$ such that for any $S>0$ there is $\gamma_0=\gamma_0(S,Q,C_1,\dots,C_\mu)$ such that for $|s|\leqslant S$, $|\nu|\geqslant\nu_0$ and $\gamma<-R|\nu|^q$ the problem*

$$(2')\qquad (P(y,D)+Q(y,D))u=f,$$

$$(3')\qquad (B_j(y,D)+C_j(y,D))u|_{\overline{x}_1=0}=g_j,\quad j=1,\dots,\mu,$$

with $f\in H^{(0,s+1|q)}_{[\gamma,\nu N]+}$ and $g_j\in H^{(r+s-\beta_j|q)}_{[\gamma]+}$, $j=1,\dots,\mu$, possesses a unique solution $u\in H^{(r,s|q)}_{[\gamma,\nu N]+}$ for which the inequality

$$(7)\qquad \begin{aligned}&|\nu|^{1/2}\|u\|_{(r,s|q)[\gamma,\nu N]}+\sum_{j=0}^{r-1}\langle D^j_{\overline{x}_1}u\rangle_{(r+s-j|q)[\gamma]}\\ &\leqslant C\Big(|\nu|^{-1/2}\|f\|_{(0,s+1|q)[\gamma,\nu N]}+\sum_{j=1}^{\mu}\langle g_j\rangle_{(r+s-\beta_j|q)[\gamma]}\Big)\end{aligned}$$

holds.

(II) *Conditions* (A), (B), (C), *and* (D) *hold.*

4.5. First we prove the implication (II)⇒(I). In the case of odd $q>1$ the desired assertion is contained in Theorem 2.5. For $q=1,\ 2b$ the corresponding assertion can be derived from Theorem 1.5. We will confine ourselves to the derivation of the inequality

$$(8)\qquad \begin{aligned}&|\gamma|^{1/2q}\|u\|_{(r,s|q),[\gamma,\nu N]}+\sum_{j=0}^{r-1}\langle D^j_{\overline{x}_1}u\rangle_{(r+s-j|q),[\gamma,\nu N]}\\ &\leqslant C\Big(|\gamma|^{-1/2q}\|f\|_{(0,s+1|q),[\gamma,\nu N]}+\sum_{j=1}^{\mu}\langle g_j\rangle_{(r+s-\beta_j|q),[\gamma,\nu N]}\Big)\end{aligned}$$

whose trivial consequence for $|\nu| \leqslant R^{-1/q}|\gamma|$ is inequality (7). After the transformation $u = \exp(-\nu\langle N, x\rangle)v$ inequality (8) takes the form

$$
\tag{9}
\begin{aligned}
&|\gamma|^{1/2q}\|v\|_{(r,s|q),[\gamma]} + \sum_{j=0}^{r-1}\langle (D^j_{\overline{x}_1} + i\nu)^j v\rangle_{(r+s-j|q),[\gamma]} \\
&\qquad \leqslant C\Big(|\gamma|^{-1/2q}\|F_\nu\|_{(0,s+1|q),[\gamma]} + \sum_{j=1}^{\mu}\langle G\nu_j\rangle_{(r+s-\beta_j|q),[\gamma]}\Big),
\end{aligned}
$$

where we set

$$
\tag{10}
F_\nu = (P+Q)(y, D_x + i\nu N, D_t)v,
$$

$$
\tag{10'}
G_{\nu_j} = (B_j + C_j)(y, D_x + i\nu N, D_t)v|_{\overline{x}_1=0}.
$$

By (1.14), inequality (9) holds for $\nu = 0$. It implies

$$
\tag{11}
\begin{aligned}
&|\gamma|^{1/2q}\|v\|_{(r,s|q),[\gamma]} + \sum_{j=0}^{r-1}\langle (D_{\overline{x}_1} + i\nu)^j v\rangle_{(r+s-j|q),[\gamma]} \\
&\qquad \leqslant C\Big(|\gamma|^{-1/2q}\|F_\nu\|_{(0,s+1|q),[\gamma]} + \sum_{j=1}^{\nu}\langle G_j\nu\rangle_{(r+s-\beta_j|q),[\gamma]}\Big) + J_1 + J_2,
\end{aligned}
$$

where

$$
\tag{12}
J_1 = C|\gamma|^{-1/2q}\|F_\nu - F_0\|_{(0,s+1|q),[\gamma]},
$$

$$
\tag{12'}
J_2 = \sum_{j=0}^{r-1}\langle (D_{\overline{x}_1} + i\nu)^j v\rangle_{(r+s-j|q),[\gamma]} - \sum_{j=0}^{r-1}\langle D^j_{\overline{x}_1} v\rangle_{(r+s-j|q),[\gamma]}.
$$

Expanding (7) in powers of ν and applying the inequality $|\nu| \leqslant R^{-1/q}|\gamma|^{1/q}$ we obtain

$$
\begin{aligned}
J_1 &\leqslant C_1|\gamma|^{-1/2q}\sum_{j=1}^{r} R^{-j/q}|\gamma|^{j/2q}\|v\|_{(r-j,s+1|q),[\gamma]} \\
&\leqslant C_1' R^{-1/q}|\gamma|^{1/2q}\|v\|_{(r-1,s+1|q),[\gamma]} \\
&\leqslant C_1' R^{-1/q}|\gamma|^{1/2q}\|v\|_{(r,s|q),[\gamma]}.
\end{aligned}
$$

Similarly, using the inequality

$$
\begin{aligned}
-\langle D^j_{\overline{x}_1} v\rangle_{(r+s-j|q),[\gamma]} \leqslant &- \langle (D_{\overline{x}_1} + i\nu)^j v\rangle_{(r+s-j|q),[\gamma]} \\
&+ \sum_{j=1}^{k}\binom{j}{k}|\nu|^k\langle (D_{\overline{x}_1} + i\nu)^{j-k} v\rangle_{(r+s-j|q),[\gamma]},
\end{aligned}
$$

we derive

$$
\begin{aligned}
J_2 &\leqslant C_2\sum_{j=0}^{r-2}\sum_{k=1}^{j}\binom{j}{k} R^{-k/q}|\gamma|^{k/q}\langle (D_{\overline{x}_1} + i\nu)^{j-k} v\rangle_{(r+s-j|q),[\gamma]} \\
&\leqslant C_2' R^{-1/q}\sum_{j=0}^{r-2}\langle (D_{\overline{x}_1} + i\nu)^j v\rangle_{(r+s-j|q),[\gamma]}.
\end{aligned}
$$

Hence, if R is sufficiently large, then $J_1 + J_2$ is less than half the left-hand side of (11), which implies (9).

4.6. We make some remarks on the necessity of conditions (A), (B), (C), and (D). In the situation under consideration analogs of the assertion in Proposition 2.6 and its corollary hold, where the variables (ξ_1, ξ') should be replaced by $(\overline{\xi}_1, \overline{\xi}')$. In particular, the inequality (cf. (2.16))

$$(13_q)\quad |v|(|\overline{\xi}_1 + iv| + |\overline{\xi}'| + |\sigma + i\gamma|^{1/q})^r \leqslant C|P_0(\overline{\xi}_1 + iv, \overline{\xi}', \sigma + i\gamma)|(|\xi'| + |\sigma + i\gamma|^{1/q})$$

holds. According to Theorem I.4.3, it follows that q is an integer and that condition (A) holds. In the case of odd $q > 1$ the necessity of (B), (C), and (D) follows from the results of §2. Let us consider the cases of even q and $q = 1$. For even q the necessity of (B) is obvious. In the case $q = 1$ (cf. III.2.3) we replace the variables $(\overline{\xi}_1 + iv, \overline{\xi}', \sigma + i\gamma)$ in (13_1) with $(\rho\overline{\xi}_1 + i\rho^{\varepsilon}v, \overline{\xi}', \sigma - iR|v|\rho^{\varepsilon})$, $0 < \varepsilon < 1$, $|v| \neq 0$, divide both sides of (13_1) by $\rho^{r+\varepsilon}$, and pass to the limit as $\rho \to +\infty$. This proves the necessity of (B). The necessity of (C) and (D) is proved by repeating literally the arguments in 2.6.

References

[A] S. Agmon, *Problèmes mixtes pour les équations hyperboliques d'ordre supérieure*, Les Équations aux Derivées Partielles (Paris, 1962), C.N.R.S., Paris, 1963, pp. 13–18.

[A1] M. S. Agranovich, *Boundary value problems for systems of pseudodifferential operators of the 1st order*, Uspekhi Mat. Nauk **24** (1969), no. 1, 61–125; English transl. in Russian Math. Surveys **24** (1969).

[A2] ———, *Boundary value problems for systems with parameter*, Mat. Sb. **84** (1971), no. 1, 27–65; English transl. in Math. USSR-Sb. **13** (1971).

[A3] ———, *A problem on a matrix depending on a parameter and its application to hyperbolic systems*, Funktsional. Anal. i Prilozhen. **6** (1972), no. 2, 1–11; English transl. in Functional Anal. Appl. **6** (1972).

[AV] M. S. Agranovich and M. I. Vishik, *Elliptic problems with parameter and general type parabolic problems*, Uspekhi. Mat. Nauk **19** (1964), no. 3, 53–161; English transl. in Russian Math. Surveys **19** (1964).

[B] T. Balaban, *On the mixed problem for a hyperbolic equation*, Mem. Amer. Math. Soc. (1971), no. 112.

[Bo] V. M. Borok, *Numerical characteristics of systems in the sense of Petrovskiĭ*, Izv. Vyssh. Uchebn. Zaved. Mat. **1959**, no. 1(8), 16–22. (Russian)

[DK] Yu. L. Daletskiĭ and M. G. Krein, *Stability of solutions to differential equations in Banach's space*, "Nauka", Moscow, 1970; English transl., Amer. Math. Soc., Providence, RI, 1974.

[DN] A. Douglis and L. Nirenberg, *Interior estimates for elliptic systems of partial differential equations*, Comm. Pure Appl. Math. **8** (1955), 503–538.

[D] A. F. Duff, *Hyperbolic differential equations and waves*, Boundary Value Problems for Linear Evolution Partial Differential Equations (H. A. Garnir, ed.), NATO Advanced Study Institute Series, vol. 29, Reidel, Dordrecht, 1977, pp. 27–156.

[E] S. D. Eidel′man, *Parabolic systems*, "Nauka", Moscow, 1964; English transl., North-Holland, Amsterdam, 1969.

[Es1] G. Eskin, *General initial-boundary value problems for second order hyperbolic equations. Singularities in boundary value problems*, Proc. NATO Adv. Study Inst., Maratea, 1980, Reidel, Boston, 1981, pp. 19–34.

[Es2] ———, *Boundary value problems for elliptic pseudodifferential equations*, Amer. Math. Soc., Providence, RI, 1981.

[F] K. O. Friedrichs, *Symmetric positive linear differential equations*, Comm. Pure Appl. Math. **11** (1958), 345–392.

[FL1] K. O. Friedrichs and P. D. Lax, *Boundary value problems for first order operators*, Comm. Pure Appl. Math. **18** (1959), 355–388.

[FL2] ———, *On symmetrizable differential operators*, Singular Integrals (Proc. Sympos. Pure Math., Chicago, 1966), Amer. Math Soc., Providence, RI, 1967, pp. 128–137.

[G] L. Gårding, *Cauchy's problem for hyperbolic equations*, Treizième Congres de Mathématicians Scandinaves (Helsinki 18-23, août 1957), Mercators Tryckeri, Helsinki, 1958, pp. 104–109.

[Gi1] S. G. Gindikin, *On a generalization of parabolic differential operators to the case of multidimensional time*, Dokl. Akad. Nauk SSSR **173** (1967), no. 3, 499–502; English transl. in Soviet Math. Dokl. **8** (1967).

[Gi2] ———, *Differential operator with real characteristics of higher order*, Dokl. Akad. Nauk SSSR **204** (1972), no. 5, 1037–1040; English transl. in Soviet Math. Dokl. **13** (1972).

[Gi3] ———, *Tube domains and the Cauchy problem*, Transl. Math. Monographs, vol. 111, Amer. Math. Soc., Providence, RI, 1992.

[GV1] S. G. Gindikin and L. R. Volevich, *Distributions and convolution equations*, Gordon and Breach, Philadelphia, 1992.

[GV2] ———, *The method of Newton's polyhedron in the theory of partial differential equations*, Math. Appl. (Soviet Ser.), vol. 86, Kluwer Academic, Dordrecht, 1992.

[GV3] ———, *Cauchy's problem for pluriparabolic differential equations*. I, II, Mat. Sb. **75** (1968), no. 1, 64–105; English transl. in Math. USSR-Sb. **4** (1968).

[GV4] ———, *Energy estimates method in the mixed problem*, Uspekhi Mat. Nauk **35** (1980), no. 5, 53–120; English transl. in Russian Math. Surveys **35** (1980).

[GV5] ———, *Sur le problème mixte pour les opérateurs differentiels q-hyperboliques*, C. R. Acad. Sci. Paris Sér. I Math. **289** (1978), 437-439.

[GV6] ———, *Sur le problème mixte pour des opérateurs differentiels a partie principale quasi homogéne*, C. R. Acad. Sci. Paris Sér. I Math. **291** (1980), 107–110.

[GV7] ———, *A mixed problem for q-hyperbolic equations*, Trudy Mosk. Mat. Obshch. **43** (1984), 215–283; English transl. in Trans. Moscow Math. Soc. **43** (1983).

[GV8] ———, *The Cauchy problem*, Encyclopaedia of Math. Sci., vol. 32 (Partial Differential Equations, III), Springer-Verlag, Berlin, 1990.

[GR] R. C. Gunning and H. Rossi, *Analytic functions of several complex variables*, Prentice-Hall, Englewood Cliffs, NJ, 1965.

[H1] L. Hörmander, *The analysis of linear partial differential operators*. II. *Differential operators with constant coefficients*, Grundlehren Math. Wiss., vol. 257, Springer-Verlag, Berlin, 1983.

[H2] ———, *Linear partial differential equations*, Grundlehren Math. Wiss., vol. 116, Springer-Verlag, Berlin, 1963.

[I] V. Ya. Ivriĭ, *Linear hyperbolic equations*, Encyclopaedia of Math. Sci., vol. 33 (Partial Differential Equations, IV), Springer-Verlag, Berlin, 1993.

[IP] V. Ya. Ivriĭ and V. M. Petkov, *Necessary conditions for correctness of Cauchy's problem for nonstrict hyperbolic equations*, Uspekhi Mat. Nauk **29** (1974), no. 2, 3–70; English transl. in Russian Math. Surveys **29** (1974).

[K] T. Kato, *Perturbation theory for linear operators*, Springer-Verlag, Berlin, 1966.

[Kr] H. O. Kreiss, *Initial boundary value problems for hyperbolic systems*, Comm. Pure Appl. Math. **23** (1970), 277–298.

[KS] M. Kryzanski and F. Shauder, *Quasilineare Differentialgleichungen zweiter Ordnung vom hyperbolishen Typus*, Gemischte Bandwertaufgaben Stud. Math. **6** (1933), 162–189.

[L] O. A. Ladyzhenskaya, *The mixed problem for a hyperbolic equation*, GITTL, Moscow, 1953. (Russian)

[Le] J. Leray, *Hyperbolic differential equations*, The Institute for Advanced Study, Princeton, NJ, 1953.

[M] D. Michelson, *Initial-boundary value problems for incomplete singular perturbations of hyperbolic systems*, J. Analyse Math. **53** (1989), 1–138.

[P1] I. G. Petrovskiĭ, *On Cauchy's problem for systems of partial linear differential equations in the domain of non-analytic functions*, Selected Works. Systems of Partial Differential Equations. Algebraic Geometry, "Nauka", Moscow, 1986, pp. 98–168. (Russian)

[P2] ———, *On Cauchy's problem for systems of partial differential equations*, Selected Works. Systems of Partial Differential Equations. Algebraic Geometry, "Nauka", Moscow, 1986, pp. 34–97. (Russian)

[R] F. V. Ralston, *Note on a paper of Kreiss*, Comm. Pure. Appl. Math. **24** (1971), 759–762.

[Ra] F. Rauch, *L_2 is a continuable initial condition for Kreiss mixed problems*, Comm. Pure. Appl. Math. **25** (1972), 277–298.

[S1] R. Sakamoto, *Mixed problems for hyperbolic equations*. I, II, J. Math. Kyoto Univ. **10** (1970), no. 3, 349–373; 403–417.

[S2] ———, *Iterated hyperbolic mixed problems*, Publ. Res. Inst. Math. Sci. **6** (1970), no. 1, 1–42.

[So] V. A. Solonnikov, *On boundary value problems for general type linear parabolic systems*, Trudy Mat. Inst. Steklov. **83** (1965); English transl. in Proc. Steklov Inst. Math. **83** (1967).

[V1] L. R. Volevich, *Energy method in Cauchy's problem for differential operators correct in Petrovskiĭ's sense*, Trudy Moskov. Mat. Obshch. **31** (1974), 147–187; English transl. in Trans. Moscow Math. Soc. **31** (1976).

[V2] ———, *On general systems of differential equations*, Dokl. Akad. Nauk SSSR **132** (1960), no. 1, 20–23; English transl. in Soviet Math. Dokl. **1** (1960).

[V3] ———, *On a linear programming problem arising in partial differential equations*, Uspekhi Mat. Nauk **18** (1983), no. 3, 155–162; English transl. in Russian Math. Surveys **18** (1983).

[V4] ———, *The Cauchy problem for systems of partial differential equations with real principal part*, submitted to Dokl. Akad. Nauk SSSR. (Russian)

[V5] ———, *A mixed problem for evolution equations*, Analysis on Manifolds with Singularities (Breitenbrunn, 1990), Teubner-Texte Math., vol. 131, Teubner, Stuttgart, 1992, pp. 298–303.

[VI] L. R. Volevich and V. Ya. Ivriĭ, *Hyperbolic equations*, I. G. Petrovskiĭ, Selected Works. Systems of Partial Differential Equations. Algebraic Geometry, "Nauka", Moscow, 1986, pp. 395–417. (Russian)

Notation Index

Subject Index

Other Titles in This Series

(Continued from the front of this publication)

105 **D. R. Yafaev,** Mathematical scattering theory: General theory, 1992

104 **R. L. Dobrushin, R. Kotecký, and S. Shlosman,** Wulff construction: A global shape from local interaction, 1992

103 **A. K. Tsikh,** Multidimensional residues and their applications, 1992

102 **A. M. Il′in,** Matching of asymptotic expansions of solutions of boundary value problems, 1992

101 **Zhang Zhi-fen, Ding Tong-ren, Huang Wen-zao, and Dong Zhen-xi,** Qualitative theory of differential equations, 1992

100 **V. L. Popov,** Groups, generators, syzygies, and orbits in invariant theory, 1992

99 **Norio Shimakura,** Partial differential operators of elliptic type, 1992

98 **V. A. Vassiliev,** Complements of discriminants of smooth maps: Topology and applications, 1992 (revised edition, 1994)

97 **Itiro Tamura,** Topology of foliations: An introduction, 1992

96 **A. I. Markushevich,** Introduction to the classical theory of Abelian functions, 1992

95 **Guangchang Dong,** Nonlinear partial differential equations of second order, 1991

94 **Yu. S. Il′yashenko,** Finiteness theorems for limit cycles, 1991

93 **A. T. Fomenko and A. A. Tuzhilin,** Elements of the geometry and topology of minimal surfaces in three-dimensional space, 1991

92 **E. M. Nikishin and V. N. Sorokin,** Rational approximations and orthogonality, 1991

91 **Mamoru Mimura and Hirosi Toda,** Topology of Lie groups, I and II, 1991

90 **S. L. Sobolev,** Some applications of functional analysis in mathematical physics, third edition, 1991

89 **Valeriĭ V. Kozlov and Dmitriĭ V. Treshchëv,** Billiards: A genetic introduction to the dynamics of systems with impacts, 1991

88 **A. G. Khovanskiĭ,** Fewnomials, 1991

87 **Aleksandr Robertovich Kemer,** Ideals of identities of associative algebras, 1991

86 **V. M. Kadets and M. I. Kadets,** Rearrangements of series in Banach spaces, 1991

85 **Mikio Ise and Masaru Takeuchi,** Lie groups I, II, 1991

84 **Đáo Trọng Thi and A. T. Fomenko,** Minimal surfaces, stratified multivarifolds, and the Plateau problem, 1991

83 **N. I. Portenko,** Generalized diffusion processes, 1990

82 **Yasutaka Sibuya,** Linear differential equations in the complex domain: Problems of analytic continuation, 1990

81 **I. M. Gelfand and S. G. Gindikin, Editors,** Mathematical problems of tomography, 1990

80 **Junjiro Noguchi and Takushiro Ochiai,** Geometric function theory in several complex variables, 1990

79 **N. I. Akhiezer,** Elements of the theory of elliptic functions, 1990

78 **A. V. Skorokhod,** Asymptotic methods of the theory of stochastic differential equations, 1989

77 **V. M. Filippov,** Variational principles for nonpotential operators, 1989

76 **Phillip A. Griffiths,** Introduction to algebraic curves, 1989

75 **B. S. Kashin and A. A. Saakyan,** Orthogonal series, 1989

74 **V. I. Yudovich,** The linearization method in hydrodynamical stability theory, 1989

73 **Yu. G. Reshetnyak,** Space mappings with bounded distortion, 1989

72 **A. V. Pogorelev,** Bendings of surfaces and stability of shells, 1988

71 **A. S. Markus,** Introduction to the spectral theory of polynomial operator pencils, 1988

70 **N. I. Akhiezer,** Lectures on integral transforms, 1988

69 **V. N. Saliĭ,** Lattices with unique complements, 1988

68 **A. G. Postnikov,** Introduction to analytic number theory, 1988

67 **A. G. Dragalin,** Mathematical intuitionism: Introduction to proof theory, 1988

66 **Ye Yan-Qian,** Theory of limit cycles, 1986

(See the AMS catalog for earlier titles)